M000086137

Distribution Family	PDF	Skew	Typical Application
Normal Chapter 21, page 317		Symmetric	Sum or average of many independent random variables with the same distribution. Common default model.
Largest Extreme Value Chapter 12, page 223		Right	Maximum value of many independent random variables. Also called Gumbel.
Smallest Extreme Value Chapter 12, page 223		Left	Minimum value of many independent random variables.
Laplace Chapter 17, page 279		Symmetric	Heavy tails and sharp peak. Also called double exponential. Excess kurtosis = 3
Logistic Chapter 18, page 287		Symmetric	Slightly heavier tails than normal. Excess kurtosis = 1.2
Student's t Chapter 25, page 365		Symmetric	Symmetric distribution with heavier tails than normal. Can match any positive value of excess kurtosis.

Six Sigma
Distribution Modeling

ABOUT THE AUTHOR

Andrew Sleeper is a Master Black Belt and General Manager of Successful Statistics, LLC. Since 1981, he has worked with product development teams as an engineer, statistician, project manager, Six Sigma Black Belt, and consultant. An experienced instructor of statistical tools for engineers, Mr. Sleeper has presented thousands of hours of training in countries around the world. Mr. Sleeper is also the author of *Design For Six Sigma Statistics: 59 Tools for Diagnosing and Solving Problems in DFSS Initiatives*, published by McGraw-Hill.

Six Sigma Distribution Modeling

Andrew D. Sleeper

Successful Statistics LLC
Fort Collins, Colorado

New York   Chicago   San Francisco   Lisbon   London   Madrid
Mexico City   Milan   New Delhi   San Juan   Seoul
Singapore   Sydney   Toronto

The **McGraw·Hill** Companies

Library of Congress Cataloging-in-Publication Data

Sleeper, Andrew D.
  Six sigma distribution modeling / Andy Sleeper.—1st ed.
    p.   cm.
  Includes bibliographical references and index.
  ISBN-13: 978-0-07-148278-3 (alk. paper)
  ISBN-10: 0-07-148278-4
  1. Production control—Data processing.   2. System analysis—
Statistical methods.   3. Distribution (Probability theory)   I. Title.

  TS156.S577   2006
  658.4'013—dc22

                                                    2006036709

Copyright © 2007 by The McGraw-Hill Companies, Inc. All rights reserved. Printed in
the United States of America. Except as permitted under the United States Copyright Act
of 1976, no part of this publication may be reproduced or distributed in any form or by
any means, or stored in a data base or retrieval system, without the prior written permission of the publisher.

1 2 3 4 5 6 7 8 9 0   DOC/DOC   0 1 3 2 1 0 9 8 7 6

ISBN-13: 978-0-07-148278-3
ISBN-10: 0-07-148278-4

*The sponsoring editor for this book was Kenneth P. McCombs and the production
supervisor was Richard C. Ruzycka. It was set in Century Schoolbook by International
Typesetting and Composition. The art director for the cover was Anthony Landi.*

Printed and bound by RR Donnelley.

This book is printed on acid-free paper.

McGraw-Hill books are available at special quantity discounts to use as premiums and sales
promotions, or for use in corporate training programs. For more information, please write
to the Director of Special Sales, Professional Publishing, McGraw-Hill, Two Penn Plaza,
New York, NY 10121-2298. Or contact your local bookstore.

Information contained in this work has been obtained by The McGraw-Hill Companies, Inc.
("McGraw-Hill") from sources believed to be reliable. However, neither McGraw-Hill nor
its authors guarantee the accuracy or completeness of any information published herein, and
neither McGraw-Hill nor its authors shall be responsible for any errors, omissions, or dam-
ages arising out of use of this information. This work is published with the understanding
that McGraw-Hill and its authors are supplying information but are not attempting to render
engineering or other professional services. If such services are required, the assistance of
an appropriate professional should be sought.

To my son, PFC Pascal N. Sleeper,
and to all the brave and unselfish people
of the United States Armed Forces

CONTENTS

PREFACE

This book combines a catalog of distribution families with a guide to selecting and using distribution models. The intended audience includes anyone who uses statistical tools to improve processes and to make better decisions. Some chapters and examples focus on Six Sigma applications, but anyone who creates and applies statistical models will find this book a useful reference.

Because of its flexibility and ease of use, Monte Carlo simulation has become a popular tool for predicting random behavior and analyzing decision scenarios. Monte Carlo simulation requires the selection of distribution models for each system input variable. Numerous examples in this book illustrate how to select and apply distribution models for simulations.

The level of mathematics has been minimized in this book. To successfully select and apply distribution models does not require sophisticated mathematics. While some of the concepts in this book involve calculus, these concepts are optional, and are included only to provide a comprehensive reference on distribution models.

How to Use this Book

This book has two parts. The first part, Chapters 1 through 4, provides a short course in the selection and application of distribution models in Six Sigma environments. The second part, starting with a glossary in Chapter 5, contains a catalog of distribution families. Most readers will begin with the first few chapters in order, and then refer to the later chapters in any order, as they need to learn more about specific distribution families.

- Chapter 1 covers the selection of distribution models using graphical and statistical tools. Since it is rare to have enough data to make a definitive choice, this chapter emphasizes the thoughtful combination of observed data, theoretical knowledge, and expert opinion.
- Chapter 2 discusses distribution modeling features of selected software products, including Crystal Ball, Excel, JMP, MINITAB, and STATGRAPHICS products. Many Excel statistical functions have been defective in past versions, and some of these defects were corrected in Excel 2003. This chapter contains recommendations on the appropriate use of Excel statistical functions.

- Chapter 3 answers questions about how to use nonnormal distributions in Six Sigma environments, or wherever normal distributions are the norm. In particular, this chapter reviews methods of assessing process stability and measuring process capability. The standard tools for these tasks all assume a normal process distribution. When the process distribution is not normal, different tools are required.
- Chapter 4 illustrates the combination of distribution models with Monte Carlo simulation and stochastic optimization to resolve complex problems easily and quickly. The three case studies in this chapter use Crystal Ball and OptQuest software.

Following a glossary in Chapter 5, each of the remaining chapters of this book is devoted to one family of distribution models, sometimes including variations of that family. These chapters include the following information tailored for each distribution family:

- Brief description of the uses and history of each family.
- Examples of that distribution family used in modeling situations.
- Relationships to other distribution families.
- Normalizing transformation, if any.
- Formulas to estimate parameter values, if simple formulas are available.
- Control chart recommendations.
- Process capability metric calculations.
- Formulas describing the distribution family.
- Characteristics of the family, including mean, median, standard deviation, skewness, kurtosis, and others.

Trademarks and Intellectual Property

- Crystal Ball® and Decisioneering® are registered trademarks of Decisioneering, Inc. CB Predictor™ is a trademark of Decisioneering, Inc. Screen shots, forms, and graphs created by Crystal Ball or OptQuest software are used in this book by permission of Decisioneering, Inc.
- MINITAB®, and all other trademarks and logos for the products of Minitab, Inc. are the exclusive property of Minitab, Inc. See *www.minitab.com* for more information.
- Microsoft® and Excel® are registered trademarks of Microsoft Corporation in the United States and in other countries.
- OptQuest® is a registered trademark of Optimization Technologies, Inc.
- Portions of the input and output contained in this book are printed with permission of Minitab, Inc.

- SAS®, JMP®, and all other SAS Institute Inc. product or service names are registered trademarks or trademarks of SAS Institute Inc. in the United States and other countries.
- STATGRAPHICS® is a registered trademark of Statistical Graphics Corporation.

Acknowledgments

I am deeply indebted to many people, who made this book possible. Here is a partial list:

- My wife Julie, for her love, support, tolerance, and good advice through 25 years.
- My son Minh, who handled many projects very well for me, while I was occupied with this project.
- Larry Goldman, of Decisioneering, who wrote most of Chapter 4, and who suggested significant improvements throughout the book.
- The authors of many excellent books, whose innovations and excellence in statistical communication are vital elements of this book. Of many references listed at the end of this book, these have been the most influential:
  - Balakrishnan and Nevzorov (2003)
  - Bothe (1997)
  - D'Agostino and Stephens (1986)
  - Johnson, Kotz, and Balakrishnan (1994 and 1995)
  - Johnson, Kemp, and Kotz (2005)
  - Kotz and Lovelace (1998)
- Many people who have reviewed portions of this manuscript, and whose suggestions made this a much better book:
  - Cathy Akritas, Minitab, Inc.
  - Sam Broderick, Colorado State University
  - Crystal Campbell, Decisioneering, Inc.
  - David Gainer, Microsoft Corporation
  - Mor Hezi, Microsoft Corporation
  - Edward Jaeck, Intel Corporation
  - Jeff Perkinson, SAS Institute, Inc.
  - Neil Polhemus, StatPoint, Inc.
- All the fine people at Decisioneering, Microsoft, Minitab, SAS Institute, and StatPoint, who have patiently answered my questions and who consistently work hard to develop better software products.

- Kenneth McCombs, Senior Acquisitions Editor at McGraw-Hill, who assembled my random thoughts into a cogent book proposal, and who guided me throughout the project.

I welcome all questions and feedback about this book. My email address is *Andy@SuccessfulStatistics.com.*

Andrew D. Sleeper
Fort Collins, Colorado

Modeling Random Behavior with Probability Distributions

Probability distributions are useful models representing random behavior. Diverse fields of human endeavor, including business, science, engineering, and many others, all use the same system of probability distribution models. Why are these models so important? Benefits of probability distribution models fall into four broad and overlapping categories:

- *Estimation.* When a sample of observations of a random variable is available, numerical characteristics describing the random variable may be estimated from the data. The mean and standard deviation are examples of these characteristics. It is common to assume a particular distribution family, such as the normal family, before estimating these characteristics. These estimates are more precise with an assumption that the random variable has a distribution from a limited family of distributions than without that assumption. Probability distribution models improve the precision of population estimates.
- *Prediction.* Following estimation, one often wants to predict future behavior based on past behavior. Using a probability distribution model, one can predict ranges of values that will occur with any given probability, or one can predict the probability that values will fall within any given range. Without using a distribution model, predictions are restricted to the specific values previously observed. Probability distribution models make predictions more flexible and versatile.
- *Simulation.* Using distribution models to represent random variables that are inputs to a system, simulation predicts how the system reacts to its inputs. The outputs of the simulated system form new random variables to be modeled, estimated, and predicted. Probability distribution models allow systems to be tested and optimized rapidly using simulation tools.

- *Communication.* Essential to all of the above is the ability to describe complex random behavior in concise terms. Probability distribution models provide a concise system of terminology for describing and communicating randomness to others.

All these functions play important roles in the Six Sigma problem solving process DMAIC: Define–Measure–Analyze–Improve–Control. In the Define and Measure phases, estimation quantifies the size of the problem in terms of parameter estimates. Prediction converts these estimates into probabilities of defects, which can be expressed in monetary terms. During the Analyze and Improve phases, simulation provides an easy way to test theories and possible improvements. During all phases, project success requires good communication to management, cross-functional team members, and stakeholders.

Distribution models make all these tasks easier. Unfortunately, many Six Sigma training programs do not equip people with the skills to select and use distribution models wisely. Often, the normal family of distributions is the only family discussed. While normal distributions are useful and important, many stable and effective processes do not follow the normal distribution. Limiting oneself to the normal distribution is like limiting a toolbox to only hammers. In a Six Sigma project, this limitation can result in significant errors in the estimation of defect rates and customer dissatisfaction.

Example 1.1

Jim is a Black Belt at a manufacturer of high-precision machined components for automotive engines. One critical new part is a valve seat that requires a flatness of less than 5 μm (one micron or micrometer is one millionth of a meter). To accomplish this feat, Jim's company installs a new lapping machine and automated flatness measurement system using monochromatic helium light.

A pilot production run of 200 parts looked very good, with C_{PK} and P_{PK} in excess of 1.50 and a stable control chart. Based on standard Six Sigma tables, this means that the seats will have a defect rate of 3.4 defects per million units (DPM), which meets the company's Six Sigma goals. Management approves the process and launches production.

Disaster follows swiftly. In the first production run of 1000 seats, the inspection system rejects six parts with flatness above 5 μm, and many more parts are close to the limit. Worse, the control chart generated by the inspection system frequently finds out of control conditions. Each time this happens, the machine stops production and e-mails the engineer and the production supervisor. People gather and investigate, but they find nothing wrong or changed in the process. This routine gets old.

The production supervisor calls Jim and assigns him a Black Belt project with this objective: "Fix this! Now!" (Since this is a family book, certain words are omitted.)

In the Measure phase of Jim's project, completed in a single afternoon, Jim uses distribution models for estimation, prediction, simulation, and communication.

Estimation. Jim pulls the data file and examines the measured flatness for a production sequence of 1000 parts. Applying the usual formulas for estimating the mean and standard deviation of a normal distribution, Jim calculates:

$$\hat{\mu} = \overline{X} = \frac{1}{n}\sum_{i=1}^{n} X_i = 1.4886$$

$$\hat{\sigma} = s = \sqrt{\frac{1}{n-1}\sum_{i=1}^{n}(X_i - \overline{X})^2} = 0.7691$$

Then, Jim estimates the long-term process capability index P_{PK}.

$$\hat{P}_{PK} = \frac{UTL - \overline{X}}{3s} = \frac{5 - 1.4886}{3 \times 0.7691} = 1.522$$

Generally, the formula for P_{PK} involves both the upper tolerance limit (UTL) and lower tolerance limit (LTL). In this case, there is only an upper tolerance limit of 5 μm.

Jim's estimates verify the estimates calculated from the pilot production run. However, the observed rate of defects, $\frac{6}{1000} \times 10^6 = 6000$ DPM is far worse than the predicted defect rate of 3.4 DPM. Why is this?

Jim creates a histogram of the 1000 measurements, shown in Figure 1-1. Jim sees why the initial prediction of 3.4 DPM is incorrect. The Six Sigma table that converts between C_{PK} or P_{PK} values and DPM assumes that the process distribution is normal. The normal distribution is symmetric and bell-shaped, but the histogram in Figure 1-1 is very asymmetric. The normal distribution is not a good model here, and the above estimates are unreliable.

Instead of the normal model, Jim decides to use a two-parameter exponential distribution family to model this data. Later in this chapter, Example 1.3 discusses why Jim chose this particular family. Using MINITAB software, Jim computes the following parameter estimates:

Scale parameter: $\hat{\beta} = 0.8946$

Threshold parameter: $\hat{\tau} = 0.594$

In Figure 1-2, Jim adds a curve to the histogram representing the two-parameter exponential model. Visually, this model appears to fit the data well.

Applying the nonnormal process capability functions in MINITAB software, Jim also estimates the process capability index to be $\hat{P}_{PK} = 0.72$. This estimate of P_{PK} is consistent with the observed rate of defects in the process.

Prediction. An important deliverable in the Measure phase is a prediction of defect rates based on the current process. Since the normal-based prediction

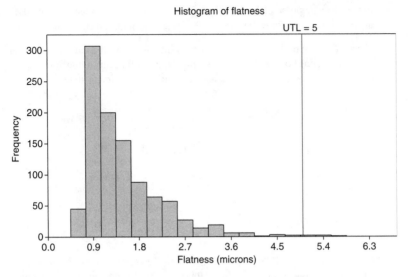

Figure 1-1   Histogram of Flatness of 1000 Valve Seats.

was incorrect, Jim must use the two-parameter exponential model. Using a formula from Section 11.1, Jim can predict the probability that any one seat will have flatness of more than 5 μm:

$$P[X > x] = R_{Exp(\beta,\tau)}(x) = e^{-(x-\tau)/\beta}$$

$$P[X > 5] = e^{-(5-0.594)/0.8946} = 0.007262$$

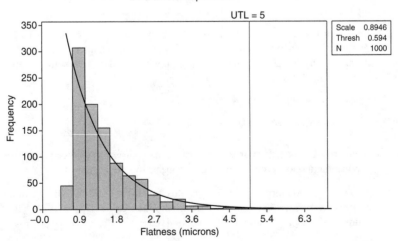

Figure 1-2   Histogram of Flatness with Two-Parameter Exponential Model.

Multiplying this probability by one million, Jim predicts that the defect rate is 7262 DPM, not far from the 6000 DPM observed in the dataset.

Simulation. Jim creates a model of the flatness random variable in a Microsoft Office Excel spreadsheet and simulates it using Crystal Ball simulation software. In a very short time, Jim produces a forecast of the long-term behavior of this process, illustrated in Figure 1-3. Later, in the Analyze and Improve phases of the project, Jim can use this simulation model to evaluate the impact of proposed changes to the process. Because a computer performs the simulations, these evaluations are quick and cost nothing more than a few minutes of Jim's time.

Communication. At the end of the day, Jim sits down with the production supervisor and reviews his findings. Jim presents the figures printed here and says, "We were fooled. We assumed that the lapper process had a normal distribution, but it seems to have an exponential distribution." By using names for these distributions, Jim can summarize complex probability concepts in a few simple words. These concepts are easier to understand when combined with a few good pictures.

From this point, Jim's Black Belt project could continue in many different ways. He might find a way to change the distribution to be closer to a normal distribution, but this is unlikely. Flatness has a physical lower boundary of zero, and usually produces skewed distributions. Jim might find a way to reduce the scale of the exponential distribution to reduce the rate of defects. Whatever happens in Jim's project, appropriate distribution models will be essential to his success.

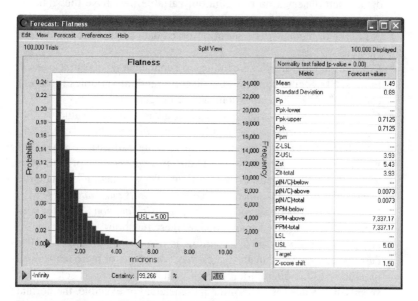

Figure 1-3   Simulated Distribution of Flatness.

This book provides guidance and reference material for selecting and using probability distribution models effectively in many applications.

- The remainder of this chapter discusses how to select and apply the most appropriate distribution models.
- Chapter 2 compares selected statistical and distribution functions of leading software products.
- Chapter 3 offers specific advice to Six Sigma practitioners on the use of nonnormal distributions in Six Sigma environments.
- Chapter 4 tells some additional application stories illustrating the uses of distribution models.
- Chapter 5 is a glossary of statistical terms used in this book.
- The remaining chapters provide detailed information about many commonly used distribution families.

1.1 Terminology to Describe Randomness

This section defines general terms used in this book to describe random variables, systems, and simulations. Chapter 5 defines additional terms describing more specific characteristics of random variables.

- An *observation* is a real number representing a measured value.
- *Observed data* are a set of observations.
- A *random variable* is any system or process that produces observed data. Each observation of a random variable may have the same or different values.
- A *sample* or a *dataset* is a set of observed data produced by a random variable.
- *Discrete* random variables are limited to a countable set of distinct values. In most Six Sigma applications, discrete random variables represent counts, which are limited to nonnegative integers $(0, 1, 2, \ldots)$.
- *Continuous* random variables may have any real number value.
- A *probability distribution*, or simply a *distribution*, is a model expressing the relative probability that a random variable will assume any set of values. Several mathematical functions express the probabilities associated with probability distributions. These functions include the *cumulative distribution function* (CDF), *probability mass function* (PMF) for discrete distributions, and *probability density function* (PDF) for continuous distributions. Chapter 5 defines these functions.
- A *parametric family* of distributions is a set of distribution models described by the same mathematical function of one or more real-valued parameters. One distribution from a parametric family is selected by specifying values for each of the parameters. For example, the normal

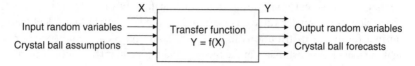

Figure 1-4   System with Transfer Function, Inputs, and Outputs.

family of distributions has two parameters, μ and σ, representing the mean and standard deviation of the random variable. Any set of values for μ and σ specifies one distribution from the normal family. In this book, a *family* of distributions refers to a parametric family of distributions.

- A *simulation* is an experiment with the objective of emulating the behavior of a physical system using a computer program. A simulation includes one or more input random variables, X, and one or more output random variables, Y, as illustrated in Figure 1-4. During the simulation, the computer calculates Y as a transfer function of X, represented as $Y = f(X)$.
- A *Monte Carlo simulation* is a simulation in which the computer selects random values for each input random variable X based on an assumed probability distribution model. For each set of random X values, the computer calculates and stores the resulting Y values. Statistical analysis and modeling applied to the stored Y values can estimate and predict the random behavior of the physical system represented by the simulation.

Using Crystal Ball software, the transfer function $Y = f(X)$ is expressed as a set of Microsoft Office Excel formulas. Elements of a Crystal Ball Monte Carlo simulation have specific terms listed below:

- An *assumption* in a Crystal Ball model is an input random variable, X. Each assumption requires a fully specified probability distribution model.
- A *forecast* in a Crystal Ball model is an output random variable, Y.
- A *trial* in a Crystal Ball simulation comprises one set of randomly generated values for all assumptions and one set of calculated values for all forecasts.
- A *simulation* is a set of trials.

1.2 Selecting a Distribution Model

Selecting a distribution model to represent the behavior of a random process requires two decisions. The first decision is to select a parametric family of distributions, for example, normal, exponential, or Weibull. The second decision is to select a single member of that family of distributions by specifying its parameter values. For example, a modeler might first select the normal family of distributions to represent some physical quantity. After making that choice, estimates of the mean μ and the standard deviation σ from the available data specify one normal distribution from the normal family.

The first decision generally requires more thought than the second decision. After selecting a family of distributions, well-established formulas or computational procedures can estimate the parameters of the distribution that best fits the available data.

The organization of this book reflects this two-step decision process. Chapters 1–4 explain and illustrate how to select the most appropriate family of distributions. Each of the following chapters describes a particular family, including formulas for estimating parameters, calculating probabilities and capability metrics, and selecting control charts.

The choice of a family of distributions to represent a random process reflects the thoughtful combination of observed data, theoretical knowledge, and expert opinion about the process. Prior information about a process distribution generally falls into these three categories:

- *Observed data.* When the process is stable, a random sample of observed data can provide the most reliable selection of a distribution family and prediction of future process behavior.
- *Theoretical knowledge.* Prior knowledge, expectations, or theory about a process may suggest that some families of distributions are more appropriate than other families.
- *Expert opinion.* When an expert has accumulated experience about a particular type of process, the expert's opinions about the distribution can be used to select a distribution family and even the parameter values within that family.

Some situations provide information in two or three of these categories. It can be challenging to combine multiple sources of information and select a single distribution family. There is no best way to do this, but here are some general guidelines. When a very large and representative sample of observed data is available, this information should outweigh theory or opinion in the decision process. In addition, theory suggesting a particular family of distributions should outweigh pure opinion. Finally, when choosing between otherwise equal alternative models, the simpler model is better.

Practical applications involve a limited quantity of observed data and limited theoretical knowledge. Figure 1-5 describes a procedure for

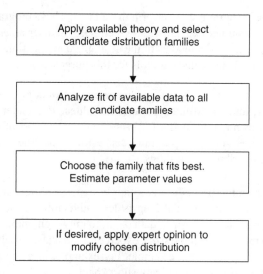

Figure 1-5   Flow Chart to Select a Distribution Using Multiple Sources of Information.

combining these sources of imperfect knowledge and arriving at the most appropriate distribution choice.

The first step in this procedure is to select a set of candidate distribution families based on a theoretical understanding of the system. This set could include one or many families. For example, in reliability analysis, failure times are always positive numbers. Using the suggestions in Section 1.3, candidate distribution families for this type of random process include exponential, gamma, Weibull, lognormal, and loglogistic families. When the random process falls into a recognized class, this knowledge generally narrows the set of candidate distributions to a manageable few.

It is a common practice to ignore the first step and to fit available data to all the distribution families offered by the fitting software. This automatic process is easy because it requires no critical thought. However, models selected by fitting data to a thoughtfully selected subset of distribution families are more credible and reliable than those selected by an automatic process. A model selected using simple theoretical arguments combined with real data is easier to explain and to defend. Even when fitting the data to all the distribution families, it is wiser to ignore families that are inconsistent with the available theory.

The second step in the procedure is to fit the available data to the candidate distributions. Crystal Ball, MINITAB, JMP, STATGRAPHICS, and many other statistical programs can perform these tasks. These programs provide either graphs or numerical statistics, which help the user assess how well the candidate distribution models fit the available data. Sections 1.4 and 1.5 describe graphical and statistical tools for this purpose.

The third step in the procedure is to choose one distribution family that fits the data best, and to estimate parameter values. Parameter estimation usually requires no additional effort. As an automatic part of distribution fitting procedures, statistical programs also estimate parameters for each of the candidate distribution families.

The final step in the procedure is to apply expert opinion to modify the selected model. This optional step is less advisable as the size of the available dataset increases, since data ought to outweigh opinion. However, with small to moderate sample sizes, expert opinion may be useful to correct for known inadequacies in the sample. For example, in many situations, a small sample represents a relatively short span of time. It is reasonable to expect shifts and drifts to increase long-term variation beyond the short-term variation seen in the sample. Six Sigma methods generally include an expert opinion that the distribution mean may shift by 1.5 standard deviations over the long term. Section 1.6 describes this and other situations calling for expert opinion.

Many real applications provide neither theory nor data about a random variable. This situation is very common in Design For Six Sigma (DFSS) applications, when designing a new product that has never been built. To apply tools such as Monte Carlo simulation to processes with no theory or data requires selection of a distribution model from opinion alone. Section 1.6 discusses approaches for selecting a distribution model in this situation.

Although mentioned earlier, one guideline for model selection is worth repeating and remembering: when choosing between otherwise equal alternative models, the simpler model is better. This is an application of the logical principle known as Occam's razor.

A good measure of simplicity of a distribution model is the number of parameters that specify a single distribution from its family of distributions. For example, a three-parameter Weibull distribution applies to certain random variables with a fixed lower bound or threshold value. One of these three parameters is the threshold value. In a two-parameter Weibull distribution, the threshold value is zero. If both three-parameter and two-parameter Weibull distributions fit the available data equally well, the two-parameter model is preferred. Also, the exponential distribution is a special case of the Weibull distribution with only one parameter. If both the exponential and Weibull distributions fit the available data equally well, the exponential model is preferred.

The best model is the simplest model that adequately fits the available data and theory.

1.3 Selecting Candidate Distributions Using Theoretical Knowledge

This section offers suggestions for distribution families that apply to random variables in many common situations. To use these guidelines effectively, think about what is causing the randomness in a process, and what external forces may be shaping the distribution of observed data. Scan the tables in this section to identify distribution families that apply to similar situations.

Here is the first question to answer: Does data produced by the random process represent a count? Examples of count data include counts of defects, customer contacts, or items sold. Count data could have many possible values or as few as two possible values. An example of a count with two possible values has the value one if a customer buys a product, or zero otherwise. This simple model, called a Bernoulli or Yes-No distribution, may represent any experiment or observation with only two possible outcomes.

Table 1-1 lists candidate distribution families for various types of count data. If the data produced by the random process is not a count, refer to Table 1-2, which lists candidate distribution families for variable data. These tables also list the lower and upper bounds for each family.

In many situations, an unbounded distribution, such as normal, is a satisfactory model for a random variable with a physical bound. When the bounds are far away from probable values, the effect of using an unbounded distribution to represent a bounded random variable is insignificant. Example 1.2 in the next section illustrates this point.

Please note that Tables 1-1 and 1-2 are not definitive or exhaustive lists. While exact theory supports some of the guidelines in the tables, others rely on approximations that are not accurate in all situations. In addition to the guidelines in the tables, many other approximate relationships between distributions can be useful to modelers. The chapters describing each distribution family list many of these relationships.

Table 1-1 Candidate Distribution Families for Count Random Variables

Situation	Examples	Candidate Distribution Families	Lower Bound	Upper Bound
Any experiment with only two possible outcomes	Flip a coin—is heads or tails on top? Does the customer buy the product or not?	Bernoulli (Yes-No)	0	1
Count of independent things that happen per unit of time, space or product	Defects per batch of material, customer calls per hour	Poisson	0	∞
In a set of n two-outcome trials (A or B), how many have outcome A?	Defective units per lot of n units	Binomial	0	n
In a set of n items selected from a finite population of N items, where each item is either type A or type B, how many type A items are in the sample?	Defective units in a sample of n units selected from a finite population of N units, including D defective units	Hypergeometric	0	n
In a series of two-outcome trials (A or B), how many outcomes A occur before the first outcome B?	How many good units are produced before the first defective unit?	Geometric, X_0 version	0	∞

In a series of two-outcome trials (*A* or *B*), how many trials happen before and including first outcome *B*?	How many total units are produced before and including the first defective unit?	Geometric, X_1 version	1	∞
In a series of two-outcome trials (*A* or *B*), how many outcomes *A* occur before the *k*th outcome of *B*?	How many good units are produced before the *k*th defective unit?	Negative binomial, X_0 version	0	∞
In a series of two-outcome trials (*A* or *B*), how many trials happen before and including the *k*th outcome *B*?	How many total units are produced before and including the *k*th defective unit?	Negative binomial, X_k version	*k*	∞
Equally likely to be any count between *A* and *B*	Roll one six-sided die—how many dots are on top?	Discrete uniform	*A*	*B*

Table 1-2 Candidate Distribution Families for Continuous Random Variables

Situation	Examples	Candidate Distribution Families	Lower Bound	Upper Bound
Sum, difference, or average of a large number independent random variables with similar standard deviations	Overall height of a stack of many similar parts with independent heights	Normal	$-\infty$	∞
Equally likely to be any value between A and B	By expert opinion, values could be anywhere between A and B	Uniform	A	B
Bounded between A and B, with one value more likely than any other value	Data or opinion suggests that values are bounded, with one value most likely	Triangular or beta	A	B
Artificial bounds imposed on an underlying distribution	Screening inspections, failure times with scheduled replacement	Truncated versions of normal, extreme value, gamma, Weibull, extreme value, or others	varies	varies
Measurements with a physical lower bound of zero	Times to failure, call times, waiting times, geometric form measurements such as flatness	Exponential, gamma, Weibull, lognormal, loglogistic	0	∞

			τ	∞
Measurements with a physical lower bound other than zero	Hours before a part wears out, when it is not possible to fail before a certain time	Two-parameter exponential, three-parameter versions of gamma, Weibull, lognormal, loglogistic	τ	∞
Right-skewed data	Times, distances, monetary quantities	Exponential, gamma, Weibull with $\alpha < 3.6$, largest extreme value, lognormal, loglogistic	varies	varies
Left-skewed data	Plating thickness or any quantity with economic or technical forces limiting higher values	Weibull with $\alpha > 3.6$, largest extreme value. Also try fitting right-skewed distributions to negated data	varies	varies
Symmetric distribution with heavier tails than a normal distribution	Any process with excess kurtosis > 0	Laplace, Student's t	$-\infty$	∞
Minimum of a large set of values with the same distribution	Material strength, fatigue life	Three-parameter Weibull or smallest extreme value	varies	∞
Maximum of a large set of values with the same distribution	Material strength, fatigue life	Largest extreme value; also, try fitting three-parameter Weibull to negated data.	$-\infty$	Varies

1.4 Selecting a Distribution Family Using Graphical Tools

When a sample of observed data is available, both graphical and statistical testing tools are available to identify the best distribution family. Graphical distribution fitting tools are best to use first, because graphs may provide insight that mere numbers cannot. A good graph reveals whether a distribution model fits, and where it does not fit. Statistical testing tools, described in the next section, quantitatively measure the lack of fit of a candidate distribution family to the available data. The best fitting distribution is consistent with available theory, appears to fit using a graphical assessment, and has insignificant lack of fit according to statistical tests.

In practice, a very large sample size is required to select one distribution family over another, especially when candidate distribution families have similar shapes. Generally, samples of 100 or fewer observations cannot definitively identify one distribution family over others. However, the graphical tools in this section may suggest certain distribution shapes, even with smaller samples. Combining these suggestions with simple theoretical arguments usually leads to an effective and useful distribution model. As additional data become available, the selected distribution model may be reviewed to measure its fit to all available data.

Three graphical tools often used to select distribution families are *histograms, probability plots*, and *quantile-quantile plots*. After brief definitions, examples will illustrate these tools.

- A *histogram* is a chart illustrating the distribution of an observed dataset. To make a histogram, the computer sorts the observed data into bins of equal width, covering the range of values. The histogram is a horizontal bar chart or a vertical column chart of the counts of data in each bin. To use the histogram to select a distribution model, compare the shape defined by the bars to the probability functions of the candidate distributions. Many programs can add overlay curves representing these probability functions to the histogram.
- A *probability plot* is a scatter plot on a special scale used to test the fit of a distribution family to the available data. Each candidate distribution family requires a separate probability plot. Probability plots include a diagonal line representing a perfect fit. When the dots on the probability plot are close to this line, the distribution family fits the data well. In a probability plot, the horizontal scale represents the measured data, and the vertical scale represents the cumulative probability predicted for each value according to the distribution model.

- A *quantile-quantile plot* or *QQ plot* is similar to a probability plot with different scales. The plot points and the diagonal line representing a perfect fit are the same on both plots. The vertical scale on a QQ plot is a linear scale representing values of the data. The horizontal scale represents quantiles which the data would follow according to the distribution model. Since the general appearance and interpretation of a QQ plot and a probability plot is the same, either plot is as effective as the other. However, interpretation of patterns on the two types of plots is different. For example, right-skewed data on a normal QQ plot curves in the opposite direction as it would on a normal probability plot.

Crystal Ball software provides a comparison chart including a histogram of the data with an overlaid probability curve for selected distributions. Many statistical packages produce histograms, probability plots, or quantile-quantile plots. MINITAB and JMP provide histograms and probability plots, while STATGRAPHICS provide histograms and quantile-quantile plots.

Example 1.2

Patricia supervises a grinding process and tracks all Critical To Quality (CTQ) characteristics on control charts. For one CTQ, the grinding process produces a mean of 10.0 mm with a standard deviation of 0.1 mm. Figure 1-6 is a histogram of 2000 observations, showing a symmetric, bell-shaped distribution. This plot, generated by Crystal Ball software, includes an overlaid curve representing the probability function of a normal distribution.

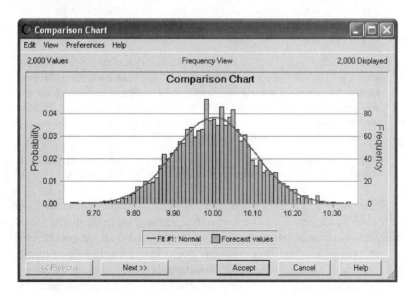

Figure 1-6   Histogram of 2000 Measurements.

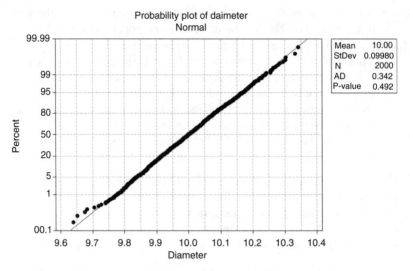

Figure 1-7   Normal Probability Plot of 2000 Measurements.

Figure 1-7 is a normal probability plot of the same data, comparing the observed distribution to a normal distribution. A probability plot is a special type of scatter plot, representing each observation by one dot on the plot. The horizontal scale of the plot represents data values. The vertical scale of the plot represents the cumulative probabilities predicted for each point by the distribution model. That is, the vertical scale represents where each data point falls on a scale from 0% at the minimum to 100% at the maximum.

Notice that the vertical scale is nonlinear. The distortion of this scale is designed so that the dots will line up on a straight diagonal line if the random variable has a normal distribution. If the random variable has some other distribution, the dots will depart from the straight line. Figure 1-7 includes a diagonal line representing the ideal fit to a normal distribution model. Except for some minor irregularities in the tails, the normal model fits this dataset well.

Since both plots indicate that the normal distribution model fits the data well, Patricia decides to accept the normal model for the grinding process.

In Example 1.2, Patricia chooses a normal distribution model, even though the process cannot be normally distributed. Physical size measurements must be positive numbers, but the normal distribution includes all positive and negative numbers as possible values. In this example, the physical bound of zero is 100 standard deviations below the mean, and the probability of observing a negative value with a normal model is negligible.

In other situations, if the physical bound were only a few standard deviations below the range of observed values, one could use a naturally bounded distribution family such as gamma, or a truncated normal model. If it is possible to generate impossible values in a simulation, it is important to select a distribution model that includes these limits.

Table 1-2 suggests several candidate distributions for processes with a lower bound of zero. In most cases, these distributions generally skew to the right, meaning that their right tail representing higher values is longer than their left tail representing lower values. This right skew naturally occurs in many processes where a physical lower bound is close to the range of observed values.

In Example 1.1, Jim examined a set of 1000 flatness measurements. This example continues below with an explanation of why Jim selected a two-parameter exponential model to represent the distribution of flatness.

Example 1.3

Jim creates a histogram and probability plot to examine whether the customary normal model fits the flatness data. The histogram in Figure 1-1 shows that the data skews to the right, and is clearly not normally distributed. Figure 1-8 is a normal probability plot, which confirms this conclusion. The curve of the dots up and to the right of the straight line is characteristic of a distribution that skews to the right more than the hypothetical model.

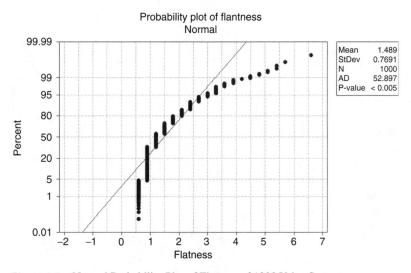

Figure 1-8   Normal Probability Plot of Flatness of 1000 Valve Seats.

The probability plot reveals another important feature of this dataset. The dots form vertical clusters occurring at regular intervals in the plot. To see why this happens, Jim looks at the data itself. Here are the first 20 observations:

0.9	0.9	1.2	0.9	1.2	1.8	1.5	0.9	0.9	0.9
0.6	1.8	1.2	0.6	1.2	0.9	1.2	1.2	0.9	1.8

Every observation is a multiple of 0.3 μm, which is the resolution of the measurement system. With the technology used to measure flatness in this example, it is not possible to record any value other than a multiple of 0.3 μm. Figure 1-9 is a histogram of the dataset with narrower bins than Figure 1-1, showing this feature of the distribution.

This is an example of what Wheeler (2003) calls "chunky data." For Jim, the chunkiness of this data is an irrelevant annoyance. The chunkiness is a result of measurement system limitations, not the manufacturing process itself. Figure 1-10 illustrates this distinction. The true values of flatness have a smooth distribution, but the measurement system produces what appears to be a discrete distribution. Jim's primary objective is to improve the manufacturing process. If the chunkiness were severe enough to make the measurements useless, it would be important to Jim, but this is not the case here.

Even so, the chunkiness of this data is annoying. The chunkiness makes it harder to use either graphical or statistical tools to select a distribution model. The clusters of points in the probability plot (Figure 1-8) make it difficult to visualize whether the distribution model fits the manufacturing process distribution.

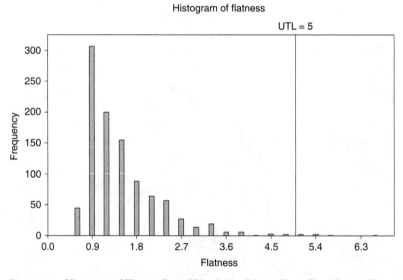

Figure 1-9   Histogram of Flatness Data, Using Many Bins to Show Chunkiness of Data.

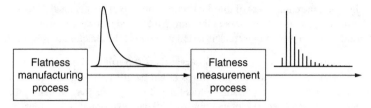

Figure 1-10   Chunkiness is a Result of Imperfect Measurements, and is Not Caused by the Underlying Process.

The histogram in Figure 1-1 uses bin widths that match the resolution of the measurement system. This particular histogram hides the chunkiness. Jim chooses this histogram to present to his boss, because Jim feels that the chunkiness is irrelevant to his project, and he decides not to draw undue attention to it.

Jim needs to select a distribution model to represent the manufacturing process. Since the flatness process has a physical lower bound of zero, Table 1-2 recommends exponential, gamma, Weibull, or lognormal models. Jim creates probability plots for these four models as shown in Figure 1-11. MINITAB software produces these graphs with its Stat ⇨ Quality Tools ⇨ Individual Distribution Identification function.

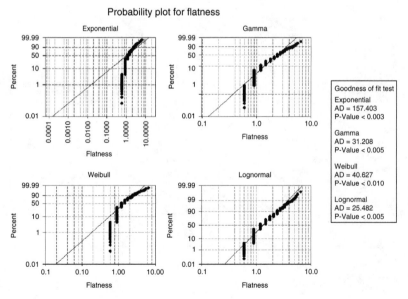

Figure 1-11   Four Probability Plots of Flatness Data with Different Distribution Families.

Ignoring the chunkiness of this data, a probability plot for the ideal distribution model would have the diagonal line through the middle of each vertical cluster of dots. None of the four models tested in Figure 1-11 are close to this ideal.

Jim notices that the minimum observation is 0.6 μm, and he wonders if the physical lower bound for this process might be greater than zero. To test this idea, he generates probability plots for the two-parameter exponential and three-parameter versions of gamma, Weibull, and lognormal. These four distributions have an additional parameter representing the physical lower bound for the process.

Figure 1-12 shows these four probability plots. Again ignoring the chunkiness, all of these plots indicate an acceptable fit to the data. Since all four of these models are equally good, Jim chooses the simplest model, which is the two-parameter exponential distribution.

Figures 1-13 and 1-14 are QQ plots of the same data used in the previous example, showing the fit of ten distribution models to that data. STATGRAPHICS software produced these plots, as part of its Analyze ⇨ Variable Data ⇨ Distribution Fitting ⇨ Fitting Uncensored Data function. A comparison of these plots to Figures 1-11 and 1-12 illustrates some of the differences between probability plots and QQ plots.

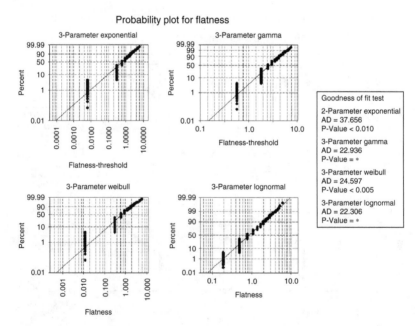

Figure 1-12   Four Probability Plots of Flatness Data with an Additional Threshold Parameter.

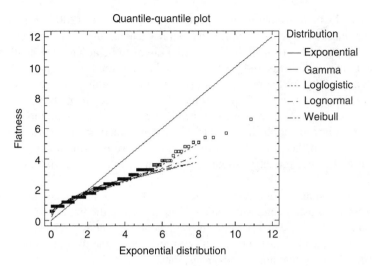

Figure 1-13   QQ plot of Flatness Data versus Five Distribution Families, All with a Lower Bound of Zero.

Measurement systems producing chunky data can cause serious problems, depending on the application. Imagine a scale that reports weight to the nearest 100 lb or to the nearest 100 kg. Such a scale might be acceptable for measuring the weight of trucks, but it would be useless for measuring the weight of people.

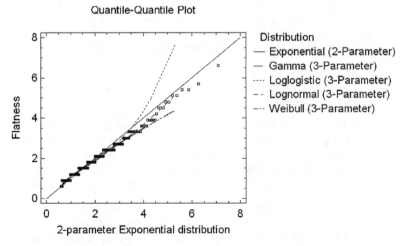

Figure 1-14   QQ plot of Flatness Data versus Five Distribution Families, All with a Variable Lower Bound Parameter.

When do measurements become too chunky? AIAG (2002) refers to chunkiness as "discrimination" and reports a general rule of thumb that "the measurement instrument discrimination ought to be at least one-tenth of the range to be measured." The range to be measured may represent the tolerance range or the process width. The process width usually refers to a range of six standard deviations. If the primary measurement objective is to separate acceptable parts from unacceptable parts, this rule of thumb means that the discrimination should be less than one-tenth of the tolerance range. Otherwise, if the primary measurement objective is to maintain process control, the discrimination should be less than 0.6 standard deviations.

As the pressures of competition, customer needs, and improvement initiatives increase, measurement systems are often pushed to the limits of available technology. Even when superior measurement technology is available, it is not economically feasible to use the best technology for all applications. No matter how far technology advances, these observations will still be true. Therefore, practitioners will always have to deal with chunky data.

Discrete or count data, illustrated in the following example, is inherently chunky. Many very useful distribution models are available for count data, especially in common Six Sigma applications.

Example 1.4

Bob is a process engineer for a company manufacturing microprocessors for personal computers. Microprocessors are manufactured on wafers before being tested and cut into individual chips. Each wafer may have several defects caused by a variety of random causes. Any defect on a chip makes that chip useless. To predict yield and plan economic order sizes, Bob needs a distribution model to represent the number of defects per wafer.

For one relatively new process, Bob collects the number of defects found on each of 500 recently manufactured wafers. Table 1-3 lists the frequency of each value observed in the dataset.

To find a candidate distribution model, Bob consults Table 1-1, which lists models for count distributions. Table 1-1 suggests the Poisson distribution for any "count of independent things that happen per unit of time, space, or product." Defects per wafer are counts of things that happen per unit of product, and they are probably independent of each other. Therefore, the Poisson is a reasonable candidate model.

To test the fit of the Poisson model, Bob enters his data into a STATGRAPHICS DataBook. Bob selects the Analyze ⇨ Variable Data ⇨ Distribution Fitting ⇨ Fitting Uncensored Data function. Bob clicks the Analysis Options button and selects the Poisson distribution from the menu of discrete and continuous distribution models.

Table 1-3 Frequency of Observations of Defects per Wafer

Defects per Wafer	Frequency
1	4
2	11
3	26
4	47
5	63
6	67
7	74
8	65
9	46
10	37
11	31
12	12
13	13
14	3
15	1

Figure 1-15 shows a histogram of Bob's data, with overlaid Poisson probability points. Closeness of these points to the histogram bars indicates a good fit for the Poisson model.

Figure 1-16 is a quantile-quantile or QQ plot showing how well the Poisson model fits this data. On this graph, the closeness of the data points to the diagonal line indicates a good fit. Based on these two graphs and the theoretical justification, Bob accepts the Poisson model as an acceptable fit for this data.

For a variety of technical reasons, many statistical tasks involving discrete distributions are more difficult than the same tasks with continuous distributions. Among the software products mentioned in this book, STATGRAPHICS software is the only product offering distribution fitting tools for several discrete distributions. The MINITAB Stat ⇨ Basic Statistics menu includes a Goodness-of-Fit Test for Poisson, which is illustrated in the next section.

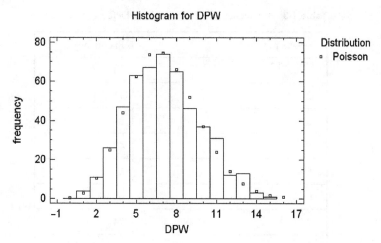

Figure 1-15 Histogram of Defects per Wafer with Poisson Probability Model.

1.5   Selecting a Distribution Family Using Statistical Tools

Numerous statistical tools are available to help modelers select a distribution that best fits a dataset. Collectively, these tools are called *goodness-of-fit tests.* A more accurate description would be *badness-of-fit tests,* since each test measures how badly one hypothetical distribution model fits the available data.

This section offers general advice on selecting, applying, and interpreting a selected few of these tools. At the end of this section are formulas and technical information about selected tools, but this is not intended to be a

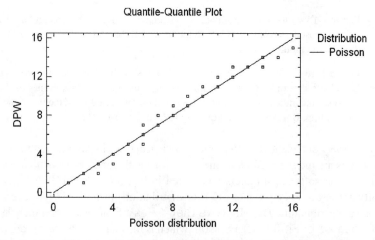

Figure 1-16 QQ plot of Defects per Wafer with Poisson Probability Model.

comprehensive reference. To learn more about the huge variety of goodness-of-fit tests, refer to D'Agostino and Stephens (1986) or Huber-Carol (2002). On this and many other statistical topics, the NIST/SEMATECH e-handbook is a very useful online resource.

The advice in this section is intended for typical Six Sigma and other modeling applications, with the following general observations:

- The modeler is more interested in selecting the most useful distribution family from several choices than in proving or disproving that one distribution family fits the data.
- The modeler estimates parameter values that select a particular distribution from a distribution family before testing whether that distribution fits the data. The alternative to this procedure is when the parameters are fixed and are not estimated from the data. In practice, it is unusual to know parameter values in advance.

All statistical tools are sensitive to sample size. Goodness-of-fit tests are particularly sensitive to the quantity of data, in ways that may be surprising. With small samples, smaller than approximately 100 observations, goodness-of-fit tests may not find enough lack of fit to reject any distribution model tested. In these cases, it is especially important to test only candidate distribution models that are consistent with available theory. After testing a select few models, one can accept the model with the least lack of fit.

Paradoxically, very large samples also cause problems for goodness-of-fit tests. These tools measure the lack of fit between the data and a hypothetical distribution model. Samples of more than 1000 observations frequently have distributions that cause every known distribution family to be rejected for relatively minor deviations from the model. These minor distribution characteristics may have no significant effect on decisions that have to be made. When every known distribution family is a bad fit to the data, the least bad fit is probably good enough.

1.5.1 Selecting a Tool for Testing a Distribution Model

The three most commonly used tools for testing the fit of distribution models are the *chi-squared* (χ^2) test, the *Kolmogorov-Smirnov* (KS) test, and the *Anderson-Darling* (AD) test. Table 1-4 summarizes some of the common features and individual strengths of these tools.

Figure 1-17 summarizes these observations into a flow chart for selecting the most appropriate goodness-of-fit test. Here is an explanation of some of the decisions and tasks in the flow chart.

Table 1-4 Comparison of Three Goodness-of-Fit Test Procedures

	Chi-Squared (χ^2) Test	Kolmogorov-Smirnov (KS) Test	Anderson-Darling (AD) Test
What does the test do?	The χ^2 test sorts the data into bins and measures how well the counts of data in the bins match the counts predicted by the distribution model	The KS test measures how well the observed distribution of the data matches the distribution model	The AD test measures how well the observed distribution of the data matches the distribution model
Does the test focus more on certain parts of the distribution?	Each bin receives equal weight Observations in bins with smaller counts have more influence	Observations in the middle of the distribution have more influence than in the tails	Each observation has approximately equal weight
Does the test provide a unique answer?	No—The test statistic depends on the number and width of bins selected for the test	Yes	Yes
Does it apply to continuous distributions?	Yes	Yes	Yes
Does it apply to discrete distributions?	Yes	No	No
What does the test statistic mean?	Lower values mean better fit Higher values mean worse fit	Lower values mean better fit Higher values mean worse fit	Lower values mean better fit Higher values mean worse fit
Are critical values and P-values available?	Available for all distributions	Only available for selected distribution families	Only available for selected distribution families

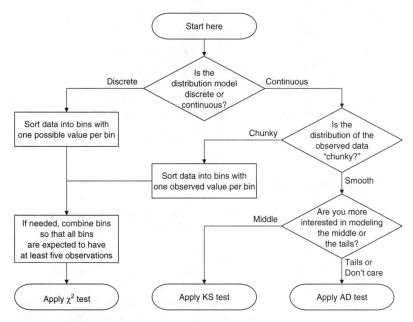

Figure 1-17   Flow Chart to Select a Goodness-of-Fit Test Procedure.

- For discrete data, the χ^2 test is the only choice. This test requires a set of bins to contain the data like a histogram, except that the bins need not have equal width. When defining bins for discrete data, start with one bin for each possible value of the distribution. Using the distribution model, calculate the expected number of observations in each bin. If the expected count of observations in any bin is fewer than 5, combine this bin with neighboring bins until the expected count of observations in all bins is at least 5.
- For continuous data, construct a histogram, dot graph, or probability plot to visualize whether the data is chunky or smooth. Chunky continuous data takes only a small number of possible values, usually spaced at regular intervals. Smooth continuous data takes a large number of possible values.
- For chunky continuous data, the χ^2 test is the best choice. This test requires a set of bins to contain the data like a histogram, except that the bins need not have equal width. When defining bins, start with one bin for each observed value in the dataset. Using the distribution model, calculate the expected number of observations in each bin. If the expected count of observations in any bin is fewer than 5, combine this bin with neighboring bins until the expected count of observations in all bins is at least 5.

- For smooth continuous data, either the KS or AD tests are good choices. To choose one tool over another, consider how the model may influence future business decisions. If the middle of the distribution is more important than its tails, use the KS test. If the tails of the distribution are more important, use the AD test. In most Six Sigma applications, quality metrics depend more on the tails than on the middle of a distribution, so the AD test is generally more appropriate.

Section 1.5.3 provides formulas and additional details about the χ^2, KS, and AD tests.

1.5.2 Interpreting Goodness-of-Fit Test Results

Here are three ways to interpret the results of a goodness-of-fit test and to make a decision about whether the model fits the data:

- *Test statistic method.* When comparing distribution models from different families, simply compare the test statistic generated by the χ^2, KS, or AD tests. For all these tests, a smaller value indicates a better fit, and a larger value indicates a worse fit. The distribution family with the lowest test statistic fits the data best. However, use caution when comparing distribution families with different numbers of parameters. Adding a parameter to the distribution will always lower the test statistic, even if the more complex model fits no better than the simpler model.
- *P-value method.* For goodness-of-fit tests, the P-value is the probability that the hypothetical distribution model could produce a dataset with at least as much lack of fit as the observed data. If the P-value is small, typically less than 0.05, this indicates that the hypothetical distribution model is a bad fit. P-values are convenient, because they always have the same range of values (0 to 1) and the same interpretation. P-values are also useful to compare distribution families with different numbers of parameters. However, P-values are not available for all distribution families.
- *Critical value method.* Published tables list critical values for specific tests and for specific distribution families. To use a critical value, compare the test statistic to the critical value. If the test statistic is greater than the critical value, then the hypothetical distribution model is a bad fit. D'Agostino and Stephens (1986) includes some of these tables. Since most people use a computer to fit distributions, and since statistical programs provide test statistics, P-values, or both, the critical value method is rarely used today.

For most Six Sigma modeling applications comparing several distribution models, the test statistic method is best, with the P-value method as a useful backup. P-values are certainly easier to remember, because the interpretation of a P-value is consistent with other types of statistical tests. However, reliance only on P-values often results in confusing dilemmas. The following examples illustrate some of the interpretation problems that may arise.

Example 1.5

In Example 1.2, Patricia examined a dataset containing 2000 observed values from a grinding process. Figures 1-6 and 1-7 show a histogram and probability plot of this data.

Patricia selected a normal distribution model, but she would like to support that choice with an appropriate goodness-of-fit test. According to the flow chart in Figure 1-17, continuous data requires either the KS or AD test. Since Patricia needs this model to predict process quality metrics, and the tails of the distribution drive these metrics, she selects the AD test.

MINITAB software, which generated Figure 1-7, automatically performs the AD test as part of any function that produces a probability plot. These functions include Stat ⇨ Quality Tools ⇨ Individual Distribution Identification and Graph ⇨ Probability Plot. The probability plot in Figure 1-7 reports the AD test results in a text box to the right of the plot. For this data and a hypothetical normal distribution model, the AD test statistic is 0.342, with a P-value of 0.492.

Since Patricia is only considering a normal distribution model, there are no alternative AD test statistics for comparison. However, the P-value provides a standardized way of interpreting the test. Since there is a high probability (P = 0.492) that a normal random variable could generate a dataset with at least as much lack of fit as this one, the normal model is a reasonable choice. If the P-value were small, usually less than 0.05, this would be reason for Patricia to look for another model.

In the probability plot, the dots depart from the line at the lower end. According to the AD test, this departure is not statistically significant.

Suppose Patricia decided to consider alternative distribution models. Figure 1-18 is another MINITAB graph testing the fit of four different distribution families to this dataset. All four families have two parameters, so the AD test statistics are directly comparable. Comparing the four AD statistics, Patricia sees that the gamma distribution has a lower AD statistic than the normal, indicating that the gamma is a better fit. The P-values for both gamma and normal are greater than 0.05. Therefore, both are acceptable fits according to statistical and graphical tools. Which should she choose?

The MINITAB Session window provides additional information to make this decision. The report includes the following table with parameter estimates for each distribution family.

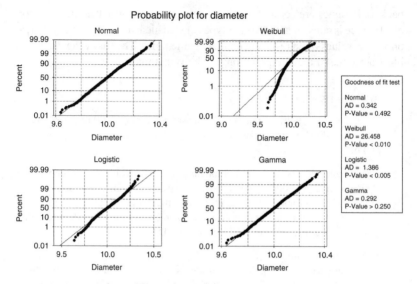

Figure 1-18   Four Probability Plots of Diameter Data.

ML Estimates of Distribution Parameters

Distribution	Location	Shape	Scale	Threshold
Normal	10.00046		0.09980	
Weibull		99.65989	10.05012	
Logistic	9.99991		0.05688	
Gamma		10048.47082	0.00100	

Based on this report, Patricia should include one of the following sentences in her report or presentation about this grinding process:

- The diameter is normally distributed with a mean of 10.00 mm and a standard deviation of 0.01 mm.
- The diameter has a gamma distribution with a shape parameter of 10,048 and a scale parameter of 0.001.

Since normal distributions are more familiar to people in her Six Sigma company, Patricia chooses the normal model to make communication easier.

As the scale parameter of a gamma distribution gets large, the shape of a gamma random variable converges to the same shape as a normal random variable. This fact further supports Patricia's choice of a normal model in the above example.

Example 1.6

In Example 1.4, Bob collected 500 observations of defects per wafer. As a visual analysis, the histogram in Figure 1-15 and the QQ plot in Figure 1-16 suggest that the Poisson distribution is a reasonable fit. The χ^2 test is recommended for testing the fit of discrete distribution models. In MINITAB, this test is provided by the Stat ⇨ Basic Statistics ⇨ Goodness-of-Fit Test for Poisson function. Figure 1-19 lists the report of this function in the Session window.

Near the bottom of Figure 1-19, MINITAB reports that the χ^2 test statistic (labeled Chi-Sq) is 10.7798, with a P-value of 0.768. Since the P-value is larger than 0.05, there is no reason to reject the Poisson distribution model for this dataset.

Example 1.7

Paula is an RF engineer for a company making antennas for satellite phones. As part of a simulation to predict the likelihood of dropped calls, she needs to select a distribution model to represent antenna gain at the carrier frequency.

Paula has gain measurements on a prototype build of 100 antennas. Despite Paula's advanced training, none of the theory available to her suggests that any distribution family is more appropriate than another. So, Paula decides to test the available data against all the available distribution models.

Goodness-of-Fit Test for Poisson Distribution

```
Data column: DPW

Poisson mean for DPW = 7.086

                     Poisson              Contribution
DPW  Observed  Probability  Expected      to Chi-Sq
  0         0    0.000837    0.4184          0.41837
  1         4    0.005929    2.9646          0.36165
  2        11    0.021007   10.5034          0.02348
  3        26    0.049618   24.8091          0.05716
  4        47    0.087899   43.9494          0.21175
  5        63    0.124570   62.2850          0.00821
  6        67    0.147117   73.5586          0.58478
  7        74    0.148925   74.4624          0.00287
  8        65    0.131910   65.9550          0.01383
  9        46    0.103857   51.9286          0.67686
 10        37    0.073593   36.7966          0.00112
 11        31    0.047407   23.7037          2.24589
 12        12    0.027994   13.9970          0.28493
 13        13    0.015259    7.6295          3.78043
 14         3    0.007723    3.8616          0.19224
 15         0    0.003648    1.8242          1.82422
 16         1    0.002706    1.3528          0.09202


  N   N*  DF   Chi-Sq   P-Value
500    0  15  10.7798     0.768


WARNING: 1 cell(s) (5.88%) with expected value(s) less than 1.  Chi-Square
         approximation probably invalid.

5 cell(s) (29.41%) with expected value(s) less than 5.
```

Figure 1-19   MINITAB Report of Goodness-of-Fit Test for Poisson Distribution.

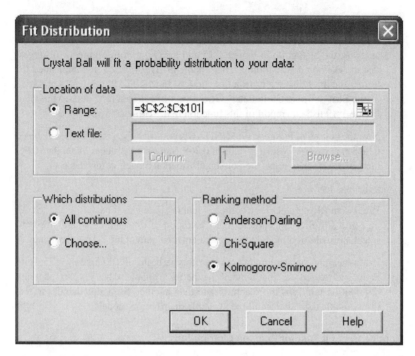

Figure 1-20   Crystal Ball Fit Distribution Form.

Crystal Ball software, which Paula is using to build her simulation model, has some convenient tools for selecting distribution models. After selecting Define ⇨ Define Assumption, Paula clicks the Fit ... button. In the Fit Distribution form shown in Figure 1-20, she can specify the location of the gain data, which distributions to fit, and which ranking method to use.

Which is the best goodness-of-fit test for this application: AD, χ^2, or KS? Paula follows the flow chart of Figure 1-17. Gain is a continuous random variable, and since the measurement system reports gain to five decimal places, the observed distribution is more smooth than chunky. This excludes the χ^2 test. Paula decides that accurately modeling the middle of the gain distribution is more important than modeling the tails, since most gain values are from the middle, and these will be more important for her simulation. Therefore, she selects the KS test.

After clicking OK, Paula sees the Comparison Chart in Figure 1-21. For this data, the distribution with the lowest KS statistic is the gamma distribution. Therefore, Paula clicks Accept to create a Crystal Ball assumption with a gamma distribution and parameters chosen to fit the observed gain data.

In Example 1.7, each of the three testing tools pointed to a different distribution family. This is a very common situation, especially with relatively small datasets. When this happens, it is important to understand the nature

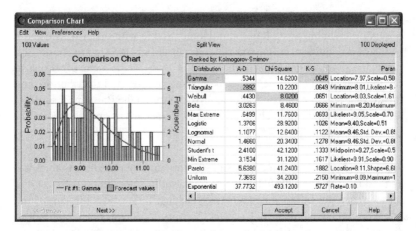

Figure 1-21   Crystal Ball Comparison Chart with Goodness-of-Fit Test Results.

of the data and the objectives of the modeling process to make a rational and defensible decision.

Chunky data, illustrated in Example 1.1 and 1.3, pose particular problems for goodness-of-fit tests. In Figures 1-11 and 1-12, the text boxes report the results of the AD test for eight distribution families. When P-values are available, the test rejects every choice. Both AD and KS tests react to the distance between the points on a probability plot and the diagonal line. In chunky data, these distances are artificially large, and it becomes impossible to use AD or KS tests to fit models to the underlying process distribution.

The χ^2 test is the best available testing choice for chunky data, but none of the software options mentioned in this book provide an easy way to do this. To apply the χ^2 test to chunky data requires control of the bin boundaries, to match the bins to the chunks. Until software improvements make this task easier, manual application of the χ^2 test formulas in the following section is the only option for chunky data.

1.5.3 Calculating Goodness-of-Fit Test Statistics

The χ^2 (chi-squared) test is probably the original goodness-of-fit test, developed by Karl Pearson (1900). In its current form, the χ^2 test uses modifications by Fisher (1922). To perform the χ^2 test, first sort the data into b bins covering the range of observed values. The test compares the observed count in each bin (o_i) with the count expected if the data were a random sample from the hypothesized distribution (e_i). If the differences between observed and expected counts are too large, the χ^2 test will reject the hypothesized distribution.

The theory behind the χ^2 test is based on the assumption that the count of observations in bin i is a Poisson random variable with mean e_i. Since the standard deviation of a Poisson random variable is the square root of the mean, the standard deviation of the count is $\sqrt{e_i}$. As the mean of a Poisson random variable gets large, its shape resembles a normal distribution, and the normal distribution is a useful approximation to the Poisson. Therefore, $z_i = \frac{o_i - e_i}{\sqrt{e_i}}$ is approximately distributed as a standard normal random variable for large values of e_i.

The sum of squares of b independent standard normal random variables is a χ^2 random variable with b degrees of freedom. In a goodness-of-fit test, the terms z_i are not independent for two reasons. First, the total sum of the counts is a fixed number N. To correct for this dependency, subtract 1 from b, the total degrees of freedom. Second, the calculations for the expected counts e_i require k parameters calculated from the same data sorted into the bins. To correct for this dependency, subtract k additional degrees of freedom. As a result, $\chi^2 = \sum_{i=1}^{b} \frac{(o_i - e_i)^2}{e_i}$ is approximately distributed as a χ^2 random variable with $b-1-k$ degrees of freedom. Therefore, $\chi^2 = \sum_{i=1}^{b} \frac{(o_i - e_i)^2}{e_i}$ is the test statistic for the χ^2 goodness-of-fit test. Table 1-5 lists the number of degrees of freedom of the χ^2 test statistic for several distribution models.

For distribution families not listed in Table 1-5, the number of degrees of freedom is $b-1-k$, where b is the number of bins and k is the number of parameters estimated from the data.

To calculate the P-value from the χ^2 test statistic, look up the probability that a χ^2 random variable with $b-1-k$ degrees of freedom will have a value greater than the test statistic. After calculating the test statistic χ^2, the Excel formula to calculate the P-value is =CHIDIST(χ^2, $b-1-k$).

The χ^2 test relies on the approximation of Poisson counts by normal random variables, which requires large expected counts in each bin. As a rule of thumb, each bin should have an expected count of at least five observations. Bins may be combined with neighboring bins to increase the expected counts to five or more.

One of the disadvantages of the χ^2 test is its non-uniqueness. A different choice of bins results in a different χ^2 test statistic and P-value. The use of an automated binning algorithm without any options for user intervention addresses this issue by providing reliable and objective test statistics. Even

Table 1-5   Degrees of Freedom for the χ^2 Goodness-of-Fit Test Statistic

Distribution Family	Parameters Estimated	DF for χ^2 Test Statistic
Beta	α, β only; A and B fixed	$b - 3$
	α, β, A, and B estimated	$b - 5$
Binomial	p (probability); n is fixed	$b - 2$
Exponential	λ (rate) or μ (mean)	$b - 2$
Two-parameter exponential	β (scale) and τ (threshold)	$b - 3$
Gamma	α (shape) and β (scale)	$b - 3$
Three-parameter gamma	α (shape), β (scale) and τ (threshold)	$b - 4$
Laplace	β (scale) and τ (threshold)	$b - 3$
Largest extreme value	η (mode) and β (scale)	$b - 3$
Lognormal	μ and σ	$b - 3$
Three-parameter lognormal	μ, σ, and τ (threshold)	$b - 4$
Normal	μ (mean) and σ (standard deviation)	$b - 3$
Poisson	λ (rate)	$b - 2$
Smallest extreme value	η (mode) and β (scale)	$b - 3$
Weibull	α (shape) and β (scale)	$b - 3$
Three-parameter Weibull	α (shape), β (scale) and τ (threshold)	$b - 4$

so, different software products use different algorithms, and may calculate different χ^2 test statistics and P-values for the same data.

Crystal Ball software uses an automated binning algorithm for the χ^2 test, which assures that the expected count of observations in each bin is at least five. In this algorithm, the number of bins b is the integer part of $1.88N^{0.4}$. For very small samples, b uses an alternate formula. After selecting the number of bins, the bin boundaries are calculated from the hypothesized distribution so that the expected counts of values in each bin are constant. This process results in bins of unequal width, but of roughly equal counts.

Figure 1-22 illustrates the difference between a histogram (top) and the Crystal Ball binning algorithm (bottom) both created from the same data. The bottom graph would be inappropriate and unethical if presented as a histogram, because people perceive area in a graph to represent relative probability. Histograms must have equal width bins. However, all the bins illustrated in the lower graph have the same expected counts of observations, which makes the χ^2 test work very well.

As noted above, an automatic binning algorithm for the χ^2 test provides consistent results any time that software is used. However, the χ^2 test results will not be the same as those calculated by different software with a different binning algorithm. Also, for certain situations like chunky data, it would be helpful to have user control over bin boundaries.

Kolmogorov (1933), Smirnov (1939) and others proposed a goodness-of-fit test which compares the empirical cumulative distribution function (ECDF) of the dataset to the hypothesized distribution. The ECDF will be defined

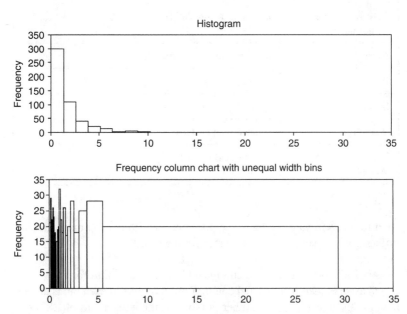

Figure 1-22   Two Charts from the Same Data, with Equal and Unequal Bin Widths.

shortly. Massey (1951) compared the KS test to the χ^2 test and found it superior in some cases. Many classify the KS test as a nonparametric procedure, although those who really care about labels might argue with that.

The KS test was the first of many goodness-of-fit tests based on the ECDF. Anderson and Darling (1952 and 1954) proposed a new ECDF test to address some perceived shortcomings in the KS test. The AD test has become very popular today. In many of its functions involving distribution fits, MINITAB software uses only the AD test. STATGRAPHICS software provides several options for goodness-of-fit tests, including KS, AD, and others.

For most practitioners who use data to make decisions, either the KS or the AD are very good tools. Both tests have unique results determined only by the values in the sample, thus avoiding one of the inherent weaknesses of the χ^2 test. Also, both the KS and AD tests are extensively researched and documented, and they have settled into the accepted statistical toolbox.

For any dataset, the *empirical cumulative distribution function* (ECDF) is a function of x expressing the proportion of values in a dataset that are less than or equal to x. In symbols: $E(x) = \frac{n_{\leq}(x)}{N}$, where $n_{\leq}(x)$ is the count of values which are less than or equal to x, and N is the total size of the dataset. Therefore, $E(x)$ is a nondecreasing function with a stairstep form, increasing from $E(-\infty) = 0$ to $E(+\infty) = 1$.

For any random variable X, the *cumulative distribution function* (CDF) is a function of x expressing the probability of observing values less than or equal to x. In symbols: $F_X(x) = P[X \leq x]$. Therefore, $F_X(x)$ is a nondecreasing function, increasing from $F_X(-\infty) = 0$ to $F_X(+\infty) = 1$.

If a dataset is a sample from a random variable X, it follows that the ECDF $E(x)$, representing the observed distribution, should be fairly close to the CDF $F_X(x)$, representing the hypothetical distribution. This is the basic idea behind the KS test, the AD test, and many others.

Figure 1-23 shows four graphs, all showing the ECDF of a dataset of 30 observations. These graphs are alternatives to probability plots as a visual analysis of distribution fits. In these graphs, the ECDF is drawn with a bold line. For comparison, each graph also shows the CDF of a distribution chosen from a different family, with parameters estimated from the data. The CDF that fits the data best will be closest to the ECDF. Looking only at Figure 1-23, which model would you choose? Take a moment to select one model, using the ECDF graphs.

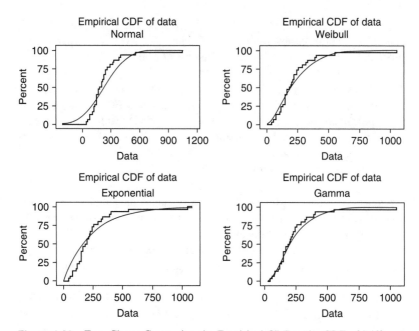

Figure 1-23   Four Charts Comparing the Empirical CDF to the CDF of Different Distribution Families.

Now look at Figure 1-24. This figure contains four probability plots and AD test results. Did you select the same model from both figures? Which style of graph is easier to use? Probability plots are a more effective tool for visual analysis, because humans can see departures from a 45° line easier than from lines at other angles.

The KS test statistic is the maximum vertical distance between any point on the ECDF and the corresponding value of the CDF. In practice, the first step of the KS test is to estimate the parameters for the random variable X from the selected family. Once $F_X(x)$ is known, here are the formulas to calculate the KS test statistic D:

$$D^+ = \max_{1 \leq i \leq N}\left[\frac{i}{N} - F_X(x_i)\right]$$

$$D^- = \max_{1 \leq i \leq N}\left[F_X(x_i) - \frac{i-1}{N}\right]$$

$$D = \max\{D^+, D^-\}$$

For many common distribution families, the distribution of D is known, and tables of P-values or critical values are available. Many of these are in

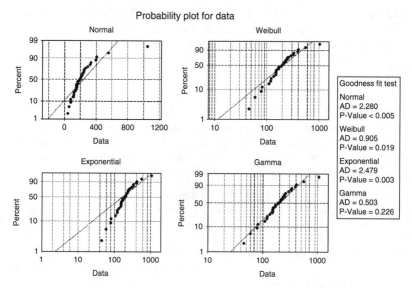

Figure 1-24   Four Probability Plots of the Same Data and Families Shown in Figure 1-23.

D'Agostino and Stephens (1986). Statistical software that automates the KS test has these tables built in, so they are not included in this book.

Most distributions have a section with high relative probability, such as the middle section of the normal distribution. At these points, the CDF curve is steepest. When the slope of the CDF is high, the differences between the CDF and ECDF tend to be larger. As a result, the KS test is more sensitive to distribution deviations in the highly probable sections, and less sensitive to deviations in the less probable sections, such as the tails. When the tails are of more interest than the middle section, this is a disadvantage for the KS test.

The AD test uses a more complex test statistic, given by this formula:

$$A^2 = -n - \frac{1}{n}\sum_{i=1}^{n}(2i - 1)[\ln F_X(x_i) + \ln(1 - F_X(x_{n+1-i}))]$$

This formula gives every observation an equal opportunity to influence the test statistic A^2. P-values and critical values for A^2 depend heavily on the distribution family. For many common families, researchers have compiled tables and derived formulas to calculate P-values and critical values by Monte Carlo simulation and regression analysis. The formulas used to calculate P-values for the AD normality test are given in Sleeper (2006) and D'Agostino and Stephens (1986).

Stephens (1974) studied the performance of the KS test, the AD test, and three other tests based on the ECDF. This is a good reference for those interested in the power of these tests to discriminate between distributions from different families. The work of Stephens leaves little doubt that there is no single best goodness-of-fit test.

Example 1.8

Max is a mechanical engineer designing automotive door latch systems. One purchased component of this system is the latch assembly. This assembly, mounted in the door, latches the door to the striker bar on the pillar. When the occupant pulls the handle to open the door, a cable releases the latch.

One critical characteristic of the latch assembly is release effort, which is the force the cable must exert to release the latch. Release effort has a tolerance of 20–35 N. For reasons unknown to Max, release effort tends to run high and the distribution is skewed to the left. Since the latch is purchased, Max is unable to investigate or improve the latch manufacturing process. To analyze and optimize the design of the handle assembly, Max needs a reliable distribution model for latch release effort.

Figure 1-25 is a histogram of measurements of release effort of 852 latches in four lots. From these histograms, Max concludes that the general shape, scale, and location of the effort distribution is stable from lot to lot.

Max has no theoretical basis to choose one distribution family over another. Max enters the measured data into a STATGRAPHICS DataBook and selects

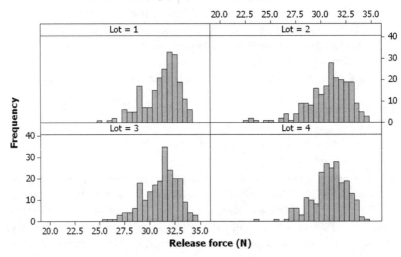

Figure 1-25   Histograms of Latch Release Effort over Four Lots.

Quantile-Quantile Plot

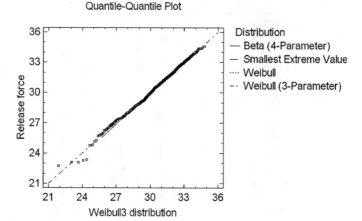

Figure 1-26   Quantile-Quantile Plot of Release Effort Data and Four Candidate Distribution Models.

Analyze ⇨ Variable Data ⇨ Distribution Fitting ⇨ Fitting Uncensored Data. Max selects several distributions to fit to the data, and he finds four families with low AD test statistics. Table 1-6 lists the parameters and AD test results for these families. Figure 1-26 is a quantile-quantile plot showing the fit of this data to these four distribution models.

There are fewer distribution models for left-skewed data than there are for right-skewed data. Perhaps left-skewed data occurs less often than right-skewed data, but it does happen. Some distribution families, such as beta and Weibull, include both left-skewed and right-skewed distributions. The right-skewed largest extreme value distribution has left-skewed mirror image, the smallest extreme value distribution.

Modelers working with left-skewed data may find it easier to fit a distribution to the negative value of the data, which reverses the direction of the skew. Many more distribution families are available to fit right-skewed data.

Example 1.8 (Continued)

Max attempts to find a right-skewed model for the negative value of release effort. In the STATGRAPHICS distribution fitting form, Max simply enters "-" before the variable name to do this. Max finds that the four-parameter beta and largest extreme value distributions fit the negated data well, but these are simply mirror images of the models listed in Table 1-6. Table 1-7 lists three additional distribution families that fit the observed distribution of negative release effort.

Table 1-6 Distribution Families Fitting the Release Effort Data

Model	Distribution Family	Parameters	Anderson-Darling Test	
			Test Statistic	P-Value
1	Beta (four-parameter)	shape $\hat{\alpha} = 770$, $\hat{\beta} = 4.97$ threshold $\hat{A} = -619$, $\hat{B} = 35.1$	0.333	≥ 0.1
2	Smallest extreme value	mode $\hat{\eta} = 31.8107$ scale $\hat{\beta} = 1.49332$	0.326	≥ 0.1
3	Weibull	shape $\hat{\alpha} = 20.9394$ scale $\hat{\beta} = 31.7716$	0.390	≥ 0.1
4	Three-parameter Weibull	shape $\hat{\alpha} = 44.1683$ scale $\hat{\beta} = 66.4477$ threshold $\hat{\tau} = -34.6553$	0.295	≥ 0.1

Table 1-7 Distribution Families Fitting Negative Release Effort Data

Model	Distribution Family	Parameters	Anderson-Darling Test	
			Test Statistic	**P-Value**
5	Gamma (three-parameter)	shape $\hat{\alpha}$ = 5.16007 scale $\hat{\beta}$ = 1.22017 threshold $\hat{\tau}$ = −35.1832	0.313	≥ 0.1
6	Loglogistic (three-parameter)	median $\hat{\mu}$ = 4.87859 shape $\hat{\alpha}$ = 0.204874 threshold $\hat{\tau}$ = −36.1328	0.906	≥ 0.1
7	Three-parameter Weibull	shape $\hat{\alpha}$ = 2.05246 scale $\hat{\beta}$ = 4.12702 threshold $\hat{\tau}$ = −34.6101	1.819	≥ 0.1

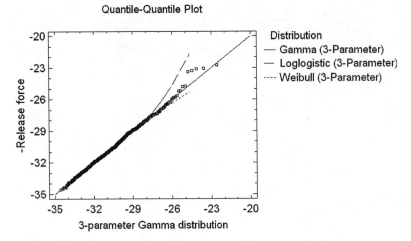

Figure 1-27 Quantile-Quantile Plot of Negated Release Effort Data and Three Candidate Distribution Models.

Figure 1-27 is a quantile-quantile plot illustrating the fit of these three distribution models to negative release effort data.

Max now has seven options for distribution models, all of which are acceptable, according to the P-value for the AD test. Which model should he choose?

A different kind of test for distribution models is to see whether the tails of the distribution model, beyond the ends of the histogram, predict plausible probability values. A good way to do this is to calculate equivalent capability metrics for that model and see whether the values are reasonable when compared to historical experience with the process. If the capability values are not plausible, neither is the model.

This example continues in Chapter 3, to illustrate the calculation and interpretation of nonnormal capability metrics.

1.5.4   Understanding Goodness-of-Fit Tests

Among statistical tests, goodness-of-fit tests pose particular challenges. To understand these challenges, compare a goodness-of-fit test to another procedure often taught to Six Sigma practitioners, the one-sample t-test.

A one-sample t-test can detect whether the mean of a normal distribution is different from a specified value, for example, 10. In statistical terms, the *null hypothesis*, representing no difference, is expressed as H_0: $\mu = 10$. The *alternative hypothesis*, representing a difference, is expressed as H_A: $\mu \neq 10$.

The one-sample t-test procedure will prove the alternative hypothesis, $\mu \neq 10$, if there is sufficient evidence to do so based on a reasonable risk level. If there is insufficient evidence to prove the alternative hypothesis, this does not prove that $\mu = 10$. In this event, the experimenter might conclude that $\mu = 10$, and act accordingly. The experimenter might also decide to collect more data, in the hope of detecting a smaller difference from 10.

Statistical tests are similar to a criminal trial, which may prove guilt, but never proves innocence. Unlike a criminal trial, an experimenter always has the option to gather more data and try again.

In a goodness-of-fit test, the null hypothesis H_0 is that the random variable follows one specific distribution, for example, normal with a given mean and standard deviation. The alternative hypothesis H_A is that the random variable follows some other distribution. If there is sufficient lack of fit, the test procedure will prove H_A and conclude that the one specific distribution model does not fit the data. But if there is insufficient lack of fit, the goodness-of-fit test proves nothing. In this event, the practitioner may conclude that hypothetical model fits the data, and act accordingly.

For statisticians, goodness-of-fit tests are problematic because the alternative hypothesis is so big. In a normality test, the null hypothesis is that the data has one specific normal distribution. The alternative hypothesis includes all other normal distributions, all other continuous distributions, all discrete distributions, all distributions which are both continuous and discrete, and countless universes of distributions that have yet to be described. It is not possible to determine if one normal distribution model is better than the countless alternative models.

In the simpler problem of testing H_0: $\mu = 10$ versus H_A: $\mu \neq 10$ for a normal distribution, the one-sample t-test is the "best" test by specific statistical criteria. However, among goodness-of-fit tests, there can never be a "best" test for all applications because of the immense variety of alternative distributions.

Within limitations, such as when testing one distribution against selected alternatives, one test procedure can be proven by analysis or simulation to be better than others. For example, the test proposed by Shapiro and Wilk (1965) has certain advantages for testing the fit of a normal distribution. JMP software applies the Shapiro-Wilk procedure to test the fit of normal distributions in certain situations. For different distribution models, JMP applies different tests. If the most important question is to decide whether one specific distribution fits the data or not, one should select the test best suited to answering that question.

However, this is not the usual question facing modelers. When building a model to represent a physical system, one must select a distribution from among several choices to represent each input variable. The modeler must choose one model from among a limited set of models.

This problem is less like a criminal trial and more like a beauty contest. A beauty contest will not send everyone away if all the contestants are less than beautiful. Or, if all contestants are equally beautiful, the contest will still pick a single winner. Every beauty contest has close decisions among many contestants, all of whom may be beautiful, but none of whom is perfect. This is the same decision facing modelers.

At its foundation, the theory of statistical inference is a set of tools for disproving hypotheses. These methods are poorly suited to judging a beauty contest. This mismatch results in confusing and useless results such as Example 1.3, in which the AD test rejects every distribution tested. There are other cases where tests reject none of the distribution models. In either case, the modeler must ignore the statistical tests and pick a winner anyway.

In time, smart researchers will develop better tools for selecting distribution models. Until then, here is some general advice about goodness-of-fit tests:

- Always plot the data. With the aid of intelligent, thinking viewers, graphical tools are more flexible and powerful than statistical tools.
- Chunky data renders most goodness-of-fit tests useless. The χ^2 test may be adapted for chunky data, but today's software does not provide automated tools for this procedure. Perhaps none of today's goodness-of-fit procedures can be effective with chunky data.
- Pay more attention to goodness-of-fit test statistics, and less attention to P-values. Be aware of the fact that distributions with more parameters will have smaller test statistics, indicating a better fit, even if the fit is no better than a simpler alternative.
- Always look for the simplest model that adequately fits the data and is consistent with available theory about the process.

1.6   Selecting a Distribution Model with Expert Opinion

In Figure 1-5, the flow chart for selecting distribution models, the last step is to apply expert opinion to modify the chosen distribution. Sometimes, when neither theory nor data exist, expert opinion is the first and only step. Monte Carlo simulation and other computerized models of complex processes require full specifications of input variables, often including unknown and unknowable parameters. While data and established theory ought to outweigh pure opinion, opinion will always be an important source of modeling knowledge.

Here are some general guidelines for modelers in the use of expert opinion:

- *Understand the objective of the project.* There are many ways to interpret and apply opinion. Incorporate opinion into the model in a manner consistent with the overall business objective. Consider two general types of projects involving modeling and simulation.
  - The objective of some projects is to predict what is likely to happen. Examples of this are a weather forecast or a prediction of interest rates. For this type of project, the most likely outcomes are more important to the business than the less likely extreme outcomes. When using expert opinion in a model like this, focus on the most likely values of each variable predicted by opinion. It is not necessary to bias the opinion one way or another.
  - The objective of other projects is to predict what might go wrong. One example is a simulation of a business plan, where investors may be particularly interested in the downside risk. Another example in DFSS applications is tolerance analysis. This type of analysis predicts the probability of failure, either a failure to assemble or a failure to function. In these projects, use conservative or safe opinions in the analysis, intentionally biasing opinions in a way that may increase the predicted probability of failure. If data later replaces opinion, the use of conservative opinions minimizes the likelihood of unhappy surprises.
- *Recognize that relatively few variables are important in a simulation.* Because of the Pareto principle, only a vital few variables in a simulation will dominate the results, while the trivial many have insignificant effect. Often modelers do not know in advance which variables are vital. To minimize wasted effort, start the simulation with very basic, simple guesses for all input variables. Then perform a sensitivity analysis to determine which inputs are vital, and gather better information for those few inputs.
- *Avoid biasing the project with the modeler's own opinion.* When opinion really matters to the outcome of a project, perform a careful survey of multiple acknowledged experts in the field. For greater credibility, these experts should not include people with a vested interest in the success or failure of the project.

This section discusses four common areas where modelers use expert opinion to select or modify distribution models in a variety of projects.

1.6.1   Truncating a Distribution Model

Truncating a distribution model is setting limits beyond which the probability of values is zero. For example, Figure 1-28 illustrates a truncated normal distribution. A random variable with a normal distribution could have any value between $-\infty$ and $+\infty$. The distribution in the figure is truncated at 4.00 and 6.00. A simulation of this distribution will only produce values between 4.00 and 6.00.

Truncation not only eliminates the probability of values outside the limits, but also increases the probability of values inside the limits. As a result, truncation alters the mean, standard deviation, and other distribution characteristics. In Figure 1-28, the untruncated normal distribution has a mean of 5.30 and a standard deviation of 0.50, but truncation changes these distribution parameters significantly.

Truncation is usually the wrong way to solve a modeling problem. The following two examples illustrate the perils of two common applications of truncation.

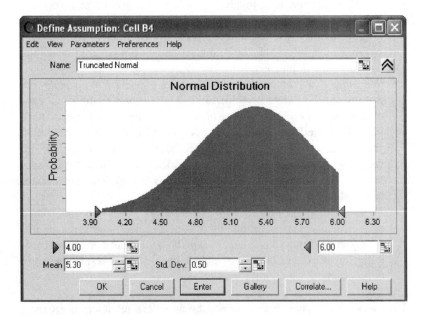

Figure 1-28   Truncated Normal Distribution.

Example 1.9

For his Six Sigma Black Belt training project, Fred is studying the flatness of vinyl floor tiles manufactured by his company. Many customers have complained about warped tiles, which create installation problems. Flatness is a measure of the amount of warp in a tile. Fred performs a temperature test and measures flatness on a sample of tiles after the test. He estimates that the mean flatness is $\hat{\mu} = 0.44$ mm and the standard deviation of flatness is $\hat{\sigma} = 0.42$ mm. Following the teachings of Fred's Six Sigma training, he assumes that flatness has a normal distribution and calculates that the probability that flatness exceeds the specification limit of 1.00 mm is 8.9% or 89,000 DPM.

Fred also observes that 14.5% of the normal distribution fit to this data is less than zero. This cannot be true, since the measured flatness must be a positive number.

In the example, Fred should not ignore the fact that the normal distribution model predicts that 14.5% of the measurements are impossible, negative numbers. A common reaction to this problem is to truncate the normal distribution at zero, as shown in Figure 1-29.

Truncation solves the problem with impossible values, but introduces another serious problem. The mean and standard deviation of the truncated distribution are no longer the same as the values Fred estimated from the sample. If a normal random variable with mean μ and standard deviation σ is

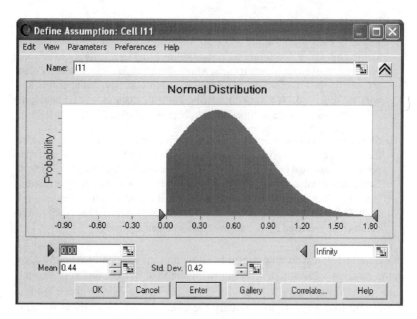

Figure 1-29   Normal Distribution Fit to the Data, with Impossible Values Truncated.

truncated on the left at A, the mean and standard deviation are given by these formulas from Section 21.2 of this book:

$$E[X] = \mu + \sigma\left[\frac{\phi\left(\frac{A - \mu}{\sigma}\right)}{1 - \Phi\left(\frac{A - \mu}{\sigma}\right)}\right]$$

$$SD[X] = \sigma\sqrt{1 + \frac{\left(\frac{A - \mu}{\sigma}\right)\phi\left(\frac{A - \mu}{\sigma}\right)}{1 - \Phi\left(\frac{A - \mu}{\sigma}\right)} - \left[\frac{\phi\left(\frac{A - \mu}{\sigma}\right)}{1 - \Phi\left(\frac{A - \mu}{\sigma}\right)}\right]^2}$$

In the above formulas, $\phi(x)$ represents the standard normal probability density function (PDF), and $\Phi(x)$ represents the standard normal cumulative density function (CDF). Excel calculates either $\phi(x)$ or $\Phi(x)$ using the =NORMDIST function. Substituting $\mu = 0.44$, $\sigma = 0.42$, and $A = 0$ into the above formulas, the truncated normal random variable has a mean of $E[X] = 0.55$ and a standard deviation of $SD[X] = 0.34$. These values are far off the mean and standard deviation of Fred's sample.

The normal distribution family is the wrong choice to model this data, and truncating off the impossible negative values only made the model worse.

Example 1.9 (continued)

Instead of using the obviously wrong normal distribution model, Fred applies the distribution fitting functions of Crystal Ball or MINITAB software to select a better model. Fred finds three models, which all have acceptable probability plots and nearly equal AD test statistics. These models are an exponential distribution with mean $\mu = 0.44$, a Weibull distribution with shape $\alpha = 1.03$ and scale $\beta = 0.45$, and a gamma distribution with shape $\alpha = 1.03$ and scale $\beta = 0.43$.

Since all three models appear equally good to Fred, he chooses the simplest model, the one-parameter exponential distribution. To analyze capability of an exponentially distributed process in MINITAB, Fred uses the Stat ⇨ Quality Tools ⇨ Capability Analysis ⇨ Nonnormal function, and specifies an exponential distribution. This function produces the graphical report shown in Figure 1-30. MINITAB predicts that the defect rate is 102,907 DPM, which is nearly 15% more defects than predicted by the incorrect normal capability analysis.

In example 1.9, it would be useful to know that the exponential distribution is the same as a gamma distribution with shape $\alpha = 1.00$, and it is also the same as a Weibull distribution with shape $\alpha = 1.00$. So, the three models Fred identified were almost exactly the same distribution. Even without knowing these facts, Fred chose the exponential model because it is the simplest model of the three.

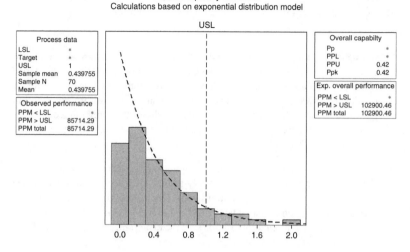

Figure 1-30   Exponential Process Capability Analysis.

The wide variety of distribution models provides great freedom for Six Sigma practitioners to choose models consistent with known physical boundaries for a system. Many systems have physical boundaries on the lower side, the upper side, or on both. When a physical boundary affects the shape of a distribution, modelers should select a bounded distribution rather than truncating an unbounded distribution.

Inspection is another process commonly modeled by truncation. Many processes are simply incapable of meeting tolerance requirements, and screening inspections are required to prevent the sale of bad parts.

When the intent of an inspection process is to remove parts outside certain limits, a modeler may choose to truncate a distribution model to represent this process. However, using truncation to represent inspection is naïve.

Inspection processes are subject to errors of many kinds, including measurement errors, inspection errors, and others. Measurement errors include the lack of accuracy and the lack of precision. According to Chapter 23 of Juran (1999), human inspection errors include technique errors, inadvertent errors, and conscious errors. For more information on measurement systems analysis, see Chapter 5 of Sleeper (2006) or AIAG (2002).

The bottom line is that all measurements are wrong, and no inspection system is perfect. For these reasons, prevention is far superior to inspection as a way of assuring product quality.

Nevertheless, many processes still rely on inspection systems, even in "Six Sigma" companies. This creates challenges for modelers.

Example 1.10

Rhoda is designing the measurement portion of an automated test system. Since measurement accuracy is Critical To Quality (CTQ), she performs tolerance analysis on all analog circuits. Following the recommendations in Chapter 11 of Sleeper (2006), she models every component with a uniform distribution between tolerance limits. Sensitivity analysis shows that only one component dominates accuracy, and this is a bandgap voltage reference, with a tolerance of 2.000 V ± 0.002 V.

Rhoda decides to measure the voltage on a sample of these references. Since the manufacturer of the reference loudly trumpets its Six Sigma program, she expects to see a normal distribution, hopefully with a good C_{PK} value. She measures voltage on 95 parts, which happen to be in stock. Instead of the high capability normal distribution, Rhoda sees the histogram in Figure 1-31. This is disturbing for at least two reasons:

- Of the 95 parts measured by Rhoda, five parts are outside the tolerance limit of 1.998–2.002 V.
- The sample contains very few parts in the middle of the distribution, closer to 2.000 V.

By viewing the histogram, Rhoda arrives at these conclusions about the manufacturing process for this part:

- The manufacturer relies on measurement to screen out unacceptable references.

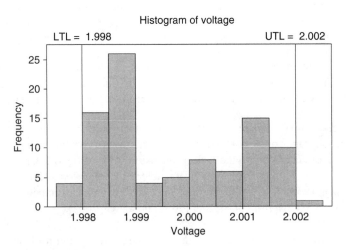

Figure 1-31   Histogram of Reference Voltage Dataset.

- Either the manufacturer's measurement system or Rhoda's measurement system is imprecise or inaccurate.
- The manufacturer appears to be selling the parts in the 2.000 V ± 0.001 V range to someone else. This option is not on the datasheet, but a quick call to the sales representative confirms that the tighter tolerance part is available, at four times the price.

Before deciding whether to buy the premium part, Rhoda must create a reliable model of the current part. There is no simple distribution family for bimodal distributions. The beta family includes some bimodal distributions, but these do not fit Rhoda's data well. Rhoda decides to create a compound distribution model in Crystal Ball software, illustrated in Figure 1-32.

The first part of the compound distribution is a uniform distribution between 1.998 and 2.002, excluding the middle section, from 1.999 to 2.001. This represents the distribution of acceptable parts, assuming that the manufacturer sells all the parts in the middle section to a higher bidder.

The second part of the compound distribution represents measurement error at the manufacturer. Rhoda has asked for gage repeatability and reproducibility (gage R&R) statistics, but the manufacturer refuses to provide them. So, Rhoda assumes that a standard metric, $GRR_{\%Tol}$, is at the limits of acceptability, which is 30%. Applying the definition of $GRR_{\%Tol}$ from Sleeper (2006), page 287, Rhoda calculates the maximum standard deviation of measurement system precision:

$$GRR_{\%Tol} = \frac{5.15\,\sigma_{GRR}}{UTL - LTL} \times 100\% = 30\%$$

$$\sigma_{GRR} = \frac{GRR_{\%Tol}(UTL - LTL)}{5.15 \times 100\%} = \frac{30\%(0.004)}{5.15 \times 100\%} = 0.000233\,V$$

In her Excel worksheet with the simulation model, Rhoda sums the bimodal uniform distribution with the normal measurement system error, and estimates a model for reference voltage as shown in Figure 1-32. With this model, Rhoda can evaluate the business case for buying the reference with a tighter tolerance. She might also consider looking for another supplier with better process control.

In distribution modeling, truncation is more often misused than used correctly. Truncation is certainly the wrong way to account for physical boundaries in a process. Many distribution models are available with bounds on the lower end, the upper end, or both. These are more reasonable and useful models than a boundless distribution with artificial truncation.

Truncation is a reasonable model for an ideal screening or inspection process. Unfortunately, no screening or inspection process is ideal. When possible in the design of new products, choose components and processes

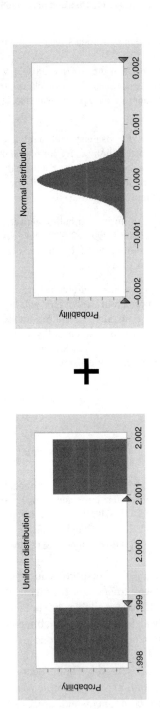

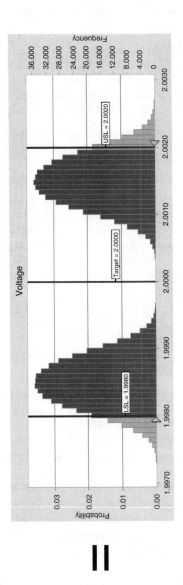

Figure 1-32   Adding Two Distributions to Simulate the Effect of Measurement Error.

with demonstrated capability to meet tolerance requirements without inspection. Because of the risk of errors, inspection is always more costly than it seems to be.

1.6.2 Modeling the Effects of Long-Term Variation

The objective of most distribution models is to represent variation over a long time, perhaps over the entire life of a product or process. In most cases, only short-term data is available to identify a distribution model. Long-term variation is always greater than short-term variation, for a variety of unpredictable reasons. Over time, the short-term distribution of any process may exhibit changes in its average, variation, or shape. Any adjustment to a distribution model to account for unknown changes is justifiable only as expert opinion.

If long-term data is available, no adjustment is necessary. Further, estimating the short-term variation within long-term data is quite important. The difference between short-term and long-term variation indicates how such processes typically change over a long time. This knowledge is useful in adjusting future short-term models into long-term models. Capability analysis performed by MINITAB and other software automatically estimates both short-term (C_P and C_{PK}) and long-term (P_P and P_{PK}) capability metrics.

When the only data available for a new process represents a short-term sample, what is the most appropriate way to adjust the model into a long-term model? There are four general opinions as to how to do this:

- **Make no adjustment.** Practitioners and companies have many reasons for choosing this approach. Many people do not know or do not believe that long-term distributions are different from short-term distributions. Other companies which once applied the 1.5-σ mean shift rule advocated in Six Sigma initiatives have discarded it. The net effect of this choice is that predictions of defect rates are always biased low, guaranteeing unhappy surprises. Long-term variation is always worse than short-term variation.
- **Shift the mean.** In most Six Sigma initiatives, long-term predictions of defect rates assume that the mean (of a normal distribution) shifts by as much as 1.5 standard deviations in either direction. This approach works for estimating long-term defect rates, but it is generally impractical for simulation, since the direction of the mean shift is unknown. The particular choice of 1.5, and not some other number, remains controversial. Bothe (2002) and Harry (2003) articulate some theoretical arguments to support this number, but it remains essentially an expert opinion.

- **Expand the Variation.** As an alternative to mean shifting, multiply the short-term standard deviation of a distribution model by a constant c. The net effect of this change is to increase the predicted long-term defect rate. Unlike mean-shifted distributions, the expanded distribution is easy to incorporate into any Monte Carlo simulation. Suggested by Evans (1975) and others before him, this method predates Six Sigma. By coincidence, the value $c = 1.5$ applied to a normal distribution of typical capability results in similar increases in defect rates to a mean shift of $Z_{shift} = 1.5$ standard deviations. Page 373 of Sleeper (2006) provides a formula to convert between c and Z_{shift} for a normal distribution, holding defects constant.
- **Follow the Data.** Analyzing long-term data from a similar process is the best way to develop a reasonable and credible adjustment for long-term variation. A similar process uses similar methods, technology, and process control techniques. Working from long-term data, JMP, MINITAB, STATGRAPHICS, or any SPC program can estimate the short-term standard deviation $\hat{\sigma}_{ST}$ and the long-term standard deviation $\hat{\sigma}_{LT}$. The estimated variation expansion coefficient for this data is $\hat{c} = \hat{\sigma}_{LT}/\hat{\sigma}_{ST}$.

Example 1.11

Todd is a mechanical engineer designing valves. In his current project, Todd is designing a larger version of a product his company now manufactures. One CTQ of the new valve is a ground inner diameter, with a target value of 7.500mm. From a pilot production run of 35 valves, Todd estimates that the mean diameter is $\hat{\mu} = 7.5008$mm with a short-term standard deviation of $\hat{\sigma}_{ST} = 0.0061$mm. Like all pilot production runs, Todd's data represents only short-term variation.

To understand the effects of long-term variation, Todd refers to the current manufacturing data on a smaller valve. The current production valve has a similar ground inner diameter with a target value of 5.000 mm. Since this is a CTQ characteristic, the manufacturing team maintains process control data. Twice each day, the team pulls a subgroup of five consecutive parts and measures the CTQ diameter using an air gage. An $\overline{X}$, s control chart from this data alerts the team when the process changes.

Todd consults the company SPC database and pulls this process data for 200 subgroups over the previous 100 days. Todd expects this duration to be sufficient that the sample will include all long-term sources of variation. Analyzing this data in MINITAB, Todd produces a capability analysis graph shown in Figure 1-33.

In Figure 1-33, the text box to the left of the graph reports the estimates of short-term and long-term variation. The short-term standard deviation, labeled StDev(Within), is $\hat{\sigma}_{ST} = 0.00510$. The long-term standard deviation, labeled StDev(Overall), is $\hat{\sigma}_{LT} = 0.00644$. From this data, Todd estimates the variation expansion coefficient to be $\hat{c} = \hat{\sigma}_{LT}/\hat{\sigma}_{ST} = 1.26$.

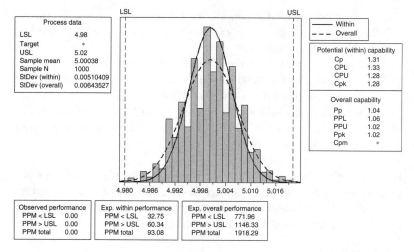

Figure 1-33 Capability Analysis with Short-Term (Within) and Long-Term (Overall) Statistics.

Finally, Todd applies this knowledge to the short-term sample of the new part, and estimates that $\hat{\sigma}_{LT} = c\hat{\sigma}_{ST} = 1.26 \times 0.0061 = 0.0077$

1.6.3 Applying Opinions in the Absence of Data and Theory

This chapter has illustrated many ways to select distribution models from observed datasets or by applying known theory about a process. When neither data nor theory is available, expert opinion is the only way to select a distribution model.

Very often, data exists, but it is not cost-effective to collect and process the data. For example, a tolerance analysis of a complex system may involve thousands of individual part characteristics. Even if this data were available, an engineer may choose to assume that every characteristic has a uniform distribution between its tolerance limits. After completing an initial analysis, the engineer sees that only a few characteristics are important to the outcome of the analysis. If needed, the engineer can then gather data on those vital few characteristics. This is an example of intelligently balancing workload, collecting data only where it matters and using opinion where it does not matter.

When opinion does matter to the outcome of a project, it deserves careful collection and handling. Observe the general guidelines listed at the start of this section. Here are specific questions to ask when eliciting opinions for distribution modeling.

- Think about the random values that we might observe in the future. Recall any relevant data or experience you have accumulated. Then answer these questions:
  - Is any one value more likely than all other values? (Yes or No)
  - If Yes, which value is most likely to occur?
  - What is the maximum value that may occur?
  - What is the minimum value that may occur?
  - Please describe any of your data or experience which supports your answers to these questions.

If one value is more likely than another value, a triangular distribution is a reasonable choice. If no value is more likely than another value, a uniform distribution is a reasonable choice. Without any data to support a more complicated distribution model, simpler models are preferred.

Example 1.12

Jill, a marketing engineer, is building a marketing analysis for a proposed new product. A critical assumption in her analysis is the market share the product will achieve two years after introduction. Jill can estimate market share from the experience of past products, but since each product and each market segment is different, these estimates do not apply to the new product. Any estimate is essentially a matter of opinion.

Jill has her opinion, but for credibility, she decides to survey her colleagues. She sends an email survey to three other marketing engineers, Paul, Raul, and Saul. In the email, she provides her plan, including product features, pricing, distribution plans, advertising plans, market segmentation data, and many other details. Then she asks each colleague for an opinion about the distribution of market share, using the questions listed above. Jill's plan is to combine all four opinions, three of her colleagues plus her own, by giving each opinion equal weight.

Before reading anyone else's answers, she answers the questions herself. Table 1-8 summarizes the responses.

Table 1-8   Responses to Jill's Market Share Survey

Question	Jill	Paul	Raul	Saul
One value more likely?	Yes	No	No	Yes
Most likely value?	10%	N/A	N/A	15%
Highest value?	25%	10%	20%	30%
Lowest value?	5%	2%	15%	0%

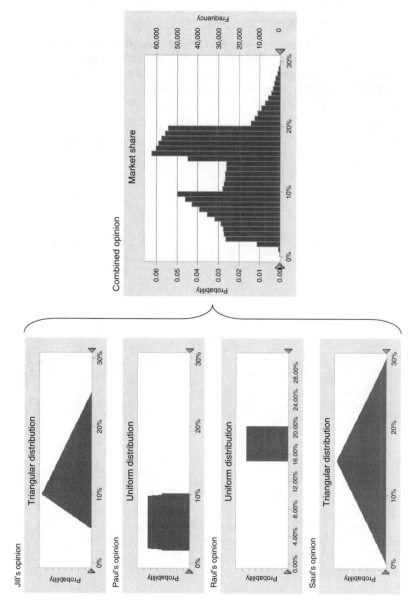

Figure 1-34 Forming a Distribution Representing Combined Opinions by Randomly Selecting One of Four Opinions.

In her Crystal Ball model, Jill defines one assumption variable for each opinion. Then, she defines a discrete uniform assumption equally likely to be 1, 2, 3, or 4. Finally, she computes market share by selecting one of the four opinions based on the discrete uniform value. She does this with an Excel VLOOKUP function, although nested IF functions would also work.

Figure 1-34 illustrates the process of combining these four opinions into one joint opinion. The resulting distribution represents the opinions of four experts, combined in a fair manner. Because of the process, the result is more credible and reliable than if it relied upon any single opinion.

Selecting Statistical Software Tools for Six Sigma Practitioners

Statistical software is an essential component of every Six Sigma initiative. Although statistical software has been an important business tool for decades, until recently, most users have had specialized statistical training and skills. Because of Six Sigma and other improvement initiatives, businesses now expect employees at all levels to use software for data visualization and analysis, and to make better decisions by using statistical tools. Statistical software is now a required tool for diverse users, including many with very little statistical training, and even some who dislike statistical methods.

Because of these challenges, the selection of statistical software is a crucial task for companies deploying Six Sigma and other data-driven initiatives. Although some companies choose to buy whatever software is sold or recommended by their Six Sigma consultant, this may not be the best choice. The management team of each company is the ultimate customer of the chosen statistical software, because they will see the graphs, review the analysis, and make the decisions supported by the software. Choosing the wrong software causes employee frustration, wasted time, and possibly incorrect decisions. Management must be actively involved in the software selection process.

The objectives of statistical software will vary between companies, but usually include these:

- *Cost.* Cost includes both the license fees and training costs. Consider ongoing costs for support and upgrades in future years. Support and upgrades are not optional. No software product is perfect, and the best ones are improved in each release.
- *Ease of use.* How easy is it for the least-trained user to perform simple tasks such as creating a histogram or a control chart? How easy is it for advanced users to apply the software in their jobs? How easy is it to update reports as new data becomes available? How easy is it to understand the

reports and graphs produced by the software? These are only a few usability issues that deserve careful consideration.

- *Input compatibility*. Input compatibility is the ability to accept raw data. Ideal statistical software would link to any source data, in whatever location or format it is maintained, with automatic updates as new data becomes available. The usual alternative to this ideal is to import data through copy and paste operations or by other means.
- *Output compatibility*. Output compatibility is the ability to produce charts and reports in any format required. Most statistical software accomplishes this by copying text and images to the clipboard. The user can then paste them into Microsoft Office PowerPoint, or into any other program. Ideal statistical software would create useful reports and presentations without any other software.
- *Software functionality*. Every software product provides a limited set of features, with strengths and weaknesses. With today's technology, companies require different tools for online, offline, and specialized types of data analysis. Very likely, there will never be *one* statistical software tool for all purposes.
- *Software quality*. The quality of statistical software is often overlooked in the decision process. One reason for this is that many people do not know how to assess software quality, or what questions to ask. This chapter provides some guidance, and highlights specific areas where today's software offerings are lacking. Another reason for the lack of focus on software quality is that many users are more tolerant of defects in statistical software than in other types of software. If a word processing program produced error messages whenever a user enters certain unusual words, this would be unacceptable. Yet, statistical software that produces error messages or incorrect answers from valid input data is widely accepted for general use. The quality of statistical software ought to be extremely important.

This chapter provides information about selected capabilities and quality of a few statistical software products that are popular among Six Sigma practitioners. To the extent possible, statements about each software product are believed to be accurate, and each software manufacturer has had the opportunity to review and correct these statements. The inclusion of any product here is not an endorsement, nor is the exclusion of any product to be interpreted negatively.

The quality of statistical software is often difficult to measure. All software contains defects, of varying levels of seriousness. Annoying defects result in error messages instead of correct answers. Dangerous defects result in incorrect answers with no error message. The most insidious defects are defective requirements. When software users expect a particular function to behave in a certain way, but the function was designed to behave differently, everyone loses. A different kind of defective requirement occurs when functions are not well defined. Some statistical functions have multiple versions with the same name. Without knowing which version a program calculates, one can be surprised and embarrassed by incorrectly interpreting the results.

Readers of this book are likely to personally experience the defects of statistical software. Even seemingly simple calculations, like the sample standard deviation, are difficult to program so that results are accurate for all practical datasets. Distribution functions referenced in this book are among the most difficult to implement accurately over the full range of possible input values.

This chapter lists differences between major statistical software products for Six Sigma practitioners, with emphasis on descriptive statistics and distribution functions. Also included are lists of known defects and limitations of Microsoft Office Excel and other programs that rely on the Excel user interface.

Since makers of successful software always work to improve their products, newer and better versions of all products listed will soon be available. One should always use the newest version of statistical software, with all available updates.

For most Six Sigma applications, Microsoft Office Excel 2003 (or 2004 for Mac) is acceptable, despite its limitations and defects noted in Section 2.3. Prior versions of Excel, up to and including Excel XP (2002) are not recommended for statistical calculations. Microsoft significantly improved many Excel statistical functions in the 2003 release. However, the Analysis Toolpak (ATP) provided with Excel is not recommended in any version.

Every company using statistical tools for Six Sigma or other purposes also needs a comprehensive statistical program such as JMP, MINITAB, or STATGRAPHICS. These programs not only provide a comprehensive, flexible, and powerful set of statistical tools, but they have been extensively tested and verified.

There are many popular add-ins for Excel providing a wide variety of statistical functions for Six Sigma and other applications. These add-ins range in complexity from fully featured attempts to clone MINITAB in Excel menus to customized templates created by in-house VBA programmers. Similarly, the quality of such software and the thoroughness of verification vary widely.

Excel add-ins either use Excel's built-in functions for statistical calculations, or they use their own algorithms. It is usually safer for an add-in to use Excel's built-in functions, thereby relying on Microsoft's verification procedures. If an add-in uses its own statistical algorithms, those functions require exhaustive testing, which requires a lot of work. Most companies making add-ins do not have the resources to perform sufficient testing on their statistical functions. Some makers of add-ins advertise that they are better than Excel functions because they do not use Excel functions. Years ago, this point had merit, but compared to Excel 2003, it is much less significant.

Before accepting any Excel add-in for widespread use inside a company, it is wise to question the thoroughness of verification procedures. The creation, verification, and support of high quality statistical software is a job for more than a handful of people.

Crystal Ball simulation software is not an add-in, but rather an application using Excel as a user interface. In its 7.x release, Crystal Ball software incorporates new routines for all its statistical functions. Decisioneering, the maker of Crystal Ball software, hired independent experts to validate its code. According to these tests, Crystal Ball's statistical routines are on a par with MINITAB in terms of quality. For more information, read Decisioneering (2006). As with Excel and every other statistical program, it is unwise still to be using older versions of Crystal Ball.

All users of statistical software must realize any software can produce calculation errors, and many of these do not result in error messages. Be particularly cautious when calculating probabilities or quantiles in the extreme tails of any distribution. Section 2.3 explains this point in greater detail. When in doubt, or when major decisions rely upon a statistical calculation, perform the same calculation in different programs and compare answers. Always use common sense to evaluate whether answers provided by statistical software are reasonable.

2.1 Comparison of Descriptive Statistics Functions in Selected Statistical Software

This section compares selected statistical functions of several leading products, at their current revision level: Crystal Ball 7.2, Microsoft Office

Excel 2003, JMP 6, MINITAB 14, and STATGRAPHICS Centurion XV. Selected new features of MINITAB release 15 are also listed, with permission of Minitab Inc. Table 2-1 lists descriptive statistics functions provided by these programs. In some cases, different programs implement different versions of the same function, and users of these functions need to be aware of these differences.

Here are explanations of some of the entries in Table 2-1, where the same function has different versions:

- The *mode* of a sample is the value that occurs most often. Many samples do not have a unique mode. In some samples, every value occurs once. In other samples, two or more values occur the same number of times.
  - Mode version 1, implemented in Excel software, returns an error (#N/A) if every value occurs once. If two or more values occur the same number of times and more than once, the minimum value that occurs most often is returned.
  - Mode version 2, implemented in STATGRAPHICS software, returns a value only if the data contains a unique mode. If two or more values occur the same number of times and more often than other values, no value is returned.
  - Mode version 3 is implemented in MINITAB software starting with release 15. This mode is an optional output of the Stat ⇨ Basic Statistics ⇨ Display Descriptive Statistics function. If every value occurs once, no mode is listed. If two or more values occur the same number of times and more than once, all such values are listed.
- *Quartiles* and *percentiles* of a sample are not uniquely defined. Each of the programs surveyed in Table 2-1 uses a different algorithm to calculate quartiles. Since boxplots represent quartiles, boxplots created by different programs will also look different. Table 2-1 lists the values returned for the first and third quartile of the dataset {1, 2, 3, 4, 5, 6}.
- The *standard deviation* of a sample is estimated by the function $s = \sqrt{\frac{1}{n-1}\sum(X_i - \overline{X})^2}$. A different function is $s_n = \sqrt{\frac{1}{n}\sum(X_i - \overline{X})^2}$. The s_n version is used for the rare situation when the entire population of data is available. As a sample statistic to estimate standard deviation, s is more appropriate than s_n. For more information on the reasons for this, see Chapter 21 on the normal distribution.
- *Kurtosis* is a measure of the shape of a distribution. Two different coefficients of kurtosis are in common use. The "normal = 3" version is always a positive number, which assigns the value 3 to the shape of a normal distribution. The "normal = 0" version subtracts 3, assigning the value 0 to a normal distribution. This book refers to the "normal = 0"

Table 2-1 Descriptive Statistics Functions

Descriptive Statistic	Crystal Ball 7.2	Excel 2003	JMP 6	MINITAB 14	STATGRAPHICS Centurion XV
Mean	✓	AVERAGE	✓	✓	✓
Median	✓	MEDIAN	✓	✓	✓
Mode	Version 1	Version 1: MODE	—	Version 3 (in MINITAB 15)	Version 2
Quartiles of {1, 2, 3, 4, 5, 6}	—	QUARTILE: 2.25, 4.75	1.75, 5.25	1.75, 5.25	2.00, 5.00
Standard deviation	s	s: STDEV s_n: STDEVP	s	s	s
Coefficient of Variation	✓	—	✓	✓	✓
Coefficient of Skewness	✓	SKEW	✓	✓	✓
Coefficient of Kurtosis	Normal = 3 version	Normal = 0 version (excess): KURT	Normal = 0 version (excess)	Normal = 0 version (excess)	Normal = 0 version (excess)

version as *excess kurtosis*. Positive values of excess kurtosis denote fatter tails or a flatter middle than a normal distribution, while negative values denote truncated tails or a peaked middle section. For more information, see the entry for *coefficient of kurtosis* in Chapter 5.

2.2 Comparison of Distribution Functions in Selected Statistical Software

Another area where programs differ is in functions describing probability distributions. Table 2-2 lists which programs provide which functions for distribution families described in this book. For each distribution family, people may require different functions to solve different problems. Here are symbols used in Table 2-2 to represent these functions:

- *P* represents $f_x(X)$, the probability mass function (PMF) for discrete distributions, or the probability density function (PDF) for continuous distributions.
- *C* represents $F_X(x) = P[X \leq x]$, the cumulative distribution function (CDF)
- *S* represents $R_X(x) = 1 - F_X(x) = P[X > x]$, the survival function or reliability function. There are reasons, explained below, why both *C* and *S* functions may be required.
- *I* represents $F_X^{-1}(p)$, the inverse CDF, used to calculate quantiles or random numbers. In general, *p* represents the left-tail probability. Variations of this parameter are noted in footnotes.
- *R* represents random numbers generated according to the distribution.

In Table 2-2, the absence of an entry does not mean that calculations for that distribution are impossible. In most cases, simple functional relationships between distribution families can be used to calculate whatever is required. When a required function is not listed in Table 2-2, see the chapter for the desired distribution family for specific suggestions.

Since the survival function $R_X(x)$ is easily calculated from the CDF $F_X(x)$ by $R_X(x) = 1 - F_X(x)$, it may not be clear why both functions are necessary. The need for both functions is a result of the inherent limitations of digital computers in representing numbers very close to one. For many applications, these limitations have insignificant effect. However, readers of this book will require accurate probability calculations in both tails of distribution models to calculate equivalent process capability metrics. For example, consider a standard normal distribution. What is $F_{N(0,1)}(-8.5)$, the probability of observing a value less than the mean minus 8.5 standard

Table 2-2 Distribution Functions in Statistical Software

Distribution	Chapter	Crystal Ball 7.2	Excel 2003	JMP 6	MINITAB 14	STATGRAPHICS Centurion XV
Beta, two-parameter	6	R	C: BETADIST I: BETAINV	P,C,I,R	P,C,I,R	P,C,S,I,R
Beta, four-parameter	6	R	C: BETADIST I: BETAINV	—	—	P,C,S,I,R
Bernoulli	7	R	see binomial	P,C,R (binomial)	P,C,I,R (binomial)	P,C[1],S,I,R
Binomial	8	R	P,C: BINOMDIST I: CRITBINOM	P,C,R	P,C,I,R	P,C[1],S,I,R
Chi-squared	9	R (gamma)	S: CHIDIST I: CHIINV[2]	P,C,I	P,C,I,R	P,C,S,I,R
Chi	9	—	—	—	—	See half-normal and Rayleigh
Chi-squared, noncentral	9	—	—	P,C,I	C,I	P,C,S,I,R

		R		R[3] (integer)	P,C,I,R (integer)	—
Discrete uniform	10	R	—		P,C,I,R (integer)	—
Exponential	11	R	P,C: GAMMADIST I: GAMMAINV	R (Exponential) P,C,I Weibull	P,C,I,R	P,C,S,I,R
Exponential, two-parameter	11	R (Weibull)	—	—	P,C,I,R	P,C,S,I,R
F	13	—	S: FDIST I: FINV[4]	P,C,I	P,C,I,R	P,C,S,I,R
F, noncentral	13	—	—	P,C,I	C,I	P,C,S,I,R
Gamma	14	R	P,C: GAMMADIST I: GAMMAINV	P,C,I,R	P,C,I,R	P,C,S,I,R
Gamma, three-parameter	14	R		P,C,I	P,C,I,R	P,C,S,I,R
Geometric, X_0 version	15	—	P: NEGBINOMDIST	P,C,R (Negative binomial)	P,C,I,R (in release 15)	P,C[1],S,I,R
Geometric, X_1 version	15	R	—	—	P,C,I,R (in release 15)	—
Half normal	21	—	—	—	—	P,C,S,I,R

(*Continued*)

Table 2-2 Distribution Functions in Statistical Software (*Continued*)

Distribution	Chapter	Crystal Ball 7.2	Excel 2003	JMP 6	MINITAB 14	STATGRAPHICS Centurion XV
Hypergeometric	16	*R*	*P:* HYPGEOMDIST	*P,C*	*P,C,I,R*	*P,C*[1]*,S,I,R*
Laplace	17	—	—	—	*P,C,I,R*	*P,C,S,I,R*
Largest extreme value	12	*R*	—	—	*P,C,I,R*	*P,C,S,I,R*
Logistic	18	*R*	—	—	*P,C,I,R*	*P,C,S,I,R*
Loglogistic	18	—	—	—	*P,C,I,R*	*P,C,S,I,R*
Loglogistic, three-parameter	18	—	—	—	*P,C,I,R*	*P,C,S,I,R*
Lognormal	19	*R*	*C:* LOGNORMDIST *I:* LOGINV	—	*P,C,I,R*	*P,C,S,I,R*
Lognormal, three-parameter	19	—	—	—	*P,C,I,R*	*P,C,S,I,R*

				P,C,R	P,C,I,R (in release 15)	P,C^1,S,I,R
Negative binomial, X_0 version	20	—	P: NEGBINOMDIST	P,C,R	P,C,I,R (in release 15)	P,C^1,S,I,R
Negative binomial, X_k version	20	R	—	—	P,C,I,R (in release 15)	—
Normal	21	R	P,C: NORMDIST I: NORMINV	P,C,I,R	P,C,I,R	P,C,S,I,R
Pareto	22	R	—	—	—	P,C,S,I,R
Poisson	23	R	P,C: POISSON	P,C,R	P,C,I,R	P,C^1,S,I,R
Rayleigh	24	—	—	—	—	P,C,S,I,R
Smallest extreme value	12	R	—	—	P,C,I,R	P,C,S,I,R
Student's t	25	R	S: TDIST I: TINV[5]	P,C,I	P,C,I,R	P,C,S,I,R
Student's t, noncentral	25	—	—	P,C,I	C,I	P,C,S,I,R

(*Continued*)

Table 2-2 Distribution Functions in Statistical Software (*Continued*)

Distribution	Chapter	Crystal Ball 7.2	Excel 2003	JMP 6	MINITAB 14	STATGRAPHICS Centurion XV
Triangular	26	R	—	R^6	P,C,I,R	P,C,S,I,R
Uniform	27	R	R: RAND[7]	R^7	P,C,I,R	P,C,S,I,R
Weibull	28	R	C,P: WEIBULL	P,C,I	P,C,I,R	P,C,S,I,R
Weibull, three-parameter	28	R	—	P,C,I	P,C,I,R	P,C,S,I,R

[1] Instead of the CDF $F_x(x) = P[X \le x]$, STATGRAPHICS software provides $P[X < x]$. The difference between these two is this: $P[X \le x] - P[X < x] = P[X = x]$. For continuous distributions, $P[X = x] = 0$, so this difference does not matter. However, for discrete distributions, the difference may be significant. In the Cumulative Distribution pane of the Probability Distributions window, STATGRAPHICS lists both the "Lower Tail Area (<)" and the "Probability Mass (=)" for discrete distributions. Adding these two values gives $F_x(x) = P[X \le x]$.

[2] In Excel, the CHIINV function calculates the quantile for right-tail probability p.

[3] The Random Integer function in JMP requires the lower bound to be 1.

[4] In Excel, the FINV function calculates the quantile for right-tail probability p.

[5] In Excel, the TINV function calculates the quantile for twice the tail probability p. This is convenient for two-tailed t-tests, but inconvenient because of its inconsistency with other inverse CDF functions.

[6] The JMP Random Triangular function always has a lower bound of 0 and an upper bound of 1.

[7] In Excel and JMP software, random uniform numbers always have a lower bound of 0 and an upper bound of 1.

deviations? According to Excel and MINITAB functions, the answer is 9.47953×10^{-18}, a very small number. Now what is $F_{N(0,1)}(+8.5)$, the probability of observing a value less than the mean plus 8.5 standard deviations? Since the normal distribution is symmetric, we know the answer is $1 - 9.47953 \times 10^{-18} = 0.999\,999\,999\,999\,999\,990\,520$, but both MINITAB and Excel round this number to 1, because the representation of numbers in both programs is limited to 15 digits in the mantissa. Using floating-point representations, numbers very close to zero can be represented with much greater precision than numbers very close to one.

Since the normal distribution is symmetric, it is easy to work around this limitation for normal probabilities using the fact that $F_X(x - \mu) = 1 - F_X(-[x - \mu])$. However, many important families of distributions are not symmetric. For example, consider a gamma distribution with shape parameter $\alpha = 4$ and scale parameter $\beta = 1$. What is the probability of observing a value greater than 45, that is, $R_{\gamma(4,1)}(45)$? When α is an integer, an exact formula is available and is listed in Chapter 14:

$$R_{\gamma(4,1)}(45) = e^{-45} \sum_{i=0}^{3} \frac{45^i}{i!} = 4.65 \times 10^{-16}$$

However, not one of the programs discussed in this book gives this value. Excel, JMP, and MINITAB only provide left tail probabilities $F_X(x)$ for the gamma distribution. These functions all return exactly 1, suggesting that $R_{\gamma(4,1)}(45) = 0$. STATGRAPHICS provides both left-tail probabilities $F_X(x)$ and right-tail probabilities $R_X(x)$ for all distributions, but in this case, the value returned is 3.97×10^{-9}, which is incorrect. The makers of STAT-GRAPHICS are aware of this issue, and they may correct this calculation error in the near future.

Why does all this matter? A specific example will illustrate the importance of these calculations to Six Sigma practitioners. Suppose the surface texture of a valve seat is extremely critical, with a single upper tolerance limit of 45 microns. The machining process produces a surface texture in microns which has a gamma distribution with shape parameter $\alpha = 4$ and a scale parameter $\beta = 1$.

What is the capability index P_{PK} for this process? According to methods explained in Chapter 3, the most appropriate formula is this: Equivalent $P_{PK}^{\%} = -\frac{1}{3}\Phi(R_{\gamma(4,1)}(45))$. The correct answer is 2.68, since a normal process with $P_{PK} = 2.68$ would have the same probability of defects. However, none of the software products mentioned in this book can calculate this value correctly using its gamma distribution functions.

The correct calculation of capability metrics requires better functions for calculating tail probabilities than any of today's leading statistical software can provide. Whatever product can best address this shortcoming in its future releases will enjoy a competitive advantage among modelers and Six Sigma practitioners.

2.3 Defects and Limitations of Microsoft Office Excel Spreadsheet Software

Microsoft Office Excel spreadsheet software contains many useful and important statistical functions. The previous sections list many of these functions. While Excel software is not a comprehensive statistical analysis package, it contains a wide range of commonly used functions. Because of its vast user base, Excel software is surely the most widely used statistical program in the world.

However, Excel software has never enjoyed wide approval among statisticians, because of defects and limitations in its statistical functions. Microsoft has corrected many of these problems, notably in Excel 2003, but some problems remain. This section lists issues described by McCullough and Wilson (1999 and 2005) and Knüsel (1998 and 2005), combined with the personal experience of the author. Classification of each issue as a problem remaining or a problem corrected is the author's opinion.

- Problems remaining in Excel 2003:
  - The random number generator (RNG) used by the Excel RAND() function is improved from earlier versions, with a period length of approximately 2^{43}. This is a big improvement, but in the opinion of some (L'Ecuyer, 1994), modern RNG periods should be 2^{60} or more. McCullough and Wilson (2005) showed that the Excel 2003 algorithm fails six out of sixty tests for uniformity. An early version of Excel 2003 software contained a defect allowing the generation of negative numbers. Microsoft fixed this defect in a hotfix issued in February, 2004. (see Microsoft Knowledge Base article 834520)
  - The random number generator in the Excel Analysis ToolPak (ATP) performs very poorly in standardized tests, and is not recommended for any purpose.

- ○ The ATP contains numerous other problems, detailed in Microsoft Knowledge Base article 829208. Some of these problems have been improved because of improvements in underlying Excel functions.
- ○ The QUARTILE and PERCENTILE functions use an algorithm that disagrees with standard statistical texts, such as Montgomery (2005). According to Montgomery, to compute the first quartile of a set of six numbers, say $\{1, 2, 3, 4, 5, 6\}$, calculate the index of the first quartile $\frac{n+1}{4} = 1.75$. Formally, the first quartile can be any value between the first and second lowest values, 1 and 2. The conventional choice is 1.5, although various programs return different values, as listed in Table 2-1. Excel's QUARTILE and PERCENTILE functions return 2.25, which is outside the range of acceptable values.
- ○ The POISSON and BINOMDIST functions for calculating the cumulative probabilities of the Poisson and binomial random variables were improved in Excel 2003, but continue to have problems. Before Excel 2003, these functions returned exact answers in the left tail of the distribution, but no answer in the middle portion of the distribution. In Excel 2003, new algorithms for these functions produce exact answers in the middle section, but zero values in the left tail. The Excel 2003 version is better, because errors now happen at least six standard deviations away from the mean, where cumulative probabilities are very small. Nevertheless, this defect creates a problem for those wishing to calculate equivalent capability indices for Poisson or binomial processes.
- ○ The CHIINV function, which calculates quantiles of the chi-squared distribution, fails to return a value for certain degrees of freedom above 750. This is a problem for anyone calculating confidence intervals or hypothesis tests on Poisson or exponential data with large observed counts, as in reliability analysis.
- ○ The GAMMADIST function, which calculates probabilities for the gamma distribution, fails to return an answer for $x = 0.1$ and $\alpha = 0.1$, and for other values in the same region.
- ○ The BETAINV function, which calculates quantiles of the beta distribution, can be wrong in the first or second significant digit, for small probability values.
- ○ The NEGBINOMDIST function, which calculates probabilities for the negative binomial distribution expects integer values for the "numbers" parameter, k in this book. Some applications require k to be a noninteger. When k is not an integer, NEGBINOMDIST rounds the value down, and returns an incorrect value, without an error message. Chapter 20 describes a workaround for this problem.
- • Problems corrected in Excel 2003:

○ The **NORMINV** and **NORMSINV** inverse normal functions were inaccurate in the tails of the distribution. This problem was severe in Excel 97 and 2000, improved in XP and fixed in 2003.

○ Univariate statistics, including **VAR** (variance), **STDEV** (standard deviation), **PEARSON**, and **CORREL** (two ways to estimate the correlation coefficient) were significantly improved in Excel 2003. Prior to Excel 2003, **STDEV** and **PEARSON** returned values with zero correct digits for certain unusual datasets. In Excel 2003, these same values are correct to at least 8 and usually 12 digits.

○ Analysis of variance and linear regression computations have improved to the extent that McCullough and Wilson (2005) report that Excel 2003 is acceptable in these areas.

The author's evaluation of the beta version of Excel 2007 suggests that the statistical functions are unchanged from Excel 2003, and that Excel 2007 will inherit remaining statistical issues from Excel 2003. Note that Microsoft may still change Excel 2007 before its release.

Here are general recommendations for Six Sigma practitioners and anyone who wishes to use Microsoft Office Excel statistical functions:

- Do not use Excel XP, 2000, 97, or any version prior to Excel 2003 (or Excel 2004 for Mac). This point is extremely important for the many large companies that are slow to upgrade software for cost reasons. The benefits of upgrading are far more significant than the nicer user interface. Excel versions prior to 2003 have so many defects in their statistical functions that they are unacceptable for any Six Sigma or other data-driven initiative.
- For Six Sigma practitioners and anyone who wants a spreadsheet tool for statistical and other calculations, Excel 2003 and later is acceptable. It is not perfect, but it is far better than earlier versions.
- Do not use the Analysis Toolpak (ATP) provided with any version of Microsoft Office Excel. To solve problems outside the capability of the built-in Excel functions, use a major statistical package instead of the ATP.
- Statistical add-ins for Excel must be evaluated carefully before acceptance. While some of these products are quite good, others have disgracefully poor quality.
- Any company with a Six Sigma initiative must have a major statistical program, such as JMP, MINITAB, or STATGRAPHICS, available for situations calling for analysis outside of Excel's capabilities. Many Six Sigma practitioners will prefer to use these more comprehensive programs instead of Excel software.

3

Applying Nonnormal Distribution Models in Six Sigma Projects

Most Six Sigma training programs teach tools based only on normal distribution models. Some training makes the incorrect assertion that nonnormal process distributions are inherently defective or unstable. Most trainees learn how to interpret metric values. They learn that if C_{PK} (or P_{PK}) = 1.5, then defect rates are 3.4 defects per million (DPM), which is a good number. They also understand that if C_{PK} (or P_{PK}) is less than 1.0, this is a bad number. Because of the global influence of Six Sigma initiatives, we now have a generation of business leaders, including trained Green Belts, Black Belts, Master Black Belts, and Champions, who accept these teachings.

Practitioners using nonnormal distribution models in a Six Sigma environment face many challenges, and the greatest of these is communication. Readers of this book want more out of Six Sigma. People who are bold and insightful enough to explore nonnormal distributions want a variety of tools to solve diverse problems efficiently. It is fair to say that most Champions and other team members do not share this desire to understand statistical tools. While they may be happy to have a colleague who can correctly use words like "Weibull" and "hypergeometric," they do not want to study these topics personally.

The key to successful communication is to understand customer (audience) requirements. The audience could be one person, as in a one-on-one meeting with the boss, or it could be thousands of people reading an article or viewing a presentation. Requirements vary, but these are important to every audience:

- *Reassurance*. Six Sigma presentations never happen in good times. The company may be bleeding profusely, and urgent action is required. The audience should leave with less anxiety than when they arrived. To achieve this requires thorough preparation and confident, friendly delivery. Be prepared to answer more questions than anyone will ever ask. Do not resort to technical termdropping, which builds a wall between presenter and audience. Use only terms that are important to the story, and be prepared to explain new terms in plain language.

- *Familiarity.* Use presentation templates, terminology, and metrics consistent with company practice and previous Six Sigma training. Unless necessary, do not require the audience to learn new technology or to question long-held beliefs. When reporting P_{PK} for a nonnormal process, use the equivalent percentile method (explained later) so that a metric value of 1.5 means 3.4 defects per million (DPM) regardless of the shape of the distribution. While probability plots and QQ plots are superior tools for selecting a distribution model, histograms may be more familiar for a presentation.

- *Simplicity.* To present the process of solving a problem, select a few logical steps connecting problem with solution. It is unnecessary to present every graph or detail every dilemma. The audience already knows that people worked hard. The presentation is the time to make it all seem easy.

When the time comes to confront misconceptions, illustrate the point with real company examples. For example, it may be necessary to challenge a general belief that nonnormal processes are bad. Decision makers must understand the difference between an unstable process and a stable, nonnormal process. To illustrate this point, look for a real company example of a stable, nonnormal process, and show that using a properly fitted distribution model results in more meaningful capability metrics and fewer false alarms on control charts.

After communication, the biggest challenges for Six Sigma practitioners using nonnormal distributions are technical issues. Statistical experts do not agree on the best ways to accomplish basic tasks with nonnormal distributions. Here are four general approaches to this problem:

- *Option 1: Apply normal-based tools.* This is the usual Six Sigma approach. Every statistical tool involves risks of error, which are predictable if the process satisfies certain assumptions, such as normality. Tests and control charts have two risks, a risk of false detections α, and a risk of missed detections β. When applying normal-based tool to nonnormal data, these risks can be dramatically higher or lower than expected.

- *Option 2: Apply nonparametric tools.* Nonparametric tools make no assumption about the shape of a process distribution. Some of these tools are very simple and useful. Many Six Sigma courses teach Fisher's one-sample sign test or Tukey's end-count test, examples of nonparametric tools. In general, nonparametric tools are less able to detect small changes in a process than parametric tools. In addition, nonparametric predictions are limited to the range of values previously observed. To predict the small defect rates required for Six Sigma initiatives requires a model to represent the tails of a distribution beyond the range of observed values. Nonparametric methods cannot provide such a model.

- *Option 3: Select a distribution model, and apply a method designed for that distribution family.* Common examples of this approach include Shewhart control charts designed for binomial and Poisson processes. In Sleeper (2006), Chapter 8 describes tests for binomial and Poisson data, and Chapter 4 describes estimation tools for reliability distributions such as exponential, Weibull, and gamma.

- *Option 4: Transform the distribution into a normal distribution and apply normal-based tools to the transformed data.* One popular transformation family is Box-Cox, which can help to normalize skewed data. The Johnson transformation is very flexible, and can normalize both unbounded and bounded distributions. MINITAB software offers both these options in many of its analysis functions.

No single approach is best for all situations. Option 1 is simple, and while it often works for mild departures from nonnormality, the risks of this method are uncontrolled. Applying normal-based methods to nonnormal distributions can lead to serious problems, as illustrated by Example 1.1. For nonnormal distributions, Option 3 is best when distribution-specific methods are available, and Option 4 is best when they are not. Nonparametric methods are also very useful, and can solve problems that cannot be solved with distribution models, but these are topics for other books. In this book, each chapter on a specific distribution family offers techniques for parameter estimation, process control, and capability metrics for that distribution family.

The following example illustrates these approaches applied to the common problem of comparing two samples. The usual tool for this problem is a two-sample t-test, which assumes a normal distribution. In this example, there is insufficient data to select a distribution model, but previous knowledge of the process suggests a specific distribution family.

Example 3.1

Wendy would like to know whether an additional annealing process changes the lifetime of snap domes, which provide tactile feedback in switches for consumer products. She performs life tests on a sample of 30 snap domes with the extra process and 30 snap domes without the extra process. Wendy wants to know if the average snap dome life is changed by this added process.

The stem-and-leaf diagram below lists the lifetimes of the "Without" snap domes, in thousands of cycles. In a stem-and-leaf diagram, the leaves, on the right, represent the least significant digit. The stems, on the left, represent more significant digits. Therefore, the "Without" dataset includes 56, 67, 67, 75, etc.

5	6
6	77
7	568
8	4599
9	12233348
10	2478
11	128
12	0339
13	
14	1

Here are the lifetimes of the "With" snap domes, in thousands of cycles, in the form of a stem-and-leaf diagram:

5	4
6	6
7	017
8	56668
9	18
10	02444
11	1348
12	47
13	2478
14	479

With only 60 total observations, this data is insufficient to select a distribution model with precision. However, these snap domes have been extensively tested in the past. Analysis of prior test results suggests that lifetime has a Weibull distribution with a shape parameter equal to 5.

- *Option 1*: Wendy applies the two-sample *t*-test, which tests samples from two normal populations to see if the population means are different. There are different versions of the two-sample *t*-test, and Wendy chooses the version that assumes both populations have the same standard deviation. (In some books and software, this version is called the homoscedastic two-sample *t*-test.) This test computes a P-value of 0.167. Since this value is greater than 0.05, Wendy concludes that there is no significant difference between the two groups of parts.
- *Option 2*: Wendy applies the Mann-Whitney test, which is a nonparametric test of two samples to see if the two populations have different distributions, without assuming any particular distribution shape. This test computes a P-value of 0.231. Since this value is greater than 0.05, Wendy concludes that there is no significant difference.

- *Option 3*: Using the reliability analysis tools in MINITAB, Wendy tests whether the two populations are different, assuming that they both have a Weibull distribution with the same shape parameter. This test computes a P-value of 0.037. Since this value is less than 0.05, Wendy concludes that there is a significant difference in life caused by the added process step.
- *Option 4*: When the shape parameter α is known, Weibull data can be transformed into exponential data by raising it to the power α. Exponential data can be normalized by raising it to the power 0.2654. Putting these two transformations together, $(X^{\alpha})^{0.2654}$ should normalize Weibull data. Since Wendy assumes that $\alpha = 5$, the overall transformation is $X^{1.327}$. Wendy raises each measurement to the 1.327 power and performs a two-sample *t*-test, resulting in a P-value of 0.148. Since this value is greater than 0.05, Wendy concludes that there is no significant difference.

In Example 3.1, out of four different procedures applied to the same data, one found a significant difference, and three did not. What does this mean? Every statistical test is an estimate of the truth, not the truth itself. In practice, the truth can never be known with certainty. The truth about this data is either that the two groups of parts are different, or they are not different.

First, suppose there is no difference between the two groups. One test out of four had a false alarm by finding a difference where there was none. If Wendy always makes her decisions based on whether the P-value is less than 0.05, then 5% of her tests will have false alarms, even when there is no difference.

This example and others in this book illustrate different tools applied to the same data. In practice, one should select the best procedure in advance, and run it. It is inappropriate to run several tests in hopes that one will give a desired answer. At best, test-shopping is confusing, and at worst, it is unethical.

Every statistical procedure has a risk of false alarms, or Type I errors. In this case, the risk of false alarms is 5% for each test. If one takes the same set of data, with no effect present, and processes it through many different statistical procedures, each with a 5% false alarm rate, eventually one test will find an effect that is not there. To publish the one test out of many that resulted in a desired conclusion would be unethical.

The second possibility is that the two groups are truly different. Since this is a textbook, we have the luxury of knowing the truth about invented examples. The truth about the data in Example 3.1 is that the two groups are different. Each consists of random numbers from a Weibull distribution with shape parameter $\alpha = 5$ and different scale parameters. The test illustrated in Option 3 is designed for Weibull distributions, and has more power than any other test to detect the effect that actually exists in this data. This example shows the value of combining a known distribution model with a small dataset of new data.

Option 4 in Example 3.1 illustrates one of the perils of applying transformation methods to two-sample tests. The transformation $X^{1.327}$ converts Weibull(5, β) data into normally distributed data. In this example, the true difference in the scale parameter β caused a difference in both the mean and the standard deviation of the transformed data. After one small effect was diluted into two smaller effects, it could not be detected using normal-based procedures. Transformation methods are more effective with one-sample tests, where this particular problem does not arise.

The remainder of this chapter discusses how to assess process stability and how to measure process capability when the distribution is not normal. This chapter assumes familiarity with standard Six Sigma tools, and a few statistical terms. Sleeper (2006) contains detailed descriptions and recipes for a variety of Six Sigma and DFSS tools. Chapter 5 in this book lists definitions for selected statistical terms.

3.1   Assessing Process Stability

The ASQ (2005) defines a stable process as "a process that is predictable within limits; a process that exhibits only random causes (of variation)." For a process to be predictable and meet this definition of stability, it must satisfy two requirements:

- For a process to exhibit only random causes of variation, the short-term process variation must have a similar distribution at all points in time. Observing a similar distribution across several short-term samples provides some assurance that future process behavior will continue to have a similar distribution.
- For a process to be predictable, a distribution model is required to predict future process behavior. Usually, the distribution model is a member of a distribution family, with parameter values estimated from the historical process data.

Figure 3-1 illustrates histograms over time from two stable processes and two unstable processes. For each of the stable processes, the histograms are similar enough over time that one can expect future histograms of the same process to have the same location, variation, and shape. For the unstable

Stable processes   Unstable processes

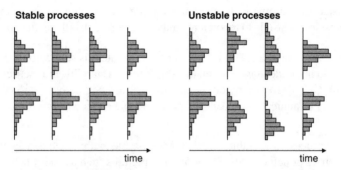

time   time

Figure 3-1 Histograms Illustrating Samples from Stable Processes and Unstable Processes.

processes, location, variation and shape change from time to time, so there is no way to predict what these processes will do in the future. Regardless of the distribution family, normal or otherwise, a process can be stable and predictable or unstable and unpredictable.

To correctly interpret capability metrics in Six Sigma applications, it is vital to understand the difference between estimates and predictions. A computer can quickly spit out estimates of C_{PK}, P_{PK}, and $DPM,$ but only a human can decide whether to predict the future with these numbers. Estimates describe the past, as represented by the data in the sample. If the process is stable, then these estimates may also serve as predictions for the near future. If the process is unstable, predictions are impossible. A thinking human must decide whether a process is stable and whether to interpret estimates as predictions or only as a snapshot of the past.

Why not plot the data on a Shewhart control chart and see if points are outside the control limits? A beagle with a computer could do this. After all, a Shewhart control chart is the stability test recommended by many excellent books on Six Sigma and SPC tools. Unfortunately, this strategy must change when stable processes might follow a variety of distribution families. Every control chart assumes that the data follows one specific family. Shewhart charts are available for normal, binomial and Poisson families. Plotting a stable process on a chart designed for the wrong family results in either too many false alarms or too little power to detect changes. With the tools available today, there is no universal control chart for all applications.

Therefore, testing a process for stability remains a task for thinking humans, and this is one of the most important decisions to make. If a process is unstable, the instability automatically becomes the biggest problem to solve, and it would be irresponsible to report any predictions before stabilizing the process.

Another important decision is how to communicate this subtle point. Many people do not understand the relationship between stability and metrics like P_{PK} or C_{PK}. Here are recommendations for Six Sigma professionals on what to communicate to stakeholders and people outside the core problem-solving team.

- If a process is unstable, say that it is unstable, but publish no capability metrics or defect rates. Show familiar pictures such as control charts or sequences of histograms that visually demonstrate the instability of the process. In the Measure phase of a Six Sigma project, it may be necessary to estimate financial impact from an unstable process. After doing this, simply state the estimated financial impact of the problem, without providing capability metrics or defect rates. If management demands these metrics, provide them only with a disclaimer that they represent past performance of an unstable process, and are not predictions of future performance.
- After stabilizing a process, it becomes harmless to publish "before" and "after" metrics to illustrate the size of the improvement. At that point, no one will presume that the "before" metrics are predictions, since the process has changed.

Every opportunity to communicate involves choices of what to communicate and what not to communicate. It is prudent to omit public statements that are likely to be misinterpreted. However, the problem-solving team must know all about the available data, so they can be informed partners in communication decisions. Team meetings provide opportunities to discuss how best to communicate findings about an unstable process.

When the distribution of a process is unknown, the best way to test it for stability is to view a sequence of probability plots. As with all statistical questions, a well-designed visual analysis is more powerful than any mathematical algorithm. Because of their design, probability plots are more effective than histograms for visualizing distribution shapes, especially over several samples. With histograms, several graphs are needed to compare the distributions of several groups. Within a single probability plot, distinctive symbols can show the distribution of many groups. The curve of points on a probability plot indicates whether the data skews more left or more right than the distribution model used by the probability plot. This information helps not only to select a good model, but also to detect whether the general shape of the distribution is stable.

In some cases, neither probability nor quantile-quantile (QQ) plots are available, and histograms must be used to visualize distribution stability. Overlaid histograms are often impractical because each histogram obscures features of others. When using histograms to test a distribution for stability, use paneled histograms, as in the example below.

When the distribution is unknown, an effective stability test requires a lot of data. Without relying on any distribution model, enough data must be available to visualize the distribution shape over several points in time. Also, the time sequence must be available. Without knowing which observations came first, it is impossible to test a dataset for stability.

Example 3.2

In Example 1.4, Bob studied the distribution of defects per wafer of micro-processor chips. He collected 500 sequential observations of defect counts, in search of a distribution model.

Before selecting a distribution model, Bob needs to test the process for stability. The c-chart, a standard Shewhart control chart, is designed for Poisson processes, but at this point, Bob does not know whether a Poisson model is appropriate for this process. If he applies a c-chart to a non-Poisson process, he cannot be sure that the control limits are meaningful.

To test the process for stability, Bob divides the data into four equal groups of 125 observations each and creates the paneled histogram shown in Figure 3-2.

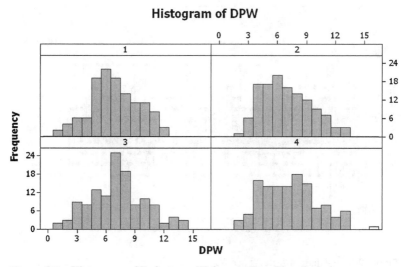

Histogram of DPW

Figure 3-2   Histograms of Defects per Wafer over Four Time Periods.

Each panel represents the distribution over a different span of time. Bob notices some differences in the shape of the distribution, but these do not seem significant to Bob. Bob concludes that the process is reasonably stable.

Next, Bob creates one histogram of the data with Poisson probability overlay, shown earlier in Figure 1-15. He concludes that the Poisson model fits the data well enough to use it for prediction.

After accepting the Poisson model, Bob now creates a c-chart, shown in Figure 3-3, as a more sensitive test of stability. This chart shows that one wafer, with 16 defects, has significantly too many defects according to the Poisson model. This wafer could have a special cause of variation, or it could be a false alarm. On a c-chart, as with all control charts, one point in a few hundred will be outside the control limits even if nothing has changed.

The sequence of actions in Example 3.2 is important. When evaluating process data for the first time, follow these steps:

- *Test the process for a stable distribution shape.* To do this, divide the data into groups, each group consisting of enough consecutive observations to evaluate the fit of a distribution model. When using histograms, each group must contain at least 100 observations. Probability plots can be effective with smaller groups. Note that this test requires

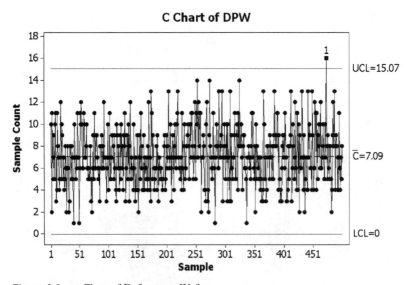

Figure 3-3   c-Chart of Defects per Wafer.

only a few groups with many observations per group. Control charts typically require at least 30 plot points before calculating control limits. When testing stability with histograms or probability plots, only a few groups are sufficient to reveal process instability.

- *If the process distribution shape is stable, select a distribution family model.* If the shape of the distribution appears similar from group to group, combine the data into a single group and create a new plot. Using methods described in Chapter 1, including theoretical knowledge, graphs, tests, and expert knowledge, select a family to represent the distribution of the data.

- *Test the process for stability using a method designed for the selected distribution family.* After selecting a distribution family, test the process again for stability using a more powerful method. This could include a control chart of the data, after dividing it into at least 30 subgroups. This could also include a statistical test, like the analysis of variance (ANOVA) which will detect if the mean of a normally distributed process is shifting.

It is often necessary to test data twice for stability. The first test is a rough visual stability test using graphs like probability plots or histograms. Then, after selecting a distribution family, use a more powerful method designed for that family with controlled risk levels.

Example 3.3

Paula works for a company that manufactures "Plung-eee" cords, which people use to tether their bodies before plunging off tall objects and saying, "Eee!" If a cord breaks, customers may become dissatisfied, which has a negative impact on repeat business.

A sample of cord material between every manufactured cord is stretched until it breaks. The burst strength recorded when each sample breaks provides assurance that each cord will withstand its intended load without breaking.

Paula is investigating problems meeting the minimum strength specification of 4500 N. She collects 30 measurements from each of five recent production weeks and creates a paneled histogram shown in Figure 3-4. From the histogram, she can see that the process mean shifts up and down from week to week. The process is unstable, and this becomes the biggest problem to solve.

Since the process is shifting, Paula cannot combine all the data into a single sample. Even so, she would like to find a distribution model for the process. With a distribution model, Paula can test the data with more powerful tools, and she can select an appropriate control chart for the process.

Histogram of Week1, Week2, Week3, Week4, Week5

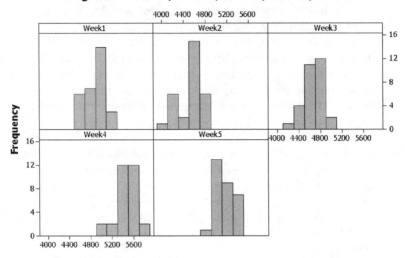

Figure 3-4   Paneled Histogram of Strength Measurements Over Five Weeks.

To test whether the process has a normal distribution with a shifting mean, Paula prepares a normal probability plot, shown in Figure 3-5. This graph shows the data from each week with distinctive plot points and a different fitted distribution model.

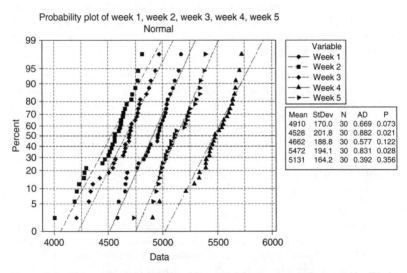

Figure 3-5   Normal Probability Plot of Strength Measurements, Separated by Week.

The P-values for the Anderson-Darling statistics are less than 0.05 for two of the five groups. This suggests either that the normal distribution family is not a good fit for this data, or that the distribution shape changes from week to week. With only 30 points per group, it is quite difficult to answer this question from the data alone.

Paula thinks about what the physical process can tell her about the distribution. When the cord stretches, one portion of the cord that happens to be weaker than the rest breaks first. Then, the rest of the cord carries the entire load and it breaks quickly. Since the weakest link breaks first, Paula decides that cord strength is really the minimum of several strengths. Based on this theoretical argument, the smallest extreme value distribution might fit the data better than a normal distribution. Also, the smallest extreme value distribution is skewed to the left, which is consistent with the curve of points on the probability plot.

Paula creates another probability plot based on the smallest extreme value distribution, shown in Figure 3-6. The plot points follow the straight lines better on this graph, and none of the P-values are less than 0.05.

Paula concludes that the shape of the strength distribution is stable and follows the smallest extreme value distribution family. The location of the distribution is unstable, increasing or decreasing from week to week.

Control charts remain standard and effective tools for evaluating process stability. When the process distribution agrees with the distribution assumed by the chart, false alarm rates and the ability to detect process shifts are reasonable.

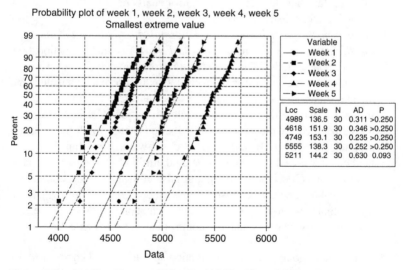

Figure 3-6   Smallest Extreme Value Probability Plot of Strength Measurements, Separated by Week.

Selecting the correct control chart requires selection of the correct distribution family. If the shape of the distribution changes from left-skewed to right-skewed to bimodal, no control chart will work well. However, if the distribution maintains a similar shape from time to time, a control chart designed for that shape will effectively detect when process parameters change, without generating too many false alarms.

As stated earlier, there is no universal control chart. However, many people have developed control charts that work well for a wide variety of distribution families. Shore (2000) identifies three general approaches to this problem:

- *Use Shewhart charts with no modifications.* This is the typical approach recommended in Six Sigma training. Because of the central limit theorem, subgroup means $\overline{X}$ tend to become normally distributed as the subgroup size increases. Therefore, the $\overline{X}$ chart tends to behave well for mildly nonnormal processes, if the subgroup size is large enough. However, the variation charts (s, R and MR) and the very popular individual X chart enjoy no such protection, and typically generate many false alarms, when applied to nonnormal data. To learn more about the impact of nonnormal data on standard Shewhart charts, read Burr (1967), Schilling and Nelson (1976), Balakrishnan and Kocherlakota (1986) or Chan *et al* (1988).
- *Use a chart designed for a specific distribution family.* The Shewhart attribute charts designed for Poisson or binomial processes are examples of this approach. Later chapters devoted to specific distribution families describe other control charts designed for processes following those distributions.
- *Transform the distribution into a normal distribution and plot the transformed data on a standard Shewhart chart.* Two transformation tools are very popular today. The Box-Cox transformation (1964) normalizes a wide range of skewed distributions. Box-Cox transformations include the log and square root transformations as special cases. The Johnson transformation, with fitting methods developed by Chou *et al* (1998) can normalize a very wide variety of distribution shapes. Both methods are available in MINITAB and other SPC software. New transformation methods developed by Shore for general attribute (2000a) and variable control charts (2000b and 2001) perform better than other transformations for lower-bounded skewed distributions. Shore's book (2005) organizes these techniques into Response Modeling Methodology (RMM). As yet, the Shore methods are not available in commercial software products.

Either of the last two approaches, distribution-specific control charts or transformation methods, is better than the first. Selecting an appropriate distribution model or transformation and using a control chart designed for that situation will always be superior to plotting nonnormal data on a normal-based control chart.

Each method has advantages. Charts designed for a specific distribution look familiar because the plot points are in the original units of measurement. However, they also look unfamiliar because of the asymmetric distribution of plot points and asymmetric control limits. Transforming the data results in a familiar chart with symmetric control limits and a symmetric data pattern, but the units of the plot points are completely unrelated to the original measurements. Practitioners can make either type of chart and decide which will work best.

The next three examples illustrate the three approaches to control charts using the flatness data collected by Six Sigma trainee Fred in Example 1.9. After viewing a histogram and fitting a variety of distribution models, Fred selected an exponential distribution to represent the variation in flatness data.

Example 3.4

Fred plots the 70 observations of flatness on an individual X, moving range chart, shown in Figure 3-7. Based on this chart, Fred concludes that the process is unstable, with several points between 43 and 50 outside the control limits.

The lower control limit of the individual X chart is -0.622, even though flatness can never have a negative value. By some control chart rules, this chart is also out of control because the plot points are unnaturally far away from the lower control limit.

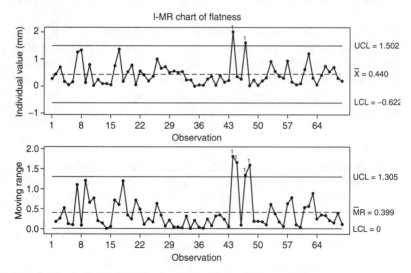

Figure 3-7   Individual X, Moving Range Chart of Flatness Measurements.

Since Fred selected an exponential model to represent the flatness distribution, he has other, more appropriate control chart options. Yang and Xie (2000) derive control limit formulas for an individual X chart for exponential data, with any false alarm risk α. For standard control charts, $\alpha = 0.0027$, corresponding to the probability of exceeding three-sigma control limits of a normal distribution. Here are the formulas for the center line and control limits, when the mean parameter μ is estimated by the sample mean $\overline{X}$:

$$UCL_X = -\overline{X}\ln(\alpha/2)$$
$$CL_X = 0.6931\overline{X}$$
$$LCL_X = -\overline{X}\ln(1 - \alpha/2)$$

These formulas are the same when estimating the rate parameter $\lambda = 1/\mu$ by $1/\overline{X}$. Yang and Xie also derived a power transformation that converts an exponential distribution into an approximately normal distribution. If $X \sim \text{EXP}(\mu_x)$, and $Y = X^{0.2654}$, then $Y \sim N(\mu_Y = 0.9034\mu_X^{0.2654}$, $\sigma = 0.2675\mu_X^{0.2654})$. The transformed exponential data may be plotted on any standard Shewhart control chart for the normal distribution.

For either option, only one chart is required to track an exponentially distributed process, instead of the two charts used to track a normal process. This is a subtle but important point. To estimate a normal distribution requires two independent statistics, $\overline{X}$ and s. Either the mean or the standard deviation could change without affecting the other statistic. Therefore, tracking a normally distributed process requires two charts, one for the mean and one for variation.

However, the exponential distribution has only a single parameter μ (or λ), which is estimated by $\overline{X}$ (or $1/\overline{X}$). A significant change in the parameter value will cause the single chart to go out of control. A needless variation chart (MR, R or s) would provide no new information, and it would add false alarms. This is yet another reason why normal-based charts are the wrong choice for exponential data.

Example 3.5

Fred decides to construct an individual X chart for the exponential distribution. He plots the data on a simple run chart and adds horizontal lines representing the center line and control limits. Here are Fred's calculations, based on the sample mean $\overline{X} = 0.4398$ and a standard false alarm rate $\alpha = 0.0027$:

$$UCL_X = -\overline{X}\ln(\alpha/2) = -0.4398 \times \ln(0.00135) = 2.906$$
$$CL_X = 0.6931\overline{X} = 0.305$$
$$LCL_X = -\overline{X}\ln(1 - \alpha/2) = -0.4398 \times \ln(0.99875) = 0.00055$$

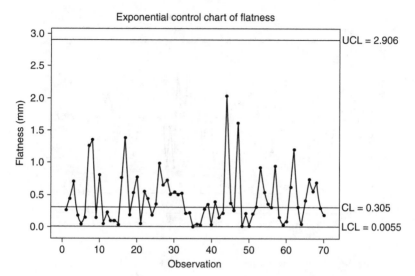

Figure 3-8 Exponential Individual X Chart of Flatness Measurements.

Figure 3-8 shows Fred's completed individual X chart for an exponential distribution. This chart shows no points outside the control limits, and a roughly equal split of points above and below the center line.

The method used in the above example can be used to generate an individual X chart for any distribution family with an inverse cumulative distribution function (inverse CDF) $F_X^{-1}(p)$. For the exponential distribution, $F_X^{-1}(p) = -\mu \ln(1 - p)$. To calculate the control limits and center line for any distribution with a false alarm rate α, use these formulas:

$$UCL_X = F_X^{-1}(1 - \alpha/2)$$
$$CL_X = F_X^{-1}(0.5)$$
$$LCL_X = F_X^{-1}(\alpha/2)$$

The inverse CDF $F_X^{-1}(p)$ is a function of the parameters of the distribution family. When estimating the parameters from the observed data, replace the parameters in $F_X^{-1}(p)$ with their estimates. The later chapters on various distribution families list inverse CDF functions and Excel formulas for calculating them, when available.

Note that this method is specifically for individual X charts, not for mean or $\overline{X}$ charts. For most distributions, the sample mean has a different distribution than individual values. In the specific case of the exponential distribution, the sample mean has a gamma distribution. Control limits for an exponential $\overline{X}$ chart would use the inverse CDF of the appropriate gamma distribution.

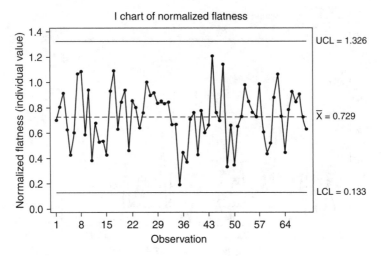

Figure 3-9   Individual X Chart of Normalized Flatness Measurements.

Example 3.6

Fred transforms his exponential data into normally distributed data by raising each observation to the power 0.2654. Then he creates a standard Shewhart individual X chart from the transformed data, shown in Figure 3-9. This chart has no points outside the control limits, and the plot points distribute as expected for a normally distributed process.

The units of measure for the points on this chart are now $mm^{0.2654}$, which is a meaningless unit. Therefore, Fred does not include units in the label for this chart.

3.2   Measuring Process Capability of a Distribution

Process capability metrics describe the relationship between process variation and tolerance (specification) limits. If the center of the process distribution is close to the target value, and if the variation is small enough that values outside the tolerance limits are unlikely to occur, the process is said to have high capability. Students of SPC and Six Sigma methods learn short-term metrics, such as C_P, C_{PK}, and C_{PM}, and long-term metrics, such as P_P, P_{PK}, and P_{PM}. Each of these metrics estimate specific aspects of short-term or long-term process capability.

All of the metrics listed above implicitly assume a normal process distribution. Experts disagree on the best way to measure capability of nonnormal distributions, but they agree that it is inappropriate to apply normal capability metrics to nonnormal processes. (See Bothe, 1997, page 431; Montgomery, 2005, page 339; Kotz and Lovelace, 1998, page 127; AIAG, 2005, page 140) The reason for this is that many people have learned relationships between defect rates and values of C_{PK} or P_{PK}. These relationships assume a normal process distribution and may be grossly high or low for other distribution families.

Before considering how to measure capability of nonnormal distributions, this section first reviews the basic types of capability indices now used by industry for normal processes. This discussion assumes that the process has a relatively stable normal distribution, with mean μ, short-term standard deviation $\sigma_{ST,}$ and long-term standard deviation σ_{LT}. Each metric considered here has a short-term (C-family) version based on μ and σ_{ST}, and a long-term (P-family) version based on μ and σ_{LT}.

The difference between σ_{LT} and σ_{ST} is in how they are estimated from the observed data. To estimate both long-term and short-term variation requires a long-term sample. Whether a dataset really represents long-term variation is a question only the practitioner can answer, based on experience with the process. A good long-term dataset includes k rational subgroups, each with n observations in each subgroup. Rational subgroups are selected to minimize variation within subgroups and to maximize variation between subgroups. The usual strategy for rational subgroups from a continuous process is to measure n consecutive units and repeat at regular intervals. If a sample represents only short-term variation, then "long-term" metrics calculated from the sample should be interpreted only as short-term metrics.

If a long-term sample from a normally distributed process contains k subgroups, each with n observations, then the most precise and unbiased estimates of population parameters are these:

$$\hat{\mu} = \frac{1}{nk} \sum_{i=1}^{k} \sum_{j=1}^{n} X_{ij}$$

$$\hat{\sigma}_{LT} = \frac{1}{c_4(nk)} \sqrt{\frac{1}{nk-1} \sum_{i=1}^{k} \sum_{j=1}^{n} (X_{ij} - \hat{\mu})^2}$$

$$\hat{\sigma}_{ST} = \frac{1}{c_4(k(n-1)+1)} \sqrt{\frac{\sum_{i=1}^{k} s_i^2}{k}}$$

In these formulas, $s_i = \sqrt{\dfrac{\sum_{j=1}^{n}(X_{ij} - \overline{X}_i)}{n-1}}$ is the sample standard deviation of

subgroup i, and $c_4(n) = \sqrt{\dfrac{2}{n-1}}\,\dfrac{\Gamma\left(\frac{n}{2}\right)}{\Gamma\left(\frac{n-1}{2}\right)}$. The values of c_4 correct for bias
in the standard deviation estimates, so that their average values are the true
standard deviation values. In Excel, calculate the gamma function $\Gamma(x)$ with
=EXP(GAMMALN(x)).

The above formula for estimating σ_{ST} is called the pooled standard
deviation. Other unbiased estimators for σ_{ST} in order or decreasing precision
are $\hat{\sigma}_{ST} = \overline{s}/c_4(n)$ based on subgroup standard deviations, $\hat{\sigma}_{ST} = \overline{R}/d_2(n)$,
based on subgroup ranges, and $\hat{\sigma}_{ST} = \overline{MR}/1.128$ based on the moving
range of two consecutive observations.

Over the years, many people have proposed new capability metrics to
correct perceived weaknesses in earlier metrics. The original metrics,
including C_P, are attributed to Juran (1974), while Kane (1986) developed
C_{PK}. Bothe (1997) and Kotz and Lovelace (1998) are both excellent references, describing both the history and current usage of capability indices.

Given upper and lower tolerance limits UTL and LTL, here are the
original five capability metrics, with short-term and long-term versions:

Short-term metrics: Long-term metrics:

$$C_P = \frac{UTL - LTL}{6\sigma_{ST}} \qquad\qquad P_P = \frac{UTL - LTL}{6\sigma_{LT}}$$

$$C_{PU} = \frac{UTL - \mu}{3\sigma_{ST}} \qquad\qquad P_{PU} = \frac{UTL - \mu}{3\sigma_{LT}}$$

$$C_{PL} = \frac{\mu - LTL}{3\sigma_{ST}} \qquad\qquad P_{PL} = \frac{\mu - LTL}{3\sigma_{LT}}$$

$$k = \frac{\left|\mu - \frac{UTL + LTL}{2}\right|}{\frac{UTL - LTL}{2}}$$

$$C_{PK} = Min\left\{\frac{UTL - \mu}{3\sigma_{ST}}, \frac{\mu - LTL}{3\sigma_{ST}}\right\} \quad P_{PK} = Min\left\{\frac{UTL - \mu}{3\sigma_{LT}}, \frac{\mu - LTL}{3\sigma_{LT}}\right\}$$

It is useful to recognize that $C_{PK} = Min\{C_{PU}, C_{PL}\}$, $C_P = (C_{PU} + C_{PL})/2$,
and $C_{PK} = C_P(1 - k)$. The same relationships apply to the P metrics.

With the exception of k, higher values of these metrics indicate better capability.

C_P and P_P measure the potential capability of the process, if it were centered between tolerance limits, but these metrics impose no penalty for being off center. The remaining metrics measure actual capability of the process, considering both the variation and centering. C_{PK} and P_{PK} are no greater than C_P and P_P, respectively, and they decrease as the mean departs from the presumed target value $T = (UTL + LTL)/2$. Although rarely used, the centering metric k provides the link between C_P and C_{PK}, and between P_P and P_{PK}. Each of these metrics may be modified for asymmetric tolerances, where $T \neq (UTL + LTL)/2$, but this case is ignored here for simplicity. Bothe (1997) describes many variations of these metrics for special situations.

Taguchi methods (Hsiang and Taguchi, 1985) popularized a new metric, C_{PM}, defined as follows, with its long-term version P_{PM}:

$$C_{PM} = \frac{UTL - LTL}{6\sqrt{\sigma_{ST}^2 + \left(\mu - \frac{UTL + LTL}{2}\right)^2}} \qquad P_{PM} = \frac{UTL - LTL}{6\sqrt{\sigma_{LT}^2 + \left(\mu - \frac{UTL + LTL}{2}\right)^2}}$$

The effect of using C_{PK} and P_{PK} is to motivate staying away from tolerance limits and minimizing defective parts. The effect of using C_{PM} and P_{PM} is to motivate staying close to the target value and minimizing variation. Figures 3-10 and 3-11 illustrate this point.

Figure 3-10 shows three normal distributions, all with $P_{PK} = 1.00$. The mean of each of these distributions is three standard deviations inside the closest tolerance limit. The defect rate is 2700 DPM for the centered distribution and 1350 DPM for the off-center distributions. For a normal process distribution, the defect rate is always between $\Phi(-3P_{PK}) \times 10^6$ DPM and $2\Phi(-3P_{PK}) \times 10^6$ DPM, so P_{PK} is a reasonable predictor of defect rates.

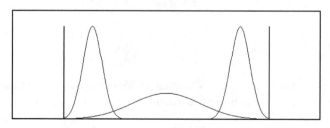

Figure 3-10 Frequency Curves of Three Normal Distributions, Each with $C_{PK} = 1.00$.

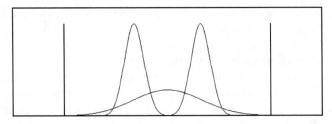

Figure 3-11   Frequency Curves of Three Normal Distributions,
Each with $C_{PM} = 1.00$.

However, P_{PK} is not a good indicator of whether values are close to the target.
The standard normal cumulative distribution function (CDF), denoted by
$\Phi(x)$, can be calculated in Excel using the formula =NORMSDIST(x).

Figure 3-11 shows three normal distributions, all with $P_{PM} = 1.00$. All
three are relatively close to the target value. In fact, $P_{PM} = 1.00$ implies
that the mean must be in the middle third of the tolerance range. In
Figure 3-11, the defect rate is 2700 DPM for the centered distribution
and 0.0000002 DPM for the off-center distributions, so there is no useful
relationship between P_{PM} and defect rates. However, since the mean
must be in the middle $1/3P_{PM}$ of the tolerance width, P_{PM} is a good
indicator of process centering.

The choice of P_{PK} versus P_{PM} depends on one's definition of quality. If quality
is the avoidance of defective units, then P_{PK} is more appropriate. If quality is
consistency around a target value, then P_{PM} is more appropriate.

To motivate both centering and defect avoidance, Pearn *et al* (1992)
proposed C_{PMK} and P_{PMK}. These metrics are defined as follows:

$$C_{PMK} = \frac{Min\{UTL - \mu, \mu - LTL\}}{3\sqrt{\sigma_{ST}^2 + (\mu - T)^2}}$$

$$P_{PMK} = \frac{Min\{UTL - \mu, \mu - LTL\}}{3\sqrt{\sigma_{LT}^2 + (\mu - T)^2}}$$

In search of a unifying theme, Vännman (1995) defined a class of capability
indices based on two nonnegative parameters u and v as follows:

$$C_P(u, v) = \frac{\frac{UTL - LTL}{2} - u\left|\mu - \frac{UTL + LTL}{2}\right|}{3\sqrt{\sigma^2 + v(\mu - T)^2}}$$

Vännman calls this a superstructure of capability indices, since it includes the four principal indices: $C_P(0, 0) = C_P$, $C_P(1, 0) = C_{PK}$, $C_P(0, 1) = C_{PM}$, and $C_P(1, 1) = C_{PMK}$, plus many others.

This is not at all a complete listing. The references listed here describe many other varieties of quality metrics, each with specific strengths and weaknesses.

Example 3.7

During a pilot production run of a new pump housing, all critical characteristics are carefully measured for $k = 30$ subgroups, each with $n = 5$ housings. A histogram and control chart of one critical milled characteristic shows a stable normal distribution. The tolerance limits for this characteristic are 12.50 ± 0.05 mm. Here are the parameter estimates from this data:

$$\hat{\mu} = 12.490$$
$$\hat{\sigma}_{LT} = 0.012$$
$$\hat{\sigma}_{ST} = 0.010$$

Based on these estimates, here are estimates for the various short-term and long-term capability metrics:

$$\hat{C}_P = \frac{12.55 - 12.45}{6 \times 0.010} = 1.67 \qquad \hat{P}_P = \frac{12.55 - 12.45}{6 \times 0.012} = 1.39$$

$$\hat{C}_{PU} = \frac{12.55 - 12.49}{3 \times 0.010} = 2.00 \qquad \hat{P}_{PU} = \frac{12.55 - 12.49}{3 \times 0.012} = 1.67$$

$$\hat{C}_{PL} = \frac{12.49 - 12.45}{3 \times 0.010} = 1.33 \qquad \hat{P}_{PL} = \frac{12.49 - 12.45}{3 \times 0.012} = 1.11$$

$$\hat{k} = \frac{|12.49 - 12.50|}{0.05} = 0.2$$

$$\hat{C}_{PK} = Min\{2.00, 1.33\} = 1.33 \qquad \hat{P}_{PK} = Min\{1.67, 1.11\} = 1.11$$

$$\hat{C}_{PM} = \frac{12.55 - 12.45}{6\sqrt{0.010^2 + (12.49 - 12.50)^2}} = 1.18$$

$$\hat{P}_{PM} = \frac{12.55 - 12.45}{6\sqrt{0.012^2 + (12.49 - 12.50)^2}} = 1.07$$

$$\hat{C}_{PMK} = \frac{Min\{12.55 - 12.49, 12.49 - 12.45\}}{3\sqrt{0.010^2 + (12.49 - 12.50)^2}} = 0.94$$

$$\hat{P}_{PMK} = \frac{Min\{12.55 - 12.49, 12.49 - 12.45\}}{3\sqrt{0.012^2 + (12.49 - 12.50)^2}} = 0.85$$

Nonnormal process distributions pose capability measurement challenges, which today's tools have not fully met. Here are three approaches to measuring process capability for nonnormal distributions:

- Modify normal capability metrics into equivalent versions for nonnormal distributions.
- Transform the distribution into normal, and then apply normal capability metrics.
- Develop a new capability metric for nonnormal distributions.

First, consider the approach of using a modified normal capability metric. If this can be done, then all the people who have already learned to use C_{PK} and P_{PK} will immediately know how to use their nonnormal versions. Because of this immediate acceptance, this approach has great advantages. Here are a few of the difficulties with this approach:

- The process mean and standard deviation completely specify a normal distribution, but without an assumed distribution family, these two parameters may provide very little information about process capability. If one knows the mean and the standard deviation, but not the distribution family, it is impossible to make a meaningful statement about defect rates or centering. According to the Chebychev inequality, as much as $1/k^2$ of a population may occur outside $\mu \pm k\sigma$ limits, for any distribution shape. This means that a "Six Sigma" process, for which $k = 4.5$, could have as much as 49,383 DPM (4.9383%) at or outside its $\mu \pm 4.5\sigma$ limits.
- The concept of short-term variation is undefined for nonnormal process distributions. The difference between short-term and long-term variation is always important, but it is not clear how to define short-term variation for distributions other than normal. Methods that may work for one family of distributions do not work for others.
- The concept of centering on a target value is also undefined for nonnormal distributions. Does centering mean that the distribution mode, median, or mean is at the target value? For a normal distribution, all three values are the same, but for other distributions, they can be quite different. For example, consider a process with an exponential distribution. How would that process be "centered" on a target value? Without a clear definition of centering, one cannot create a metric that measures the degree of centering.

Because of the above facts, only a few normal capability metrics can be modified for nonnormal processes. Without a consistent way to estimate short-term variation, all the short-term C-family metrics are unavailable. Also, the lack of a definition of centering means that P_{PM} and P_{PMK} are unavailable. The only remaining normal metrics are P_P, P_{PU}, P_{PL}, and P_{PK}. A further limitation is that the mean and standard deviation statistics are not useful predictors of tail probabilities for nonnormal distributions. This means that any useful modification of these metrics for nonnormal applications must use information other than the mean and standard deviation.

Two different methods to calculate equivalent P_P, equivalent P_{PU}, equivalent P_{PL}, and equivalent P_{PK} for nonnormal distributions are widely accepted today:

- *The ISO method.* First suggested by Clements (1989), the ISO Technical Committee 69 on Applications for Statistical Methods has proposed this method, in Document N13 of Working Group 6, Subcommittee 4. Bothe (1997) describes this method in some detail. AIAG (2005) also lists this method. To apply the ISO method after selecting a distribution model, determine the median $X_{0.5}$, the 0.135 percentile $X_{0.00135}$, and the 99.865 percentile $X_{0.99865}$. Then calculate equivalent metrics as follows:

$$\text{Equivalent } P_P = \frac{UTL - LTL}{X_{0.99865} - X_{0.00135}}$$

$$\text{Equivalent } P_{PU} = \frac{UTL - X_{0.5}}{X_{0.99865} - X_{0.5}}$$

$$\text{Equivalent } P_{PL} = \frac{X_{0.5} - LTL}{X_{0.5} - X_{0.00135}}$$

$$\text{Equivalent } P_{PK} = Min\{\text{Equivalent } P_{PL}, \text{Equivalent } P_{PU}\}$$

The rationale behind the ISO method is that a nonnormal process with equivalent $P_{PK} = 1$ should have the same defect rate as a normal process with $P_{PK} = 1$. While the metric achieves this goal, one and zero are the only values where defect rates are equal for normal and nonnormal distributions.

- *Bothe's percentage method.* Bothe (1992) proposed this method and developed it further in his 1997 book. To apply the Bothe percentage method after selecting a distribution model, with CDF $F_X(x)$ and survival function $R_X(x) = 1 - F_X(x)$, calculate equivalent percentage metrics as follows:

$$\text{Equivalent } P_{PL}^{\%} = \frac{-1}{3}\Phi^{-1}(F_X(LTL))$$

$$\text{Equivalent } P_{PU}^{\%} = \frac{-1}{3}\Phi^{-1}(R_X(UTL))$$

$$\text{Equivalent } P_P^{\%} = \frac{1}{2}(\text{Equivalent } P_{PL}^{\%} + \text{Equivalent } P_{PU}^{\%})$$

$$\text{Equivalent } P_{PK}^{\%} = Min\{\text{Equivalent } P_{PL}^{\%}, \text{Equivalent } P_{PU}^{\%}\}$$

In these formulas, $\Phi^{-1}(p)$ represents the inverse CDF of the standard normal distribution, calculated by the Excel formula =NORMSINV(p). The rationale for the percentage method is to maintain the same relationship between metric values and defect probabilities, regardless of the distribution model. The reason for using $F_X(LTL)$ and $R_X(UTL)$ is so that these numbers are both close to 0 instead of close to 1. Having these values closer to 0 improves the accuracy of the calculation for higher values of the metrics.

Both ISO and Bothe methods should be modified for discrete distributions. When X has a discrete distribution, and LTL is an integer, replace LTL by $LTL - 1$ in the above formulas.

Software products vary widely in how they calculate capability metrics for nonnormal distributions. Here is a brief survey:

- Crystal Ball simulation software (Version 7.2 or later) tests the distribution for normality. If it fails the normality test, Crystal Ball calculates capability metrics using the ISO method. Crystal Ball provides goodness-of-fit statistics to help the user select a distribution model, but Crystal Ball does not select a model automatically. The user has options to control the capability analysis functions for nonnormal data.
- JMP software provides normal or nonnormal capability analysis as part of its **Analyze** ⇨ **Distribution** function. The **Capability Analysis** option shows only normal-based capability metrics. To display nonnormal capability metrics, first use the **Fit Distribution** option to select one of 11 distribution models. After doing this, select and specify **Spec Limits** to enable the nonnormal capability analysis, which uses the ISO method. One of the distribution options is labeled "extreme value" option, but this is actually a two-parameter Weibull distribution with different parameters, and the capability metrics for "extreme value" and Weibull options are the same. The JMP "extreme value" distribution is not the same as the Gumbel extreme value distribution described in Chapter 12.
- MINITAB software includes nonnormal capability calculations for any of 13 distribution families. The user of MINITAB may choose either the ISO method or the Bothe method, which is labeled the "Minitab" method.
- STATGRAPHICS software provides either ISO or Bothe method, plus two other methods covered later, for any of 27 distribution families. In the **Capability** tab of the **Edit** ⇨ **Preferences** form, users may select either **Use distance between percentiles** (ISO method) or **Use corresponding Z-scores** (Bothe method).

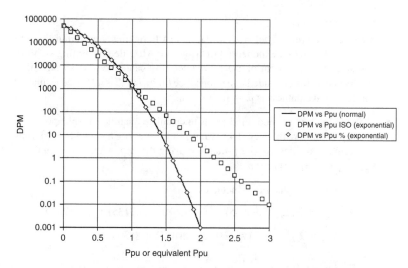

Figure 3-12   Graph of the Relationship Between P_{PU} and DPM for Three Cases: P_{PU} of a Normal Distribution; Equivalent P_{PU} (ISO Method) for an Exponential Distribution; Equivalent $P_{PU}^{\%}$ (Bothe Method) for an Exponential Distribution.

Figure 3-12 illustrates the important difference between the ISO method and the Bothe method. The bold line in the graph shows the relationship between P_{PU} and DPM for a normally distributed process with a single upper tolerance limit. The two sets of markers show the relationship between equivalent P_{PU} and DPM for an exponentially distributed process with a single upper tolerance limit, using the ISO method and the percentage method. The diamond-shaped markers representing DPM versus equivalent $P_{PU}^{\%}$, follow the normal distribution line exactly.

Using the percentage method, the defect rate associated with an equivalent capability metric is always the same as it would be for a normal process with the same metric value. The ISO method satisfies this important property only at two points, where the metric equals 1 and 0.

A normally distributed process with a single tolerance limit achieves a defect rate of 3.4 DPM when $P_{PK} = 1.5$. For an exponential process to achieve 3.4 DPM requires either equivalent $P_{PK}^{\%} = 1.5$ or equivalent $P_{PK} = 2.0$. The familiar and consistent interpretation of metric values over all distributions gives the percentage method a tremendous advantage over the ISO method. For this reason, the Bothe equivalent percentage metrics are the best nonnormal capability metrics for Six Sigma practitioners.

Example 3.8

In earlier examples, Fred studied flatness of floor tiles, and found that the process has a stable, exponential distribution. After demonstrating stability with an exponential control chart, Fred needs to calculate process capability with respect to the one-sided upper tolerance limit, $UTL = 1.0$. Fred estimated that the average flatness is $\hat{\mu} = 0.4398$ mm.

- Applying the ISO method, Fred needs to calculate quantiles of an exponential random variable with mean 0.4398 mm. Referring to Chapter 11, Fred finds this formula to calculate the p-quantile using the exponential inverse CDF: $F_{Exp(\mu)}^{-1}(p) = -\mu \ln(1 - p)$

Applying this formula, Fred calculates:

$$X_{0.99865} = F_{Exp(\mu)}^{-1}(0.99865) = -0.4398 \ln(0.00135) = 2.906$$

$$X_{0.5} = F_{Exp(\mu)}^{-1}(0.5) = -0.4398 \ln(0.5) = 0.305$$

$$\text{Equivalent } P_{PU} = \frac{UTL - X_{0.5}}{X_{0.99865} - X_{0.5}} = \frac{1 - 0.308}{2.906 - 0.308} = 0.267$$

$$\text{Equivalent } P_{PK} = P_{PU} = 0.267$$

- Applying the percentile method, Fred needs the survival function for the exponential distribution, which is $R_{Exp(\mu)}(x) = e^{-x/\mu}$. Since $UTL = 1$, $R_{Exp(0.4398)}(1) = e^{-1/0.4398} = 0.1029$. Using the Excel =NORMSINV function for $\Phi^{-1}(x)$, Fred calculates:

$$\text{Equivalent } P_{PU}^{\%} = \frac{-1}{3}\Phi^{-1}(0.1029) = 0.4217$$

$$\text{Equivalent } P_{PK}^{\%} = \text{Equivalent } P_{PU}^{\%} = 0.4217$$

Fred also needs to calculate the defect rate using the exponential model. The defect rate is $P[X > 1] = R_{Exp(0.4398)}(1) = e^{-1/0.4398} = 0.1029$. Multiplying this by 10^6, Fred predicts that the defect rate is 102,900 DPM.

Example 3.9

This example continues Example 1.8 from Chapter 1, in which Max studied the release effort of automotive door latches. After some investigation, Max found seven candidate distribution models, all of which have acceptable goodness-of-fit test results. Four of these models are left-skewed distributions fitted to the data. Three additional models are right-skewed distributions fitted to the negative of the data.

To help Max decide which models are plausible, he calculates equivalent capability metrics according to the Bothe percentage method, using all seven candidate models. Table 3-1 summarizes these calculations.

Max notices that model 7, the three-parameter Weibull distribution fitted to the negative of the data, has an upper bound at 34.61 N. This results in zero predicted

Table 3-1 Equivalent Long-Term Capability Metrics for Release Effort, Using Seven Candidate Models

Model	P[X < 20]	Equiv. $P_{PL}^\%$	P[X > 35]	Equiv. $P_{PU}^\%$	Equiv. $P_{PK}^\%$
1: β(770, 4.97, −619, 35.1)	0.00007868	1.260	1.942×10^{-7}	1.692	1.260
2: SEV (31.81, 1.493)	0.0003669	1.126	0.0002095	1.176	1.126
3: Weibull (20.94, 31.78)	0.00006144	1.280	0.0005285	1.092	1.092
4: Weibull (44.17, 66.45, −34.66)	0.0001792	1.190	0.0003240	1.137	1.137
5: -γ(5.16, 1.22, −35.18)	0.00007037	1.269	0.00000211	1.533	1.269
6: -Loglogistic(4.879, 0.2049, −36.13)	0.00000209	1.534	0.0007931	1.053	1.053
7: -Weibull(2.052, 4.127, −34.61)	0.00000154	1.555	0	∞	1.555

probability of values above the tolerance limit of 35 N, and an infinite capability metric. The histogram of the data, in Figure 1-22, shows a left-skewed distribution quite close to the upper limit of 35 N. When data is relatively close to a tolerance limit, it is not appropriate to use a distribution model that predicts infinite process capability at that tolerance limit. Therefore, model 7 should be rejected.

In the end, Max decides to use model 2, the smallest extreme value distribution, for these reasons:

- With only two parameters, models 2 and 3 are simpler than the other models.
- Model 2 has a smaller Anderson-Darling test statistic than model 3, as listed in Table 1-6. Other models with three-parameters have lower AD test statistics, but these models are more complex.
- The capability predicted by model 2, equivalent $P^{\%}_{PK} = 1.126$, is neither the highest nor the lowest of the seven models, and it is a plausible value to represent the histogram of measured results.

Transformation methods are the second of three ways to calculate nonnormal capability metrics. If the process distribution can be transformed into a normal distribution, then all the familiar normal capability metrics can be applied to the transformed data. To use this approach, the specification or tolerance limits are also transformed before applying normal capability metric formulas. Kotz and Johnson (2002) describe several transformation methods for capability measurement, with numerous references. Two transformation methods are now in common use:

- *The Box-Cox transformation.* Proposed by Box and Cox (1964), this family of power series transformations includes popular log, reciprocal, and square root transformations. Many distributions that skew to the left or the right can be transformed into a symmetric, approximately normal distribution using this method. This method is available in JMP, MINITAB, STATGRAPHICS and many other statistical programs.
- *The Johnson transformation.* Johnson (1949) defined a diverse family of distributions with the common property that each can be transformed into a normal distribution. Chou *et al* (1998) developed methods for selecting the member of the Johnson family that best fits the distribution of a dataset. MINITAB has built this algorithm into many of its functions for nonnormal data.

Transformation methods for capability analysis are very appealing, because a fast, computerized algorithm can select a distribution model and calculate metrics with no human thought required. Rodriguez (1992) articulated several valid statistical concerns about transformation methods for capability analysis. Without a very large dataset, such models can be wildly imprecise, and confidence intervals are usually unavailable. Aside from these statistical issues, it is always risky to use an automated algorithm to select a distribution model. The computer cannot know that some models are theoretically more appropriate than others; nor can the computer integrate expert opinion into the analysis[1]. As Rodriguez observed, "the black box approach tends to displace basic practices, such as plotting the data and applying simple normalizing transformations, that provide genuine understanding of the process."

Example 3.10

Fred applies transformation methods to compute capability metrics for the exponentially distributed flatness data.

* To apply the Box-Cox transformation in MINITAB, Fred selects Stat ⇨ Quality Tools ⇨ Capability Analysis ⇨ Normal. In the Capability Analysis form, Fred clicks the Box-Cox button. He selects Box-Cox power transformation ($W = X$**lambda) and Use optimal lambda. Figure 3-13 shows the graph produced by this analysis. This report includes histograms of the original and the transformed data, plus capability metrics. The report lists $P_{PK} = 0.42$, matching the equivalent $P_{PK}^{\%}$ computed in the previous example. Yang and Xie (2000)

---

[1]Bayesian statistics is a set of statistical tools that combine observed data with prior information. Prior information could represent expert opinion. Bayesian tools for process capability analysis are not fully developed, and no major software product provides these calculations. Because of the potential value of combining expert opinion with limited data in Six Sigma applications, it is likely that better tools will soon be available.

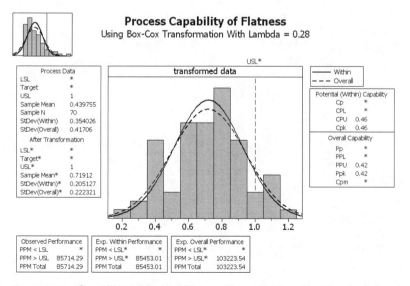

Figure 3-13 Capability Analysis of Flatness Data After Box-Cox Transformation.

showed that if X is exponentially distributed, then $X^{0.2654}$ is approximately normally distributed. The optimal power of 0.28 selected from the data by MINITAB is very close to the theoretical power that normalizes an exponential distribution. As part of the analysis, MINITAB also computes short-term capability metrics based on moving ranges in the transformed data.

- To apply the Johnson transformation in MINITAB, Fred selects Stat ⇨ Quality Tools ⇨ Capability Analysis ⇨ Nonnormal. In the Capability Analysis form, Fred selects the Johnson transformation button. Figure 3-14 shows the resulting report, with histograms before and after transformation. MINITAB selected $Y = 1.408 + 0.822 \ln\left(\frac{X + 0.037}{2.327 - X}\right)$ as the one that best normalizes the observed data. Based on this transformation, MINITAB estimates that $P_{PK} = 0.41$.

When a simple technique works, more complex techniques are unnecessary. In the above example, Box-Cox is the simplest of transformation methods, producing a simple model, $Y = X^{0.28}$. In this case, since Box-Cox works, there is no need to use the more complex Johnson transformation.

The third approach to calculating nonnormal capability metrics includes new metrics invented specifically for nonnormal distributions. From many published techniques, here are three important examples of new capability metrics:

- *The Clements method.* Clements (1989) devised a method of selecting a distribution from the Pearson distribution family by matching the

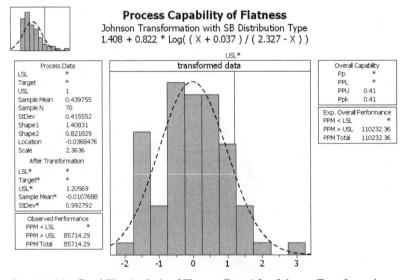

Figure 3-14   Capability Analysis of Flatness Data After Johnson Transformation.

mean, standard deviation, skewness, and kurtosis of an observed dataset. After selecting the appropriate Pearson distribution, the percentiles $X_{0.5}$, $X_{0.00135}$, and $X_{0.99865}$ can be calculated from Pearson formulas or looked up in tables provided by Clements. With these percentile values, apply what has become known as the ISO method described above. The resulting metrics are labeled $P_{P(Q)}$ and $P_{PK(Q)}$. For complete instructions on calculating these metrics, see Clements (1989) or Kotz and Lovelace (1998). Since STATRAPHICS software and some computer-aided design (CAD) software now incorporate the Clements method, it is one of the most frequently used methods of nonnormal capability analysis.

- Yeh and Bhattacharya (1998) define C_F as follows, based on a targeted defect rate α:

$$C_F = Min\left\{ \frac{\alpha/2}{F_X(LTL)}, \frac{\alpha/2}{R_X(UTL)} \right\}$$

Yeh and Bhattacharya illustrate this index using $\alpha = 0.0027$, so that for a normal distribution, $C_F = C_{PK}$ only when $C_{PK} = C_P = 1$.

- Luceño (1996) introduced C_{PC}, defined as follows:

$$C_{PC} = \frac{UTL - LTL}{6\sqrt{\frac{\pi}{2}}E\left| X - \frac{UTL + LTL}{2} \right|}$$

C_{PC} is only defined for bilateral tolerances. In cases when the target value is not the average of the tolerance limits, replace $\frac{UTL + LTL}{2}$ by the target value. The second C in the subscript stands for "confidence," since Luceño provides a confidence interval for this index. The $6\sqrt{\frac{\pi}{2}}$ factor in the denominator causes $C_{PC} = C_P$ when the distribution is normal. To estimate C_{PC}, compute $c_i = \left| X_i - \frac{UTL + LTL}{2} \right|$ for each observation. Then $\bar{c}$ and s_c are the usual sample mean and standard deviation formulas applied to the c_i. The estimate of C_{PC} is $\hat{C}_{PC} = \frac{UTL - LTL}{6\sqrt{\frac{\pi}{2}}\bar{c}}$ with a

$100(1-\alpha)\%$ confidence interval of $\dfrac{\hat{C}_{PC}}{1 \pm t_{\alpha/2,\, n-1}\dfrac{s_c}{\bar{c}\sqrt{n}}}$. In this formula,

$t_{\alpha/2,\, n-1}$ is the $(1 - \frac{\alpha}{2})$ quantile of the t distribution with $n - 1$ degrees of freedom, which may be calculated by the Excel formula =TINV(α, $n-1$).

Capability metrics by Clements' method and Luceño's C_{PC} are calculated by STATGRAPHICS software. In the Process Capability Analysis window, enable the Non-Normal Capability Metrics pane. To calculate C_{PC} and a 95% confidence interval, specify both specification limits plus a target value.

Clements' method is important, because many software products incorporate it. Users of Clements' method ought to recognize its limitations. Its biggest limitation is the need for a very large sample to produce reliable results. Clements' method relies on the sample mean, standard deviation, skewness, and kurtosis, which are all characteristics of the middle portion of the distribution. When the distribution family is unknown, it is unreliable to predict tail probabilities based on characteristics of the middle. Even if the distribution family is known, the sample skewness and sample kurtosis have so much variation that metrics by Clements method can be wildly imprecise.

Yeh and Bhattacharya's C_F uses the simple idea of computing the ratio of targeted to actual defect rates. In practice, using C_F magnifies the apparent size of problems, and the apparent size of improvements, when compared to today's metrics. Suppose that a Six Sigma project improves the defect rate of a process from 1350 DPM ($C_{PK} = 1$) to 3.4 DPM ($C_{PK} = 1.5$). This same improvement would change C_F from 1 to 397.

Luceño's C_{PC} has been controversial partly because of the claim that it does not rely on a normal distribution. The confidence interval relies upon the central limit theorem, so its accuracy increases as sample size increases. Some have questioned Luceño's claim that the confidence interval is reliable (Kotz and Lovelace, 1998, p. 118), but it may be the only nonnormal capability index with an easily calculated confidence interval.

Unfortunately, incorrect formulas for C_{PC} have been published and quoted in multiple references. In the otherwise excellent Kotz and Lovelace (1998), the mean absolute deviation, $\bar{c}$, is inside instead of outside the radical sign. The fourth edition of Montgomery's *Introduction to Statistical Quality Control* (2000) contained the same incorrect formulas for C_{PC}, but this error was corrected in the fifth edition (2004).

Example 3.11

Fred decides to calculate all available capability metrics for his flatness data. Using STATGRAPHICS software, Fred calculates these statistics from the floor tile flatness data:

$$\overline{X} = 0.4398$$
$$s = 0.4155$$
$$Skew[X] = 1.579$$
$$ExKurt[X] = 2.788$$

$ExKurt[X]$ is the coefficient of excess kurtosis, which would be 0 for a normal distribution. This is the version of kurtosis calculated by most software products. For more information on this, see the definitions in Chapter 5.

- To apply Clements' method in STATGRAPHICS software, Fred selects Analyze ⇨ Variable Data ⇨ Capability Analysis ⇨ Individuals, and enters a single upper specification limit of 1. Then, Fred enables the Non-Normal Capability Indices pane. This analysis reports that the Pearson distribution matching the mean, standard deviation, skewness, and kurtosis of this data has a median $X_{0.5} = 0.3057$ and a 99.865 percentile $X_{0.99865} = 2.378$. Based on these values, $P_{PK(Q)} = \frac{1 - 0.3057}{2.378 - 0.3057} = 0.335$

- To calculate C_F with a single upper tolerance limit, Fred sets $\alpha = 0.0027$. Earlier, he estimated that the probability of a unit exceeding the tolerance limit of 1.00 is 0.1029. Therefore,

$$C_F = \frac{\alpha/2}{R_X(UTL)} = \frac{0.00135}{0.1029} = 0.0131$$

Since $C_F = 0.0131$, this means that the target defect rate is 1.31% of the current defect rate.

- Luceño's C_{PC} metric is undefined for unilateral tolerances, but for the purpose of illustration, let $LTL = 0$, and the target value $T = \frac{UTL + LTL}{2} = 0.5$. In STATGRAPHICS, when Fred enters the lower limit and target values, the nonnormal capability index pane reports that $C_{PC} = 0.4038$ with a 95% confidence interval of (0.3403, 0.4964).

Table 3-2 summarizes the previous examples, listing all the nonnormal capability metrics applied to Fred's flatness data.

Table 3-2 Nonnormal Capability Metrics Applied to Flatness Dataset

Capability Metric	Estimated Value
Equivalent P_{PK} (ISO method)	0.267
Equivalent $P_{PK}^{\%}$ (Percentile method)	0.4217
P_{PK} after Box-Cox transformation $Y = X^{0.28}$	0.42
P_{PK} after Johnson transformation $Y = 1.408 + 0.822 \ln\left(\frac{X + 0.037}{2.327 - X}\right)$	0.41
$P_{PK(Q)}$ by Clements method	0.335
C_F with $\alpha = 0.0027$	0.0131
C_{PC}	0.4038

Many others have proposed methods of measuring process capability for nonnormal distributions, and research continues in this area. So far, the best metrics yet devised for Six Sigma practitioners are Bothe's equivalent percentage metrics, such as equivalent $P_{PK}^{\%}$. These metrics are simple to calculate and maintain the same relationship to defect rates as the familiar P_{PK} and C_{PK} for normal processes.

Applying Distribution Models and Simulation in Six Sigma Projects

After selecting distribution models, the next step is to apply those models to solve real problems. The first two chapters described graphs, statistical tools, and software for selecting the most appropriate distribution model. Chapter 3 showed how to overcome barriers to acceptance and how to use nonnormal distributions effectively in a Six Sigma environment. This chapter illustrates how to combine distribution models with Monte Carlo simulation in Six Sigma projects.

With distribution models describing variation for each system input, Monte Carlo simulation quickly reveals the distribution of the system outputs. Monte Carlo simulation is a computerized tool that calculates the system outputs from random numbers representing the system inputs. Monte Carlo simulation is certainly the most widely used application for distribution models. Here are a few reasons for its popularity:

- *Monte Carlo simulation is easy to understand.* Everyone who has played games with cards or dice has an innate understanding of randomness. Monte Carlo simulation builds on this basic knowledge to solve real problems, without requiring the user to understand complex mathematics. Anyone who can define a simple math function representing a physical system can apply Monte Carlo simulation successfully.
- *Monte Carlo simulation is visual.* People understand graphs and pictures far more easily than tables and statistical metrics. Nothing can tell the story in a dataset more effectively than a well-designed graph. Monte Carlo simulation produces histograms to visualize distributions and bar charts to visualize sensitivities. With nothing but these two simple graphs, people can make decisions and take action with confidence.
- *Monte Carlo simulation is flexible.* Real problems are more complex than book or classroom examples. Intricacies such as dependent

assumptions and complex transfer functions are easy to simulate with the right software.

- *Monte Carlo simulation is widely available.* Numerous software products provide Monte Carlo simulation. These range from specialized modeling software with integrated Monte Carlo functions to general-purpose, standalone simulation products, such as Crystal Ball software, featured in this book.

Crystal Ball software, based on Microsoft Office Excel spreadsheet software, provides a comprehensive set of Monte Carlo simulation tools. Crystal Ball software can apply Monte Carlo simulation to any system model expressed as a set of formulas in an Excel worksheet.

This chapter begins with an overview of Monte Carlo simulation, illustrated by the Crystal Ball software. Next are three case studies, which combine appropriate distribution models with simulation and optimization tools, quickly resolving complex problems. These are the three case studies:

- *Bank loan process improvement project.* This example illustrates how simulation can help improve a transactional process.
- *Simulation with design of experiments (DOE).* This example is a Six Sigma improvement project including an experiment analyzed with MINITAB, progressing to simulation and optimization with Crystal Ball software.
- *Perishable inventory optimization.* This example illustrates the optimization of a supply chain.

These case studies include both normal and nonnormal distribution models. Each of these models derives from observed data, theoretical knowledge, and expert opinion, as explained in Chapter 1.

4.1   Understanding Monte Carlo Simulation

Simply put, Monte Carlo simulation is a computerized experiment with random inputs. Instead of testing and measuring physical objects, a Monte Carlo simulation generates random input values and calculates how the system responds to those random inputs. A Monte Carlo simulation can measure thousands or millions of virtual objects in a very short time.

The history of Monte Carlo simulation spans hundreds of years. In 1654, during an association with gambling friends and before a religious conversion, Blaise Pascal (1623–1662) developed the first probability models for games of chance. Today, Monte Carlo is a region of the Principality of Monaco famous for celebrities, Formula One racing, and casinos. If not for Pascal's probability models, Monaco might not have become as rich as it is today. Probability models for games of chance assure casino owners that they will continue to enjoy healthy profits.

Although Monte Carlo simulation today is a computerized technique, random or stochastic experiments predate the era of computers. In 1930, Enrico Fermi used random methods to predict the properties of neutrons. After the advent of computers in the 1940s, scientists at the Los Alamos National Laboratory used Monte Carlo simulation to predict the possible effects of nuclear explosions.

Today, Monte Carlo simulation is a proven, efficient technique requiring nothing more complex than a computerized random number generator (Kelton and Law, 1991). A Monte Carlo simulation uses a model, which is a mathematical expression of outputs as functions of inputs. The objective of the model is to represent the behavior of some real system, perhaps a business process, a chemical reaction, or an electronic circuit. The accuracy of a Monte Carlo simulation cannot be better than the accuracy of its model in representing the real system. During the Monte Carlo simulation, the computer generates random numbers according to selected probability distributions for each input. Practitioners can easily analyze many scenarios created by the simulation using interactive charts and tables.

With the continuing development of faster personal computers and sophisticated software products, Monte Carlo simulation is fast becoming a staple in the desktop analytic toolkit. When applied correctly, Monte Carlo simulation provides valuable insights not available through deterministic methods.

4.1.1 Recognizing Opportunities for Simulation

As with any Six Sigma tool, simulation can be effectively applied to solve a wide variety of problems, including strategic analysis, value stream analysis, tolerance design, cost estimation, financial analysis, market forecasting, and resource allocation. Six Sigma professionals most often apply Monte Carlo analysis to two broad categories of applications:

- **Project management** where inputs such as costs and potential revenue are uncertain or unknown.
- **Product and process design** where inputs such as part dimensions and task cycle times vary between units.

Table 4-1 lists some of the more common concerns and conditions within project management and design that indicate the need for Monte Carlo simulation.

Monte Carlo simulation is not the only way to predict variation in the applications listed in Table 4-1. In some limited cases, analytical formulas and procedures provide enough information to predict uncertainty. Project managers use techniques such as Program Evaluation and Review Technique (PERT) to predict, manage, and control project risk. Engineers use root-sum-square (RSS) analysis for tolerance design of systems with linear transfer functions. All of these analytical formulas require a system model meeting restrictive assumptions and sufficient data to estimate statistical characteristics of system inputs.

Monte Carlo simulation is often the best way to predict variation. Most real problems are too complex or unusual for analytical formulas to solve, and Monte Carlo simulation is easy to apply to a very broad class of problems. When measurement data for system inputs is costly, scarce, or nonexistent, Monte Carlo simulation can be applied to the available data, combined with expert opinion. When a mathematical model exists that relates system inputs and outputs, it is easy to explore this model with Monte Carlo simulation. Table 4-2 describes conditions of the available data and of the system model which make Monte Carlo simulation a good choice.

4.1.2 Defining Input Distributions and Output Variables for Crystal Ball Simulations

In a Monte Carlo simulation, each input variable requires a fully specified distribution model. Crystal Ball software offers 20 predefined distribution families plus a custom distribution for unique situations that cannot be described with the other distribution families. Example 4.3 below illustrates a custom distribution. In a Crystal Ball simulation, each system input variable and each system output variable occupies one cell in an Excel worksheet.

Models are mathematical representations of real phenomena. As George Box famously said, "Every model is wrong, but some are useful." To be useful, models need not be perfect. However, a useful model should be consistent with all available information, including measurement data, theoretical knowledge, and expert opinion.

Every Monte Carlo simulation involves two types of models. The first type of model is the probability distribution model chosen to represent each input

Table 4-1 Reasons for Using Simulation in Project Management and Product and Process Design

	Concern/Condition
Project Management	**1. Project has financial uncertainty.** When the financial returns and risks of a product or process are uncertain, and the financial estimations or forecasts can be described in a spreadsheet model (Example 4.3)
	2. Project has schedule uncertainty. When a project schedule includes a critical path with tasks that have uncertain durations or efforts.
	3. Project has cost controls. When it is prohibitively expensive to acquire data for an input or output of a process or product, and simulation can create realistic virtual data.
	4. Project is high risk. When making decisions about a high-risk project, for example one with a potentially huge negative revenue impact, simulation helps to estimate the certainty of success, magnitude of risk, and key drivers influencing success.
Product and Process Design	**1. Robust design is required.** When a product or process design must be robust (insensitive) to input variation and when validation is expensive or time-consuming. Robust design requires stochastic optimization (Example 4.2).
	2. Tolerance analysis is performed. When the variation of a system in response to input variations within tolerance bounds must be predicted.
	3. Process is relatively simple. When a business process is linear or can be viewed at a high level, it can be modeled in spreadsheet form and simulated. This helps to measure baseline performance and test the effect of process improvements prior to implementation (Example 4.1).

variable. Earlier chapters of this book discussed the process of selecting distribution models. Later chapters list features and characteristics of many families of distribution models. The second type of model is the mathematical transfer function that calculates system output values from the system input

Table 4-2   Data and Model Conditions that Require Monte Carlo Simulation

	Condition
Data on Input or Output Variables	**1. Data is infrequent.** When a dataset of sufficient size to make decisions takes too long to collect.
	2. Data is costly. When the cost of obtaining sufficient measurements is prohibitively expensive.
	3. Data does not exist. When data for a system input or output is absent, for example, in competitive benchmarking analysis.
	4. Data is estimated. When the variation of system inputs is estimated from scarce data or from expert opinion.
	5. Variation is nonnormal. When system inputs have a distribution other than normal, this can invalidate many statistical tools that assume normality.
System Model	**1. Mathematical relationship exists.** When there is a known mathematical relationship between system inputs and outputs.
	2. Spreadsheet model already exists. When a model of the system including input variables and formulas already exists in the spreadsheet form.
	3. Equations are nonlinear. When the model includes nonlinear or highly complex mathematical functions.
	4. Physical models are impractical. When physical models of the system are impractical to create or to measure, due to cost or technological limitations.

values. The examples in this chapter illustrate the development of transfer function models, but this process generally involves specific skills beyond the scope of this book.

The maxim of "Garbage In, Garbage Out" is nowhere more appropriate than with the use of simulation. Whether Monte Carlo simulation produces accurate predictions depends on the quality of input distribution models and the transfer function model. Prior to simulation, it should be verified that models are consistent with the available data and other knowledge about the process.

In a Crystal Ball simulation, input distributions are referred to as *assumptions*. Assumptions are specified by one of two methods, either by graphical menus or by Excel functions. The graphical method is more popular because of its strong visual interface. After selecting a distribution family from the Distribution Gallery, shown in Figure 4-1, the analyst identifies a specific distribution within the family by specifying its parameter values. Figure 4-2 shows the Define Assumption form for the normal family of distributions. The two parameters for this family are the mean and the standard deviation.

It is usual to assume that each assumption, or input variable, is independent of every other assumption. In some systems, this is not the case, when the distributions of some assumptions depend on the values of other assumptions. Crystal Ball functions provide two ways to introduce dependent relationships. The first way is to specify a correlation coefficient between assumptions. The correlation coefficient is a number between -1 and $+1$, with 0 indicating no relationship. From the Define Assumption form, click the Correlate . . . button to access this function. The second method of specifying dependent relationships is to use cell references, such as =B3, instead of numeric parameter values. If the referenced cell depends on another assumption, this technique changes the distribution based on the value of the referenced assumption.

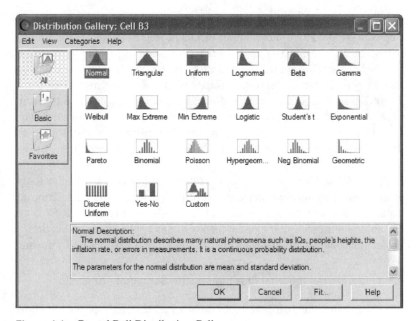

Figure 4-1   Crystal Ball Distribution Gallery.

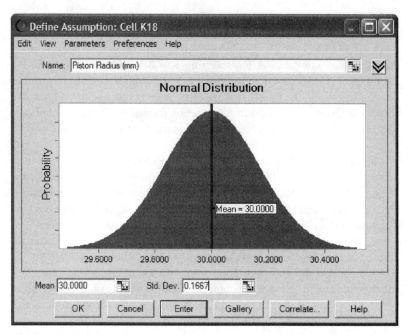

Figure 4-2   Crystal Ball Define Assumption Dialog for Normal Distribution.

After specifying distribution models for all assumptions, the analyst's next step is to define one or more output variables, referred to in Crystal Ball as *forecasts*. These forecasts are the responses or effects of the system represented by the transfer function. If the transfer function model is an equation like $Y = f(X)$, then Y represents forecasts, and X represents assumptions. Following a successful simulation, forecast results are displayed in the form of histograms, tables, or a combination of both, as shown in Figure 4-3. The Crystal Ball Forecast window can include a histogram and a table of descriptive statistics or percentiles. Chapter 5 contains definitions of the various statistics included in this display. When one or both specification limits are defined for a forecast, the Forecast window can also display long-term or short-term capability metrics of the forecast. The calculation of capability metrics was added to Crystal Ball software in version 7.2. Because capability metrics are so important for Six Sigma applications, no Six Sigma professional should still be using an earlier version of Crystal Ball software.

4.1.3   Identifying the Vital Few Inputs with Sensitivity Analysis

After predicting the distribution of output variables, analysts need to know which input variables contributed the most to variation in the output variables. There are many techniques to gain this knowledge from the data

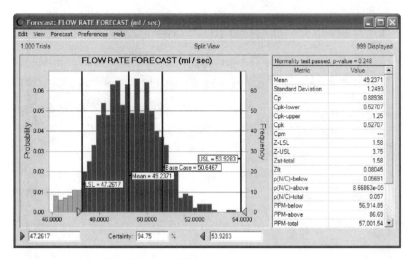

Figure 4-3 Crystal Ball Forecast Chart and Table.

collected during a Monte Carlo simulation, and these methods are collectively known as *sensitivity analysis*.

Crystal Ball software provides two forms of sensitivity analysis. The default method provides *contributions to variance* for each input variable. These values are percentages for each input variable, and their absolute values add up to 100%. The optional method provides *rank correlation coefficients* for each input variable. Rank correlation coefficients are expressed as values between −1 and 1. Crystal Ball software calculates contributions to variance by squaring the rank correlation coefficients, scaling so they sum to 100%, and preserving the sign of each correlation.

Both sensitivity metrics have the same interpretation. Assumptions with sensitivity values closest to −1 (−100%) or +1 (100%) are responsible for the largest part of variation in the forecast. Positive values indicate that increasing assumption values lead to increasing forecast values. Negative values indicate that increasing assumption values lead to decreasing forecast values. The sensitivity metrics consider both variation in the assumption distribution and also how much the variation propagates through the transfer function into the forecast. The Crystal Ball Sensitivity Chart presents these values in order, with the biggest contributor to variance on top, and others listed in order of decreasing contribution. Figure 4-4 shows an example Sensitivity Chart.

It is largely a personal choice whether to use rank correlation coefficients or contributions to variance for sensitivity analysis. Both metrics provide the same information. Contributions to variance tend to make the differences

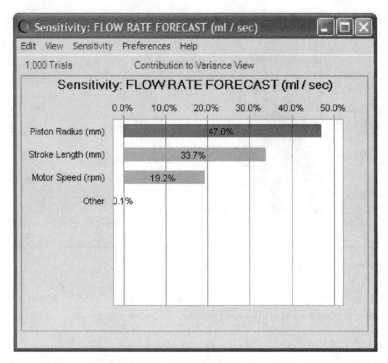

Figure 4-4 Crystal Ball Sensitivity Chart.

between sensitivities seem larger than rank correlation coefficients, and all but the top few contributions tend to have very small values. Contributions to variance are expressed in percentages, which are more familiar to many people. Since contributions to variance were added recently to Crystal Ball software, many long-time Crystal Ball users prefer rank correlation coefficients.

Sensitivity analysis helps the analyst and project team understand which inputs are responsible for the most variation and the most defects. This knowledge is vital to focus the team on the most important inputs to the system. After reducing variation on the vital few Xs, the analyst can run the analysis again to see how forecast variation and sensitivities have changed.

Equally important, sensitivity analysis tells the team which assumptions contribute the least to forecast variation. This knowledge is important for two reasons. First, it is unnecessary to study or to reduce the variation of these trivial inputs. Second, these inputs provide cost reduction opportunities. Increasing variation or tolerances in the trivial inputs will have very little impact on the variation of outputs.

4.2 Case Study: Bank Loan Process Improvement

The example in this section is a transactional process with too much variation in cycle time. By collecting data, fitting distribution models, and using Monte Carlo simulation, a Six Sigma team can find and correct the primary cause of the variation. The simulation model also helps to predict how changes in the process will improve the distribution and capability of cycle time.

Example 4.1

A financial organization wishes to make its loan process a best-in-class process. The current loan process, from the initial customer inquiry to the loan disbursement, takes an average of 92 hours. Given a performance target of 96 hours of cycle time, the process would seem to be solid. However, the loan specialists have complained (quite loudly) that completing a loan can take over 130 hours and that the process should be an improvement project. Walt, a Six Sigma Black Belt, sets up an improvement project to tackle this problem. The project charter includes these goals:

- Understand what is causing the variation around this business process
- Shift the mean cycle time to less than 96 hours for any individual loan
- Reduce the overall variation in the process
- Quantify the process capacity and Six Sigma quality level

In the Define phase, Walt and his team maps the transactional process and defines six broad steps. Figure 4-5 shows the macro map for the loan process with the average time for each major process step.

Walt's first question is whether the process is stable. By tracking 100 loans through the application process, the team collects 100 durations for each process step. Control charts and histograms of these times show relatively stable distributions. By combining observed data with theoretical knowledge and expert opinion, the team selects distribution models for each of the six task times. Table 4-3 summarizes the six steps and distribution models for each.

Walt creates a simple Excel model, shown in Figure 4-6, that sums the durations for each of the process steps. Using Crystal Ball software, Walt defines the appropriate assumption for the duration of each process step, according to the findings in Table 4-3. The cycle time formula sums these steps and represents

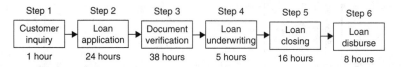

Step 1	Step 2	Step 3	Step 4	Step 5	Step 6
Customer inquiry	Loan application	Document verification	Loan underwriting	Loan closing	Loan disburse
1 hour	24 hours	38 hours	5 hours	16 hours	8 hours

Figure 4-5 Macro Map of Loan Process.

Table 4-3 Description of Steps for Loan Process

Process Step	Step Description	Probability Distribution
1. Customer Inquiry	Occurs via a phone call, office visit, the Internet, or home visit by a mortgage officer. Includes the creation of an initial rate quote for the customer.	Observed data indicates this cycle time is lognormally distributed with a mean of 1 hour and a standard deviation 0.25 hours.
2. Loan Application	With the inquiry completed, the applicant must complete all of the necessary forms. The timing around the distribution of blank forms and the collection of completed forms is difficult to estimate.	The staff who oversees this step provides an expert opinion that it takes no more than one week (40 hours), no less than one day (8 hours), and most often takes three days (24 hours). This is a triangular distribution.
3. Document Verification and Processing	Completed application is reviewed by a loan specialist, who contacts the applicant to verify the information, present the best loan alternatives, and help the applicant to decide which option is best for his or her situation. The loan specialist will obtain other information, such as credit score and history, directly from the credit bureaus, and will independently verify the information in the application.	Observed data shows that this step usually takes between two and four days (16–32 hours), but two out of ten times, due to suspended loans, the step takes between four and six days (32–48 hours). This situation is best described by a custom distribution with 80% of the values generated between 16–32 hours, and 20% of the values generated between 32–48 hours.

4. Loan Underwriting	Here the application is sent directly to an underwriter for review. The underwriter either approves it outright, approves with conditions to be met, or declines it.	Expert opinion indicates that the loan will receive a pre-approval in no more than 8 hours but no less than one hour. All values between 1 hour and 8 hours have the same likelihood of occurrence. This is described by a uniform distribution.
5. Loan Closing	The closing of the loan includes the preparation of the final documentation, the locking in of the interest rate, and the arrangement of where to deposit the funds.	By fitting distributions to the historical data, the team decides that the data is normally distributed with a mean of 16 hours and a standard deviation of 4 hours.
6. Loan Disbursement	Moving the funds to the applicant's bank usually takes 2 days (16 hours).	Disbursement time is normally distributed with a mean of 16 hours and a standard deviation of 4 hours.

A	B	C	D	E	F	G	H	I
9	Performance Target:			96				
10								
11	Process	Simulated						
12	Step	Cycle Time			Assumption Parameters			
13								
14	Step 1:	1		1	0.25		lognormal (mean, st dev)	
15								
16	Step 2:	24		8	24	40	triangular (min, likely max)	
17								
18	Step 3:	29		16	32	80%	custom (two conditions)	
19				32	48	20%		
20	Step 4:	5		1	8		uniform (min, max)	
21								
22	Step 5:	16		16	4		normal (mean, st dev)	
23								
24	Step 6:	16		16	4		normal (mean, st dev)	
25								
26	Cycle Time	91						
27								

Figure 4-6 Microsoft Excel Model of Loan Process.

the only forecast in the model. The upper tolerance limit for this forecast, labeled "Performance Target" in the spreadsheet, is 96 hours.

After running a Monte Carlo simulation of 1000 trials, the equivalent of 1000 virtually processed loans, Walt determines that the current baseline performance is as poor as the loan specialists originally described. The cycle time forecast chart in Figure 4-7 indicates that the mean cycle time is 89 hours, with

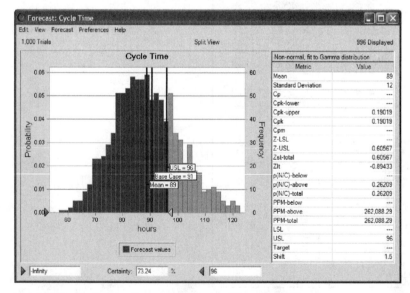

Figure 4-7 Cycle Time Forecast of Simulated Loan Process.

a standard deviation of 12 hours, a C_{PK} of 0.19, and a Sigma level of 0.61. The cycle time distribution is nonnormal, slightly skewed to the right, and fits a gamma distribution model. Chapter 14 provides more information on gamma distributions. The probability of achieving the 96 hour USL and target is roughly 73%, or 732 of the 1000 simulation trials.

Using sensitivity analysis, Walt learns that Step 3, document verification, contributes most to variation in total cycle time. Document verification is responsible for nearly 43% of total cycle time variation, as shown in Figure 4-8.

Given this result, Walt's team returns to the loan specialists to break down the sub-steps of this task and to investigate the root causes for the associated delays. By automating several of the sub-tasks, the team calculates that they could reduce the variation in the Step 3 assumption so that 95% of loans will be verified in 8–16 hours and 5% will be verified in 16–24 hours. The team knows that this one change would make a substantial improvement in the capability of the process, but implementing the change will take several weeks.

Because Walt already has a model of the process, he can use simulation to predict the effects of the proposed solution prior to implementation. Figure 4-9 shows the cycle time forecast distribution with improvements to the document verification step. This forecast distribution passes the normality test with a

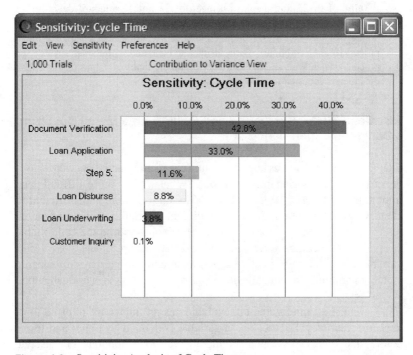

Figure 4-8   Sensitivity Analysis of Cycle Time.

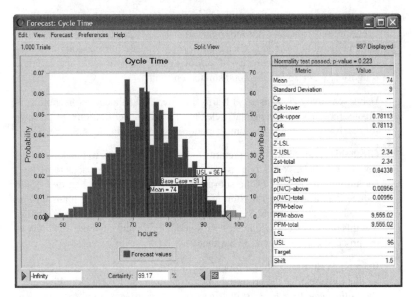

Figure 4-9   Improved Solution of Loan Process Cycle Time.

P-value of 0.223 and displays a mean of 74 hours, standard deviation of 9 hours, a C_{PK} of 0.78, and a Sigma level of 2.34. In this proposed state, only 0.83% of loans exceed the tolerance limit of 96 hours, with a maximum loan process time of 100 hours. With this data in hand and with the approval of the process owner, Walt and his team work with the loan specialists to implement the changes required to bring about the improved solution.

4.3   Case Study: Simulating and Optimizing a Model Built from a Designed Experiment

Example 4.1 demonstrates how simulation can identify and reduce the causes of variation, answering the question, "What if?" A more difficult and important question is "What is best?" Simulation alone is insufficient to identify the best solution. The combination of simulation and optimization, referred to as stochastic optimization, helps modelers find the controllable design variable settings resulting in the best statistical parameters of forecast variables. For example, stochastic optimization can find ways to design a system to minimize standard deviation or maximize C_{PK} in a critical output.

In a Crystal Ball model, the controllable design inputs are referred to as *decision variables*. Each decision variable is defined by upper and lower bounds and value type, which is either discrete or continuous. For discrete decision variables, the user also specifies the increment between potential values.

The classic application for stochastic optimization is the development of robust designs that optimize the tradeoff between the highest possible material quality and the lowest possible material cost. Example 4.2 below is a robust design example. Other optimization examples include production scheduling, project/strategy selection and prioritization, workforce queuing, and inventory optimization, illustrated in Example 4.3.

OptQuest global optimization software is an add-in created specifically for Crystal Ball software. The optimizer enhances Crystal Ball by searching for and finding optimal solutions to simulation models. A single OptQuest run comprises many Crystal Ball simulations, with perhaps 1000 trials per simulation, at different settings of the decision variables. OptQuest runs sequences of Crystal Ball simulations and uses a mixture of numerical techniques to find the right combination of decision variables for the best possible results. As the program runs, adaptive and neural network technologies help it learn from past simulations, so that it achieves better results in less time. The numerical methods used in OptQuest algorithms are collectively known as *metaheuristic optimization.* (Laguna, 1997 and 1997a) Stochastic optimization is particularly valuable for models with several decision variables and because of many of the complex conditions cited in Tables 4-1 and 4-2.

Example 4.2

A Six Sigma team, led by Linda, is working hard to improve a process that involves manufacturing plastic-injected molded parts. These parts have to meet length specifications (LSL of 59.0 and USL of 67.0), but variation in the process has resulted in incomplete filling of the mold and different part lengths.

Three process inputs that provide significant control over the process are mold temperature, cycle time, and hold pressure. Controls in the process adjust the mean value of these parameters to any desired value, but each has a wide variation around that average value. Since the equipment is old, it is difficult to troubleshoot and reduce variation in the control of these parameters. Linda and her team would like to find settings at which the variation in the control factors has less influence on the variation of the parts. In other words, the team wants to make the process more robust to variation in the control factors.

In the Measure phase of the project, the team collects measurements of mold temperature, cycle time, and hold pressure for 150 parts. Each of these datasets passes a normality test, and the mean of each is reasonably close to the control setting. The standard deviation of mold temperature and cycle time is 10 units, and the standard deviation of hold pressure is 5 units.

Table 4-4   Input Factors and Levels for the Experiment

Input Factor	Low	High
Mold Temperature ($X1$)	100	200
Cycle Time ($X2$)	60	140
Hold Pressure ($X3$)	120	140

Linda's plan is to use design of experiments (DOE) to build a transfer function model representing how part length changes as a function of the three control factors. Then, she will apply Monte Carlo simulation and stochastic optimization to look for ways to make the system more robust.

Using MINITAB software, the team defines, conducts, and analyzes an experiment with eight runs in a full factorial 2^3 treatment structure. This experiment includes five replicates of the eight runs for a total of 40 trials. Table 4-4 lists the three control factors, with two levels for each factor.

In their analysis, the team determines that all of the main effects have significant effects on part length, as does the interaction between mold temperature and cycle time. Figure 4-10 shows the MINITAB Session window listing the reduced model and final analysis of the experiment. The bottom section of the report lists the coefficients for the transfer function for part length. Using these coefficients, the model for the system as determined by the experiment is this: Part Length = −9.18 + 0.05755*MoldTemp + 0.09025*CycleTime + 0.4565*HoldPres − 0.0002825*(MoldTemp*CycleTime).

Now that the experiment has produced a transfer function representing the system, Linda and her team need to use that transfer function to learn more about the system. Since the three control factors vary, the team must predict how much part length varies in response to the three control factors. From this information, they can calculate the process capability and the probability of noncompliance. Monte Carlo simulation provides a quick way of answering these questions without producing or measuring any additional parts.

In an Excel worksheet, Linda recreates the transfer equation, as shown in Figure 4-11. Then, Linda defines three Crystal Ball assumptions, one for each of the three control variables, MoldTemp, CycleTime, and HoldPres. Each of these three assumptions has a normal distribution with mean and standard deviation referring to cells in the worksheet. In the worksheet, part length is expressed as a function of the three control variables. Linda runs a Crystal Ball simulation of 1000 trials, the virtual equivalent of 1000 plastic parts.

Analyzing the results, Linda determines that the part length process passes a normality test with a P-value of 0.752, that the defect rate is 2.35%, and that the

Factorial Fit: Length versus MoldTemp, CycleTime, HoldPres

Estimated Effects and Coefficients for Length (coded units)

Term	Effect	Coef	SE Coef	T	P
Constant		63.5850	0.2039	311.87	0.000
MoldTemp	2.9300	1.4650	0.2039	7.19	0.000
CycleTime	3.8300	1.9150	0.2039	9.39	0.000
HoldPres	9.1300	4.5650	0.2039	22.39	0.000
MoldTemp*CycleTime	-1.1300	-0.5650	0.2039	-2.77	0.009

S = 1.28946   R-Sq = 94.88%   R-Sq(adj) = 94.30%

Analysis of Variance for Length (coded units)

Source	DF	Seq SS	Adj SS	Adj MS	F	P
Main Effects	3	1066.11	1066.11	355.369	213.73	0.000
2-Way Interactions	1	12.77	12.77	12.769	7.68	0.009
Residual Error	35	58.20	58.20	1.663		
Lack of Fit	3	5.75	5.75	1.916	1.17	0.337
Pure Error	32	52.45	52.45	1.639		
Total	39	1137.07				

Estimated Coefficients for Length using data in uncoded units

Term	Coef
Constant	-9.18000
MoldTemp	0.0575500
CycleTime	0.0902500
HoldPres	0.456500
MoldTemp*CycleTime	-2.82500E-04

Figure 4-10   Reduced Model for Injection Molded Plastic Part Lengths.

sigma level is 1.95. The process, shown in Figure 4-12, is off center to the right, with a C_{PK} of 0.65. This indicates that more parts were longer than average instead of shorter than average.

Sensitivity analysis reveals that part length variation is greatly influenced (~50%) by the variation in the mold temperature. After research into mold temperature, Linda finds that she can reduce the standard deviation by half, from 10 degrees to 5 degrees. A subsequent simulation using this new parameter for mold temperature results in a sigma level of 2.5, with less than 1% defects.

With this improved model, Linda's team is now ready to finally address the optimization question: can the process settings be configured to achieve a minimum quality goal while at the same time reducing the process cost per part? By analyzing the relationship between cost and process parameters, Linda finds that the energy consumed by molding equipment is proportional to the

	C11	▼		f_x =B20*C7+B21*C8+B22*C9+B23*(C7*C8)+B19		
	A	B	C	D	E	F

1 Simulation with Design of Experiments

2 Length Specification

3	**LSL**	**USL**	**Target**			
4	59.0	67.0	63.0			
5						
6	**Factor**	**Name**	**Value**	**Nominal Value**	**Std Dev**	**Dist.**
7	1	MoldTemp	155	155	10	*Normal*
8	2	CycleTime	100	100	5	*Normal*
9	3	HoldPres	130	130	3	*Normal*
10						
11	**Response:**		Length	63.732		
12						

13 Transfer Function:

14 Length = 0.05755MoldTemp + 0.09025CycleTime + 0.4565HoldPres + -0.0002825MoldTemp*CycleTime + -9.18

15

16 Results from Minitab

17 Estimated Coefficients for Length using data in uncoded units

	Term	**Coef**
18		
19	Constant	-9.18
20	MoldTemp	0.05755
21	CycleTime	0.09025
22	HoldPres	0.4565
23	MoldTemp*CycleTime	-2.83E-04
24		

Figure 4-11   Microsoft Excel Model for Injection Molded Plastic Part Lengths.

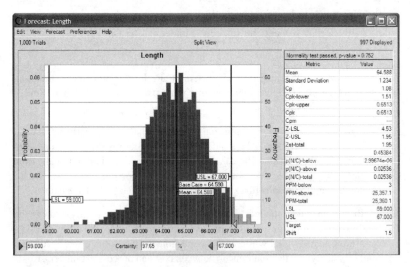

Figure 4-12   Initial Simulation Forecast of Part Length.

product of cycle time and mold temperature. She also finds that the labor cost to run molding equipment is proportional to cycle time. She builds a process cost equation into her Excel model where $Cost = K1 \times Time \times Temp + K2 \times Time$. She defines the cost as a second Crystal Ball forecast. The cost per part of the non-optimized design is $2.03.

Linda assumes that she can control the mean value of the three control variables, so she defines the mean of each of the three inputs as a decision variable. The range of each decision variable is based on the low and high levels from the designed experiment, listed in Table 4-4. In OptQuest, Linda sets the primary goal to be minimizing process cost, with a requirement that the sigma level (Zst) of part length be at least 4.

After 800 OptQuest runs (each representing 1000 Crystal Ball simulations), the computer identifies a new, more robust design with a process cost of $1.16 per part and part length with a sigma level of 4. Figure 4-13 graphs the lowest cost per part found by OptQuest during the optimization run.

For additional assurance, Linda repeats the simulation of the optimal settings at a greater number of trials to verify that the final design solution meets performance requirements with a lower cost than the first-cut design. For comparison, Table 4-5 lists the settings and results for the original design, after reducing variation in temperature control, and after the optimization process.

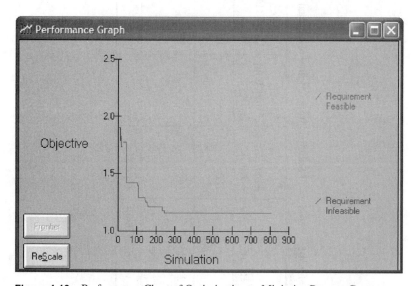

Figure 4-13   Performance Chart of Optimization to Minimize Process Cost.

Table 4-5 Iterations of Solutions from Original to Optimal State

Iteration #	Mold Temp Mean	Mold Temp StDev	Cycle Time Mean	Cycle Time StDev	Hold Press Mean	Hold Press StDev	Sigma Level of Part Length	Process Cost
1 (original settings)	160	10	100	10	130	5	1.94	$2.03
2 (reduced MoldTemp StDev)	160	5	100	10	130	5	2.53	$2.03
3 (optimization)	150	5	61	10	140	5	4.01	$1.16

4.4 Case Study: Perishable Inventory Optimization

The case study in this section illustrates the use of Monte Carlo simulation and stochastic optimization to improve the performance of a complex inventory system.

Example 4.3

In this project, a company that sells perishable inventory uses Monte Carlo simulation within the Six Sigma process to create and maintain an optimal balance between lost sales and wasted inventory. In the recent past, the company has been turning away customers with increasing frequency due to unanticipated demand and a lack of inventory. This problem is identified and delegated to Marco, a Six Sigma Black Belt, and his project team for resolution.

The first step in the Six Sigma project is to clearly define the defect and project objectives. A defect is defined as any instance in which the company turns a customer away because the company does not have the materials required to complete the customer's order. By tracking the number of declined orders for the past ten years, Marco's team calculates that this statistic had been increasing annually, with a record 1631 refused orders in just the past year!

Using historical sales data and time-series forecasting, the team forecasts an expected value of 2113 lost sales for the current year. Marco performs this analysis with CB Predictor, a time-series forecasting and multiple linear regression tool, included in Crystal Ball Professional Edition. The team then creates a simple Excel-based sales model and, with Monte Carlo simulation, predicts an expected loss of $211,000 this year due to lost sales, with a minimum loss of $166,000 and a maximum loss of $258,000.

Marco's team wants to reduce the cost of lost sales, but they must balance this goal with the need to control the cost of perishable materials discarded due to expiration. At the end of the Define phase, the team defines an overall project goal of reducing the defect rate by 75% without increasing inventory costs, which were $169,000 the previous year. Inventory costs include cost of orders, cost of discarded materials, and cost of lost sales. For this project to succeed, the company needs no more than 408 lost sales for the year and total inventory costs of no more than $169,000.

In the Measure phase, Marco's team develops and validates an Excel-based inventory process model, shown in Figure 4-14. This model describes inventory costs over a period of 52 weeks. Using several years of historical data, they use Crystal Ball to define two sets of similar assumptions for each week:

- The weekly demand has a Poisson distribution with a rate of 150 units per week. Figure 4-15 shows the frequency plot for this assumption, and Chapter 23 lists more information on the Poisson family of distribution.
- Lead time has a custom distribution shown in Figure 4-16.

Inventory Simulation With Lost Sales

Order Quantity	200	units
Reorder Point	800	units
Initial Inventory	400	units

Order Cost	$ 50	
Cost of Waste per Unit	$ 50	
Lost Sales Cost	$ 100	

Optimize order quantity and reorder point to maximize adjusted profit

Total Cost (Including Lost Sales): $ 7,050 $ 2,100 $ -

Total Cost $ 9,150

Week	Beg Inv Pos	1 Week Old Inv	2 Week Old Inv	Beg Inv	Order Rec'd	Units Rec'd	Dmd	Waste Units	End Inv	Lost Sales	Order Placed?	Ending Inv Pos	Week Due	Waste Cost	Order Cost	Short Cost	Total Cost
1	400	0	0	0		400	156	0	244	0	YES	444	3		$ 50	-	$ 50
2	444	244	0	244		0	143	0	101	0	YES	501	4		$ 50	-	$ 50
3	501	0	101	101	YES	200	138	0	163	0	YES	563	6		$ 50	-	$ 50
4	563	163	0	163	YES	200	175	0	188	0	YES	588	7		$ 50	-	$ 50
5	588	188	0	188		0	158	0	30	0	YES	630	7		$ 50	-	$ 50
6	630	30	0	30	YES	200	165	0	65	0	YES	665	9		$ 50	-	$ 50
7	665	65	0	65	YES	400	144	0	321	0	YES	721	11		$ 50	-	$ 50
8	721	321	0	321		0	130	0	191	0	YES	791	12		$ 50	-	$ 50
9	791	65	126	191	YES	200	150	0	241	0	YES	841	13		$ 50	-	$ 50
10	841	241	0	241		0	159	0	82	0	YES	882	14		$ 50	-	$ 50
11	882	65	17	82	YES	200	138	0	144	0	YES	944	13		$ 50	-	$ 50
12	944	144	0	144	YES	200	158	0	186	0	YES	986	15		$ 50	-	$ 50
13	986	186	0	186	YES	400	156	0	430	0	NO	830	16			-	$ 50
14	830	430	0	430	YES	200	179	0	451	0	YES	851	16		$ 50	-	$ 50
15	851	386	65	451	YES	200	145	0	506	0	YES	906	18		$ 50	-	$ 50
16	906	506	0	506	YES	200	138	0	568	0	YES	968	18		$ 50	-	$ 50
17	968	568	0	568		0	145	0	423	0	NO	823				-	$ 50
18	823	423	0	423	YES	400	151	0	672	0	YES	872	22		$ 50	-	$ 50
19	872	672	0	672		0	150	0	522	0	YES	922	22		$ 50	-	$ 50
20	922	522	0	522		0	135	0	387	0	NO	987	22			-	$ 50
21	987	387	0	387		0	150	0	237	0	YES	837	24		$ 50	-	$ 50
22	837	237	0	237	YES	600	153	0	684	0	YES	884	24		$ 50	-	$ 50
23	884	684	0	684		0	150	0	534	0	YES	934	25		$ 50	-	$ 50
24	934	534	0	534	YES	200	156	141	437	0	YES	837	27	$ 7,050.00	$ 50	-	$ 7,100
25	902	437	297		YES	200	156	0	602	0	YES		27		$ 50	-	$ 50

Figure 4-14 Microsoft Excel Model of Inventory Process.

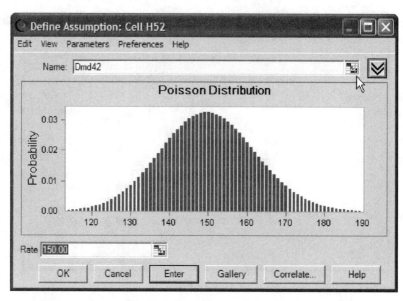

Figure 4-15   Poisson Distribution for Weekly Demand Variation.

The primary forecast is total cost, which is the sum of waste costs, order costs, and lost sales costs for the year. Marco sets order quantity = 250, reorder point = 250, and initial inventory level = 250, which reflects the current operating discipline. A simulation of the current state indicates that, over the 52-week

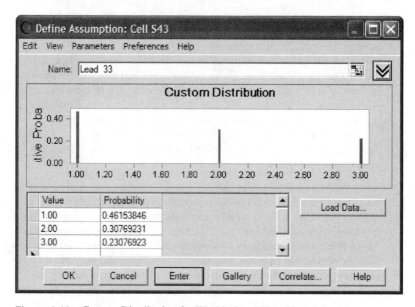

Figure 4-16   Custom Distribution for Weekly Lead Time Variation.

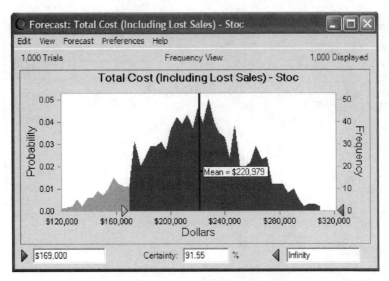

Figure 4-17   Simulation Forecast for Total Lost Sales.

period, the company would lose between 1200 and 3100 sales due to lack of inventory, with a mean value of lost sales near 2150. No doubt, the increased number of lost sales would be a major contributor to the 91.6% likelihood that the total inventory costs would exceed the previous year's value of $169,000. Figure 4-17 shows the predicted distribution of total cost.

Now in the Analyze phase, the team uses the Crystal Ball Tornado Chart tool to determine which inputs had the greatest influence on the total lost sales forecast. The tornado chart represents a one factor at a time (OFAT) test of the model, which varies each assumption and decision variable independently and on a pre-simulation basis. Using the Tornado Chart tool, Marco finds that the reorder point, order quantity, and initial inventory are the most influential on the number of lost sales. Because these inputs are controllable, Marco defines each of these as a decision variable with a lower bound of 0 units, an upper bound of 1500 units, and a discrete increment of 100 units. Standard operating procedure dictates that orders of perishable inventory are in increments of 100 units.

In the Improve phase, the team uses OptQuest to minimize the mean total lost sales and to find the optimal settings for reorder point, order quantity, and initial inventory. The team includes a requirement that the optimal solution will provide at least 90% certainty that the total inventory costs would not exceed $169,000. OptQuest eventually converges on a reorder point of 1400, an order quantity of 300, and an initial inventory of 700. The model indicates that the company can reduce the expected number of lost sales to less than 21, with a 100% chance of reaching the goal of 408 or less lost sales. Figure 4-18 shows

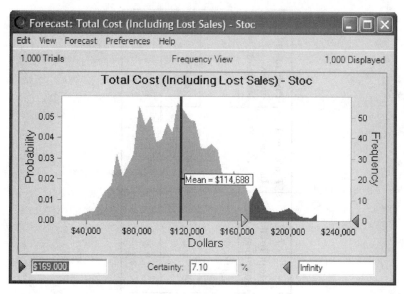

Figure 4-18   Optimized Forecast for Total Lost Sales.

the total cost forecast for this optimal solution, which will reduce the likelihood of exceeding the previous year's inventory costs from 91.6% to 7.1%.

Using this knowledge, Marco and his team address the final question of whether they might be able to reach their goal of 75% defect reduction and achieve even more substantial gains in terms of reduced costs. They run another OptQuest optimization with the new objective of minimizing the mean total cost subject to the following requirements: (1) 90% certainty that the number of lost sales will not exceed the target value of 408, and (2) 90% certainty that total costs will not exceed last year's value of $169,000.

This optimization gives the team a new solution of a reorder point of 800, an order quantity of 200, and an initial inventory of 400. Despite the side effect that this solution raises the expected number of lost sales to 65, the company can now maintain a 100% certainty of reaching the goal of 408 or fewer lost sales. This solution virtually eliminates the possibility of any increase in total inventory costs, because the total expected inventory cost is only $17,000, a $162,000 reduction from the previous year's costs.

All of these predictions depend on the validity of the original assumptions and the inventory model. Before implementing long-term change in the process, Marco's team needs to validate the performance of the optimized settings. For comparison purposes, Table 4-6 lists the settings and results for each step of the problem-solving process.

Table 4-6 Iterations of Solutions from Original to Optimal State

Iteration #	Reorder Point	Order Quantity	Initial Inventory	Mean No. Lost Sales	Mean Total Inventory Cost	Prob. Total Inv. Costs > $169,000	Prob. of 408 or less lost sales
1 (simulated original state)	250	250	250	2150	$211,000	91.6%	0%
2 (first optimization)	300	1400	700	< 21	$115,000	7.1%	100%
3 (second optimization)	200	800	400	65	$17,000	~0.0%	100%

In the Control phase of the project, the team develops a plan to ensure that the anticipated gains are achieved and maintained. They use Monte Carlo simulation to forecast the 90th percentile of the total inventory costs for each week for comparison with actual costs. Going forward, if the actual costs exceeded the 90th percentile from the forecast, the process owner would incorporate the latest demand and delivery lead times into the assumptions and repeat the optimization to determine if the optimum reorder point or order quantity had changed.

In the final project review, Marco and his team were asked if running a stochastic optimization made a difference over a purely deterministic assessment of inventory. Marco said that he also used a deterministic optimizer to analyze this problem. The deterministic optimizer recommended an optimal solution of order quantity 300, reorder point 400, and initial inventory 500, without considering the variability in demand and lead delivery time and using the average values of 149 units and 2 weeks. Comparing these settings to the final optimized settings, Marco and his team calculated that 99.6% of the time, the stochastic solution resulted in lower costs than the deterministic solution. Displaying the results in the overlay chart shown in Figure 4-19, Marco proved that the average savings per year by using stochastic optimization instead of deterministic optimization is $85,000.

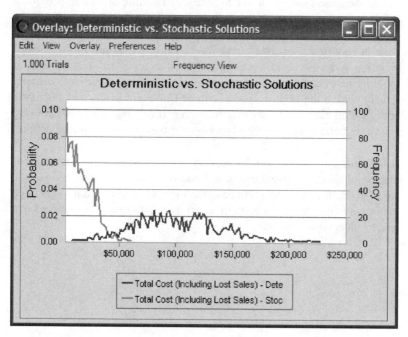

Figure 4-19   Comparison of Deterministic and Stochastic Inventory Solutions.

4.5 Benefits of Simulation and Optimization

As discussed in Chapter 2, every software tool has strengths and weaknesses. In the case of Monte Carlo simulation, the weakness often lies in the quality of the simulation model. All simulation and optimization models should be tested for accuracy and applicability, and practitioners should be able to defend their reasoning for applying specific probability distributions within a model. Simulation can be used as a testing tool to identify poor model design and flaws in defining assumptions. Poorly built models will cause a simulation tool to shut down due to calculation errors, and poorly defined assumptions will lead to inaccurate and unrealistic forecast and sensitivity charts. Inexperienced modelers and those with particularly narrow agendas can easily misrepresent and misuse simulation results.

When applied in the proper context, Monte Carlo simulation and stochastic optimization are increasingly popular tools, because their potential strengths far outweigh their potential weaknesses. For project and financial managers, for example, simulation can help to account for cost, schedule, and success uncertainty in initial stages of a project and can help to delineate the impacts of a proposed change on customer satisfaction and profitability. In the Measure phase of a Six Sigma project, simulation can provide a baseline process and capability measurement and can estimate infrequent, costly, or nonexistent data. In the Analyze phase, sensitivity analysis plays a major role in the discovery and validation of the underlying causes of variation and waste, and works equally well as a communication tool for building team consensus and gaining early approval of process owners. Finally, within the Improve phase, Monte Carlo simulation and stochastic optimization can help teams save resources, time, and money by allowing them to experiment on proposed process changes prior to implementation.

In the end, any tool that can harness the power of probability distributions to drive insights, increase quality, and improve customer satisfaction is one worth using. It is not enough to describe variation or uncertainty. It is more important to use that knowledge to reveal the solutions and innovations that lead to successful projects.

5

Glossary of Terms

The following chapters contain descriptions of many distribution families. This chapter defines symbols, functions, and terminology used throughout this book. References to other entries in this glossary are *italicized*.

^: The symbol ^ over a *parameter* denotes a *statistic* that estimates the parameter. For example, if μ is the *mean* of a *random variable*, then $\hat{\mu}$ is a statistic calculated from sample data that estimates μ.

~: The symbol ~ means "is distributed as." For example, if X is a *random variable* with *CDF F*, then $X \sim F$. Also, if X is a normal random variable with *mean* μ and *standard deviation* σ, then $X \sim N(\mu,\sigma)$.

$\overset{iid}{\sim}$**:** The abbreviation iid is shorthand for "independent and identically distributed."

$\approx$ **:** This symbol means "is approximately distributed as."

$\overset{D}{\longrightarrow}$ **:** This symbol means "is asymptotically distributed as." The arrow suggests a limit, as the sample size gets very large. The D above the arrow stands for distribution.

Average: *Average* is a synonym for *mean*. In common usage, some people confuse the terms *average*, *mean*, and *median*. To reduce misunderstanding, use of the specific terms *mean* and *median* is recommended instead of *average*.

Average Run Length: The *average run length* (ARL) of a control chart is the average number of plot points between points indicating an out of control condition. When Shewhart control charts are applied to a stable normal distribution, the ARL is approximately 370 for X and $\overline{X}$ charts, corresponding to a false alarm rate $\alpha = 1/ARL = 0.0027$. For stable data, a higher ARL indicates fewer false alarms. In general, control charts and interpretation rules require that ARL > 100 to be acceptable.

The ARL can also measure the power of a control chart to detect changes in the distribution. For normal data with the mean μ shifted by 1.5σ, an individual X chart has a ARL of 15, but an $\overline{X}$ chart with subgroup size of 4 has an ARL of 2. This means that the $\overline{X}$ chart has more power to detect smaller shifts than an individual X chart.

When normal Shewhart control charts are applied to nonnormal data, the ARL changes for both stable and unstable processes. A Monte Carlo simulation of a control chart with nonnormal data can estimate ARL to determine if the control chart is appropriate for a specific application.

Beta Function: The *beta function*, represented by $B(\alpha, \beta)$ is defined to be $B(\alpha, \beta) = \int_0^1 t^\alpha (1 - t)^\beta dt$. The beta function is easy to calculate from the *gamma function* using this formula: $B(\alpha, \beta) = \frac{\Gamma(\alpha)\Gamma(\beta)}{\Gamma(\alpha + \beta)}$. To calculate $B(\alpha, \beta)$ in an Excel worksheet, use this formula =EXP(GAMMALN(α)+GAMMALN(β)-GAMMALN($\alpha+\beta$)).

Coefficient of Excess Kurtosis: The *coefficient of excess kurtosis* of a random variable X, denoted by $ExKurt[X]$, is the *coefficient of kurtosis* minus 3. $ExKurt[X] = Kurt[X] - 3$. For any normal random variable X, $ExKurt[X] = 0$. Many software programs, including Excel, MINITAB, and JMP calculate the coefficient of excess kurtosis and simply call it "kurtosis." See the definition of *coefficient of kurtosis* for more information on interpreting values of $ExKurt[X]$.

Coefficient of Kurtosis: The *coefficient of kurtosis* of a random variable X, denoted by $Kurt[X]$, is a unitless measure of the shape of a distribution. Kurtosis represents the relationship between the probabilities of middle values and probabilities of extreme values. The definition is $Kurt[X] = \frac{E[(X - E[X])^4]}{(SD[X])^4}$. The *coefficient of excess kurtosis* $ExKurt[X]$ is related by the formula $ExKurt[X] = Kurt[X] - 3$. If X has a normal distribution, then $Kurt[X] = 3$, providing a reference for the kurtosis of other distributions. If $Kurt[X] < 3$ or $ExKurt[X] < 0$, then X is said to be *platykurtic*. Platykurtic random variables tend to have a flatter middle section or truncated tails, relative to a normal distribution. If $Kurt[X] > 3$ or $ExKurt[X] > 0$, then X is said to be *leptokurtic*. Leptokurtic random variables tend to have a sharper middle peak or heavier tails, relative to a normal distribution. If $Kurt[X] = 3$ or $ExKurt[X] = 0$, then X is said to be *mesokurtic*. Books, articles, and software may use either version of kurtosis. Material written by and for academia usually uses $Kurt[X]$, while material written for practitioners usually uses $ExKurt[X]$. Crystal Ball software defines "kurtosis" as $Kurt[X]$, while Excel, MINITAB, and JMP software all calculate $ExKurt[X]$, and call it "kurtosis." If in doubt about which version of kurtosis is used by a program, generate at least 10,000 normal random numbers and use the program to estimate the kurtosis of those numbers. If the estimate is nearly 0, the program uses $ExKurt[X]$. If the estimate is nearly 3, the program uses $Kurt[X]$.

Coefficient of Skewness: The *coefficient of skewness* of a random variable X, denoted by $Skew[X]$, is a unitless measure of the shape of a distribution. Skewness measures the lack of symmetry of a distribution around a middle value. The definition is $Skew[X] = \frac{E[(X - E[X])^3]}{(SD[X])^3}$. When $Skew[X] < 0$, X is said to be "skewed to the left," because it has a heavier tail on the left than on the right. When $Skew[X] > 0$, X is said to be "skewed to the right," because it has a heavier tail on the right than on the left. For symmetric distributions like the normal distribution, $Skew[X] = 0$. It is possible for an asymmetric distribution to have $Skew[X] = 0$. One example of this is a Weibull random variable with a shape parameter $\alpha = 3.602$. While this Weibull random variable has zero skewness and a "nearly normal" probability curve, it has a lower bound of zero, which makes it asymmetric.

Coefficient of Variation: The *coefficient of variation* of a random variable X, denoted by $CV[X]$, is the ratio of the standard deviation of X to the mean of X. That is, $CV[X] = SD[X]/E[X]$. The coefficient of variation is a unitless measure of variation, and is useful for comparing random variables with very different means, or with different units of measure. The coefficient of variation is not useful for random variables that may have zero or negative values.

Continuous: A *continuous random variable* may assume any value within a set of real numbers. A *continuous distribution* is a distribution describing the probabilities associated with a continuous random variable.

Cumulative Distribution Function (CDF):   The *cumulative distribution function* or *CDF* of a *random variable* X, denoted by $F_X(x)$, is a function representing the probability of observing a value less than or equal to x. That is, $F_X(x) = P[X \le x]$. Every CDF has the same properties that $F_X(-\infty) = 0$, $F_X(+\infty) = 1$, and it is monotone nondecreasing. The monotone nondecreasing property means that $F_X(B) \ge F_X(A)$ for every $B \ge A$, and that the CDF has an inverse function $F_X^{-1}(p)$ which is unique everywhere except on sets of zero probability. The *inverse CDF* is very useful for calculating *quantiles* and for generating random numbers. For any number p between 0 and 1, $F_X^{-1}(p)$ is the *p-quantile* of X. If U is a random number uniformly distributed between 0 and 1, then $F_X^{-1}(U)$ is a random number which follows the distribution of X. This property is called the probability integral transform.

Discrete: A *discrete random variable* is limited to a countable set of distinct values. In most Six Sigma applications, discrete random variables represent counts, and are limited to nonnegative integers $(0, 1, 2 \ldots)$. A *discrete distribution* is a distribution describing the probabilities associated with a discrete random variable.

Distribution: A *distribution* is a mathematical model expressing the relative probability that a random variable will assume any set of values. Other names for a distribution include "probability distribution" and "probability law."

Estimator: An *estimator* is a *statistic* calculated from sample data to estimate the value of a *parameter* of the distribution of a *random variable*.

Expected Value: The *expected value*, also called the expectation, of a random variable X is a function representing the *mean* of X. The expected value is denoted by $E[X]$. If X is a *discrete* random variable with *support S* and *probability mass function (PMF)* $f_X(x)$, then $E[X] = \sum_{x \in S} x f_X(x)$. If X is a *continuous* random variable with *support S* and *probability density function (PDF)* $f_X(x)$, then $E[X] = \int_S x f_X(x) dx$.

Gamma Function: The *gamma function*, denoted by $\Gamma(x)$ is shorthand notation for this quantity: $\Gamma(x) = \int_0^\infty e^{-u} u^{x-1} du$. $\Gamma(x)$ has a real-number value for all $x > 0$. The gamma function is related to factorials by this relationship: $\Gamma(x + 1) = x!$ Most statistical and mathematical software can evaluate the gamma function. Excel computes the natural log of the gamma function with its GAMMALN function. To calculate $\Gamma(x)$ in an Excel worksheet, use the formula =EXP(GAMMALN(x)).

Hazard Function: The *hazard function* of a random variable X, denoted by $h_X(x)$, is defined by $h_X(x) = f_X(x)/R_X(x)$, where $f_X(x)$ is the *probability density function* and $R_X(x)$ is the *survival function* of X. In reliability applications, the hazard function represents the instantaneous rate of failures at time x. Among all the units surviving at time x, the probability of failure in the next small unit of time is proportional to $h_X(x)$. The hazard function is the inverse of the *Mills ratio*.

Inverse Cumulative Distribution Function: The *inverse CDF* of a random variable X, denoted by $F_X^{-1}(prob)$ represents a value that is greater than or equal to X with probability $prob$. That is, $P[X \le F_X^{-1}(prob)] = prob$. Also, $F_X^{-1}(prob)$ is a *prob-quantile* of X. Since $prob$ is a probability, $F_X^{-1}(prob)$ is only defined for $prob \in [0, 1]$. Some distribution families use p to represent a parameter, which is also a probability. In the context of the inverse CDF, this book uses $prob$ to distinguish it from the parameter p. One important use for the inverse CDF is to generate random numbers representing simulated observations of X. Uniform random numbers between 0 and 1, passed through the inverse CDF of X, have the same distribution as X. That is, if $U \sim Unif(0, 1)$, then $F_X^{-1}(U) \sim X$. This property is known as the "probability integral transform." Crystal Ball and most other statistical software use this property to generate random numbers for any given distribution.

Kurtosis: See *coefficient of kurtosis*.

Leptokurtic: See *coefficient of kurtosis*.

Location Parameter: θ is a *location parameter* of a *parametric family* of *distributions* with CDF $F_{X(\theta)}(x)$ if $F_{X(0)}(x + \theta) = F_{X(\theta)}(x)$ for all x, and for all values of θ. Examples of location parameters include the mean μ of a normal distribution, the mode η of an extreme value distribution, or the threshold parameter τ of a two-parameter exponential distribution. In effect, the location parameter adds itself to the value of a random variable. When the location parameter is optional, its default value is 0. A location parameter θ may be added to any family which does not have one by replacing the value x with $x - \theta$ in all formulas.

Maximum Likelihood: *Maximum likelihood* is a method of estimating *parameters* of a *random variable* X based on a sample of observed data. Informally, a maximum likelihood estimator (MLE) is the most likely value for a parameter, given the observed data and an assumed distribution family. The method of calculating an MLE depends on which parametric family is selected for X, and on other assumptions. For some families, explicit formulas for MLEs are available, but in general, a computer must calculate MLEs using iterative numerical methods. One technical advantage to MLEs is that they tend to be normally distributed, as the sample size gets large. For this reason, approximate confidence intervals are always available for MLEs. MLEs are sometimes *unbiased*, but often they are not. When both an unbiased estimator and an MLE are available, the unbiased estimator is generally preferred.

Mean: The *mean* of a random variable X is the *expected value* $E[X]$. The mean has the same units of measure as X. The mean is analogous to the center of gravity of a system of masses. If the *PDF* or *PMF* of X is symmetric around a value μ, then μ is both the *mean* and the *median* of X. If X is "skewed to the right" by a heavy upper tail, then the mean is greater than the median. If X is "skewed to the left" by a heavy lower tail, then the mean is less than the median. The mean may also be called the "arithmetic mean" to distinguish it from the less often used geometric mean, harmonic mean, and other means.

Median: The *median* of a random variable X is the 0.5-*quantile* or the 50th *percentile* of X. Informally, the median is the value separating the lower 50% from the upper 50% of possible values. If x is a median of X, then $P[X \le x] \ge 0.5$ and $P[X \ge x] \ge 0.5$. The median has the same units of measure as X. Some random variables, particularly discrete random variables, do not have a unique median. The usual convention in these cases is to report the midpoint of the range of possible medians as the median. This is similar to the procedure used to calculate sample medians for even sample sizes.

Mesokurtic: See *coefficient of kurtosis*.

Mills ratio: The *Mills ratio* of a random variable X is the ratio of the *survival function* to the *probability density function*. $m(x) = \frac{R(x)}{f(x)}$. The Mills ratio is the inverse of the *hazard function*.

Mode: The *mode* of a *discrete* random variable X is the value with the largest probability of occurrence. The *mode* of a *continuous* random variable X is the value x at which the *PDF* $f_X(x)$ is at a local maximum. Some random variables have multiple modes. The uniform random variable has no modes.

Moment Generating Function: The *moment generating function* (MGF) of a random variable is the *expected value* of e^{tX}, that is, $M_X(t) = E[e^{tX}]$. The MGF has two major uses. First, the rth derivative of $M_X(t)$, evaluated at 0, is the rth raw moment of X, $\mu'_r = E[X^r]$. That is, $\mu'_r = \frac{\partial^r}{\partial t^r} M_X(t)|_{t=0}$. Common properties of X are calculated as functions of these raw moments. $E[X] = \mu = \mu'_1$, $V[X] = \mu'_2 - \mu^2$, $Skew[X] = \frac{\mu'_3 - 3\mu'_2\mu + 2\mu^3}{(V[X])^{3/2}}$, and $Kurt[X] = \frac{\mu'_4 - 4\mu'_3\mu + 6\mu'_2\mu^2 - 3\mu^4}{(V[X])^2}$. The second major use of the MGF is to identify the distribution of functions of random variables. There is a 1:1 relationship between the PDF and MGF of random variables. Calculating the PDF for a MGF is sometimes difficult, but if the form of the MGF matches the form of a known distribution family, then the PDF must belong to that family.

Parameter: A *parameter* is a real number describing a defined characteristic of a distribution. Each *parametric family* of distributions has a different set of parameters. For example, within the normal family of distributions, the mean μ and standard deviation σ are parameters.

Parametric Family: A set of *distributions* described by the same mathematical function of one or more *parameters* is a *parametric family*. For example, the normal distribution family comprises an infinite number of distributions with the same bell-shaped *probability density function*. An individual normal distribution from the family is identified by specifying the values of two parameters, its mean μ and standard deviation σ

Percentile: The Pth percentile of a random variable X is a value x such that X has probability $P/100$ of being less than or equal to x and probability $1\text{-}(P/100)$ of being greater than or equal to x. That is, $P[X \leq x] \geq \frac{P}{100}$ and $P[X \geq x] \geq 1 - \frac{P}{100}$. The Pth percentile is the same as the $(P/100)$-quantile. Informally, the Pth percentile separates the lower $P\%$ from the upper $(100\text{-}P)\%$ of the possible values. There may be no unique value of a percentile, especially for *discrete* random variables. The Pth percentile of any random variable is calculated by the *inverse cumulative distribution function* $F^{-1}\left(\frac{P}{100}\right)$. Mathematically, percentiles and quantiles are interchangeable, but many people find percentiles easier to understand.

Platykurtic: See *coefficient of kurtosis*.

Probability Density Function (PDF):   The *probability density function* or *PDF* of a *continuous random variable* X, denoted by $f_X(x)$, is a function representing the relative probability that X will assume any value x in the *support* of X. Note that a continuous random variable has zero probability of assuming any single value. The probability of observing any value within a range of values is determined by the area under the PDF curve. That is, $\int_A^B f_X(x)dx = P[A < X < B]$ for every real numbers A and B such that $A < B$.

Probability Mass Function (PMF):   The *probability mass function* or *PMF* of a *discrete random variable* X, denoted by $f_X(x)$, is a function representing the probability that X will assume any value x in the *support* of X. That is, $f_X(x) = P[X = x]$ for every value x in the *support* of X, and $f_X(x) = 0$ otherwise.

Quantile: The p-*quantile* of a random variable X is a value x such that X has probability p of being less than or equal to x and probability $1\text{-}p$ of being greater than or equal to x. That is, $P[X \leq x] \geq p$ and $P[X \geq x] \geq 1 - p$. Informally, the p-quantile separates the lower $100p\%$ from the upper $100(1-p)\%$ of the possible values. There may be no unique value of a p-quantile, especially for *discrete* random variables. The p-quantile of any random variable is calculated by the *inverse cumulative distribution function* $F^{-1}(p)$. *Medians*, quartiles, deciles, and *percentiles* are all specific types of quantiles.

Random Variable: Any system or process that produces observed data is a random variable. Each observation of a random variable may have the same or different values. The relative probabilities of observing different values of a random variable are described by a *distribution*.

Reliability Function: See *survival function.*

Scale Parameter: β is a *scale parameter* of a *parametric family* of *distributions* with CDF $F_{X(\beta)}(x)$ if $F_{X(\beta)}(x) = F_{X(1)}(\beta x)$ for all x, and for all values of β. If the family has both a *location parameter* θ and a scale parameter β, then $F_{X(\theta,\beta)}(x) = F_{X(0,1)}(\beta x + \theta)$ for all x, θ, and β. Examples of scale parameters include the standard deviation σ of a normal distribution and scale parameters of Weibull, gamma, and many other families. In effect, the scale parameter multiplies itself by the value of a random variable. When the scale parameter is optional, its default value is 1. A scale parameter β may be added to any family which does not have one. The PDF or PMF of a random variable $f_X(x)$ with an added scale parameter β is $f_{X(\beta)}(x) = \frac{1}{\beta} f_X(\frac{x}{\beta})$. The CDF of a random variable $F_X(x)$ with an added scale parameter β is $F_{X(\beta)}(x) = F_X(\frac{x}{\beta})$.

Shape Parameter: α is called a *shape parameter* of a *parametric family* of *distributions* if it changes the shape of the density or mass function, and it is not a scale or location parameter.

Skewness: See *coefficient of skewness.*

Standard Deviation: The *standard deviation* of a random variable X is a measure of the variation of X, with the same units of measure as X. The standard deviation, denoted by $SD[X]$ is defined as $SD[X] = \sqrt{E[(X - E[X])^2]}$. An alternate formula is $SD[X] = \sqrt{E[X^2] - (E[X])^2}$.

Statistic: A *statistic* is a numerical characteristic of a sample. Some statistics are intended to estimate *parameters* of a *random variable*. For example, the statistic $\overline{X}$ is an estimator of the mean μ. A carat ($\wedge$) over a parameter symbol denotes a statistic intended to estimate the parameter. For example, $\hat{\mu} = \overline{X}$.

Support: The *support* of a random variable X is a set of observed values of X with probability greater than zero. For example, the support of a normal random variable is the entire real number line $(-\infty, +\infty)$. The support of a Poisson random variable is the set of nonnegative integers $\{0, 1, \dots \}$.

Survival Function: The *survival function* (also called the *reliability function*) of a *random variable* X, denoted by $R_X(x)$, is a function representing the probability of observing a value greater than x. That is, $R_X(x) = P[X > x] = 1 - F_X(x)$. The survival function is typically used in reliability analysis, when X represents a time to failure.

Unbiased: An *unbiased estimator* is an estimator with the desirable property that the expected value of the estimator is equal to the parameter value. That is, if $\hat{\theta}$ is an estimator of θ, $\hat{\theta}$ is unbiased if $E[\hat{\theta}] = \theta$.

Variance: The *variance* of a random variable X is a measure of variation of X. The units of measure of variance are the square of the units of measure as X. The variance, denoted by $V[X]$ is defined as $V[X] = E[(X - E[X])^2]$. An alternate formula is $V[X] = E[X^2] - (E[X])^2$.

Chapter

6

Bernoulli (Yes-No)
Distribution Family

A Bernoulli random variable has one of the simplest of all distributions, with only two possible values, 0 and 1. In modeling, a Bernoulli random variable represents whether any event occurs or not. This application is so common that Crystal Ball software calls this distribution the "Yes-No" distribution. In the Bernoulli distribution family, the single parameter p represents the probability of observing the value 1.

The Bernoulli random variable is a basic building block of many other discrete distributions, including binomial, geometric, and negative binomial.

The name of the Bernoulli distribution honors Jakob Bernoulli (1654–1705), who developed the law of large numbers and the binomial distribution in his book *Ars Conjectandi* (*The Art of Conjecture*), published in 1713. The honor could extend to seven other Bernoullis, all related, and all celebrated mathematicians and scientists. In *Against the Odds*, Bernstein (1998) provides an enjoyable and readable account of the Bernoullis and many other pioneers of probability.

Example 6.1

Jack is simulating a business plan to predict the expected profit and the risk of losing money. He will market his product to three potentially large customers who will either award him a contract or not. In the Crystal Ball model representing profit projections for his business plan, Jack defines one Yes-No assumption for each of the three potential customers. Calculations for revenue and profit depend on the values of the three Yes-No assumptions.

Parameters: The Bernoulli or Yes-No distribution family has one parameter: p represents the probability of observing the value 1. $0 \le p \le 1$

Figure 6-1 Jakob Bernoulli.

Representation: $X \sim Bern(p)$

Support: $\{0, 1\}$

Relationships to Other Distributions:

- The sum of n independent Bernoulli random variables is a binomial random variable with number of trials n. If $X_i \overset{iid}{\sim} Bern(p)$ for $i-1, \ldots n$, then $\sum_{i=1}^{n} X_i \sim Bin(n, p)$.
- A Bernoulli random variable is a special case of a binomial random variable, with number of trials $n = 1$. If $X \sim Bin(1, p)$, then $X \sim Bern(p)$.
- Consider an experiment consisting of a series of independent $Bern(p)$ random variables, also called Bernoulli trials. In this experiment, p is the probability of outcome "A" in each trial. Several other families of random variables describe aspects of this experiment:
  - The number of outcomes "A" in a set of n trials is a binomial random variable, $Bin(n, p)$.
  - The number of outcomes "B" before the first outcome "A" is a geometric random variable, $Geom_0(p)$.
  - The number of trials up to and including the first outcome "A" is a geometric random variable, $Geom_1(p)$.
  - The number of outcomes "B" before the kth outcome "A" is a negative binomial random variable, $NB_0(k, p)$.
  - The number of trials up to and including the kth outcome "A" is a negative binomial random variable, $NB_k(k, p)$.

Estimating Parameter Values: The sample mean $\overline{X}$ is an unbiased, maximum likelihood estimator for p. $\hat{p} = \overline{X} = \frac{1}{n}\sum_{i=1}^{n} X_i$.

Probability Mass Function (PMF):

$$f_{Bern(p)}(x) = \begin{cases} p^x(1-p)^{1-x} & x \in \{0,1\} \\ 0 & \text{otherwise} \end{cases}$$

To calculate $f_{Bern(p)}(x)$ in Excel, use =BINOMDIST(x,1,p,FALSE)

Cumulative Distribution Function (CDF):

$$F_{Bern(p)}(x) = \begin{cases} 0 & x < 0 \\ 1-p & 0 \le x < 1 \\ 1 & x \ge 1 \end{cases}$$

To calculate $F_{Bern(p)}(x)$ in Excel, use =BINOMDIST(x,1,p,TRUE)

Figure 6-2 illustrates the PMF and CDF of a Bernoulli random variable.

Inverse Cumulative Distribution Function:

$$F_{Bern(p)}^{-1}(prob) = \begin{cases} 0 & prob \le 1-p \\ 1 & \text{otherwise} \end{cases}$$

To calculate $F_{Bern(p)}^{-1}(prob)$ in Excel, use =IF($prob$<=1 − p,0,1)

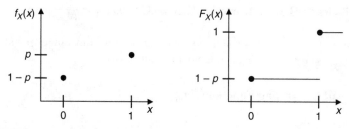

Figure 6-2   PMF and CDF of a Bernoulli or Yes-No Random Variable, with Probability Parameter p.

Random Number Generation:

To generate Bern(p) random numbers in Excel, use =IF(RAND()<=1-p,0,1)

To generate Bern(p) random numbers in Excel with Crystal Ball, use =CB.YesNo(p)

Mean (Expected Value): $E[Bern(p)] = p$

Median: The median is $\begin{cases} 0 & p < 0.5 \\ 0.5 & p = 0.5 \\ 1 & p > 0.5 \end{cases}$

When $p = 0.5$, any number between 0 and 1 is a median. By convention, the median is 0.5.

Mode: The mode is $\begin{cases} 1 & p > 0.5 \\ 0 & p < 0.5 \end{cases}$

Standard Deviation: $SD[Bern(p)] = \sqrt{p(1 - p)}$

Variance: $V[Bern(p)] = p(1 - p)$

Coefficient of Variation: $CV[Bern(p)] = \sqrt{\frac{1 - p}{p}}$

Coefficient of Skewness: $Skew[Bern(p)] = \dfrac{1 - 2p}{\sqrt{p(1 - p)}}$

A Bernoulli random variable is skewed to the right if $p < 0.5$ and to the left if $p > 0.5$

Coefficient of Kurtosis: $Kurt[Bern(p)] = \dfrac{1}{p(1 - p)} - 3$

Coefficient of Excess Kurtosis: $ExKurt[Bern(p)] = \dfrac{1}{p(1 - p)} - 6$

A Bernoulli random variable is platykurtic for mid-range p values, $\frac{1}{2} - \frac{1}{2\sqrt{3}} < p < \frac{1}{2} + \frac{1}{2\sqrt{3}}$, and leptokurtic otherwise.

Moment Generating Function: $M_{Bern(p)}(t) = 1 - p + pe^t$

7

Beta Distribution Family

A beta random variable is a continuous random variable with both upper and lower bounds. A two-parameter beta random variable is bounded between 0 and 1, and is often applied to model uncertainty in probability values. A four-parameter beta random variable can have any upper and lower bounds.

The beta family of random variables has two shape parameters, one describing the lower half and the other describing the upper half of the distribution. With two shape parameters, the beta probability curve can take a wide variety of shapes including a nearly normal distribution with truncated tails, a uniform distribution, bimodal distributions, and many others.

Example 1.8 in Chapter 1 illustrates the beta distribution as a model for a dataset that is skewed to the left.

To introduce an important application for the beta distribution, consider a different way of thinking about uncertainty. Usually, people use distribution models to describe the uncertainty of measured quantities, such as length or whether a unit passes a test. Instead, why not use distribution models to describe the uncertainty of an unknown parameter, such as *average* length or the *probability* of passing a test? Over time, as we collect data and learn more about the process, the distribution of an unknown parameter will have less variation, reflecting the benefits of accumulated experimental knowledge.

Here is one example of this approach, as applied to a series of pass-fail trials, known as a binomial experiment. A typical experiment to learn more about p, the probability of some event, consists of n independent trials, each of which may or may not produce the event of interest. If x is the number of events observed in n trials, then x has a binomial distribution, and the series of trials is called a binomial experiment.

In a binomial experiment, the unknown parameter is the probability p. The beta distribution is a natural choice to model unknown probabilities. The two-parameter beta distribution is bounded between 0 and 1, as are probabilities, and the beta distribution may assume a wide variety of shapes.

In a binomial experiment, the beta distribution is an excellent distribution model for the probability p, because the formulas are quite simple. Before a binomial experiment, suppose that the probability p has a two-parameter beta distribution, with parameters α_0 and β_0. That is, $p_0 \sim \beta(\alpha_0, \beta_0)$. Then, a binomial experiment with n trials is performed, and each trial produces either an event or a non-event. Suppose that n trials produced x events and $n - x$ non-events. After the experiment, the probability p still has a beta distribution with parameters $\alpha_0 + x$ and $\beta_0 + n - x$. Using this distribution model, the expected value of p after the experiment is $\frac{\alpha_0 + x}{\alpha_0 + \beta_0 + n}$.

By applying this technique to successive experiments, the distribution of p reflects the accumulated knowledge of all experiments. Before the first experiment, the prior distribution of p could be a uniform distribution, which is also $\beta(1, 1)$, representing a lack of information about p. Or, the prior distribution might be chosen from the beta family to reflect expert opinion.

Example 7.1

Robert works for a pharmaceutical company, which is developing a new blood test for a specific type of cancer. A blood test now in use has a false positive rate of 30%. This means that if the test is administered to people who do not have the cancer, 30% of those tests will falsely indicate the presence of cancer. One critical requirement of the new test is to have a significantly lower rate of false positives. Robert's company will verify the performance of the new test through a series of clinical trials.

Robert decides to represent the probability of false positives p by a two-parameter beta random variable. During each clinical trial, the test is administered both to people who have the cancer and to people who do not have the cancer. Robert will use the number of false positive results in the clinical trials to estimate the false positive probability p.

Before starting any clinical trials, Robert makes the conservative assumption that p is uniformly distributed between 0 and 1. Expressed as a beta distribution, $p \sim \beta(1, 1)$. Therefore, the parameters of this distribution are $\alpha_0 = \beta_0 = 1$.

The first clinical trial includes $n_1 = 50$ people without the cancer. Of these, $x_1 = 10$ tests were falsely positive. After the first clinical trial, the new distribution for p has a beta distribution with parameters $\alpha_1 = \alpha_0 + x_1 = 1 + 10 = 11$ and $\beta_1 = \beta_0 + n_1 - x_1 = 1 + 50 - 10 = 41$. That is, $p \sim \beta(11, 41)$ with an expected value of $\frac{11}{11 + 41} = 0.211$ after the first clinical trial. Based on this first clinical trial, the probability that the new test is better than the old test, with $p = 0.3$, is the probability that a $\beta(11, 41)$ random variable is less than 0.3. The Excel formula to calculate this is =BETADIST(0.3,11,41) which returns the value 0.932. Robert can be 93.2% confident that the new test is better than the old.

This is encouraging, but not definitive. Robert's company proceeds to the second clinical trial, which includes $n_2 = 300$ people without the cancer. Of these, $x_2 = 68$ tests were falsely positive. Robert combines these results with those of the first clinical trial. After both clinical trials, the new distribution for p has a beta distribution with parameters $\alpha_2 = \alpha_1 + x_2 = 11 + 68 = 79$ and $\beta_2 = \beta_1 + n_2 - x_2 = 41 + 300 - 68 = 273$. That is, $p \sim \beta(79, 273)$ with an expected value of $\frac{79}{79 + 273} = 0.224$ after the second clinical trial. At this point, the probability that the new test is better than the old test, with $p = 0.3$, is the probability that a $\beta(79, 273)$ random variable is less than 0.3. The Excel formula to calculate this is =BETADIST(0.3,79,273) which returns the value 0.99932. Robert is now 99.932% confident that the new test is better than the old.

Figure 7-1 illustrates three distributions of the false positive probability p. The three density functions in the figure represent the prior distribution before any trials, the posterior distribution after the first clinical trial, and the posterior distribution after both clinical trials.

The preceding example is a simple application of Bayesian statistics, a collection of inference tools named in honor of Thomas Bayes (1702–1761), a Presbyterian minister and mathematician who developed early forms of these techniques.

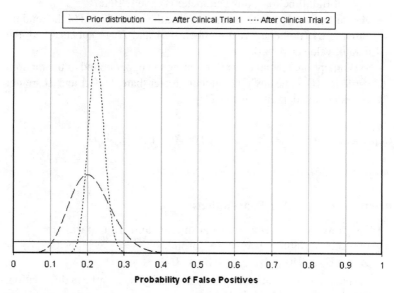

Figure 7-1 Prior and Posterior Probability Density Functions of the Probability of False Positive Test Results.

In Bayesian statistics, special terms refer to distribution models when applied to unknown parameters:

- The *prior distribution* is a distribution model representing the probability of possible values for an unknown parameter, before an experiment.
- The *posterior distribution* is a distribution representing the probability of possible values for an unknown parameter, after an experiment. The posterior distribution combines the prior distribution with the experience gained in the experiment.

In the example of a binomial experiment, a beta prior distribution for p leads to a beta posterior distribution for p. This relationship is convenient, but unusual in Bayesian methods. In general, the posterior distribution may be much more complex than the prior distribution. The beta and binomial distributions are said to be a *conjugate pair* of distributions because of this convenient relationship in Bayesian statistics.

Parameters: The beta family of random variables has two shape parameters, plus two optional parameters representing lower and upper bounds:

- α is the first shape parameter, which controls the shape of the lower half of the distribution. α can be any positive number. In both Excel and Crystal Ball functions, this parameter is called "Alpha."
- β is the second shape parameter, which controls the shape of the upper half of the distribution. β can be any positive number. In both Excel and Crystal Ball functions, this parameter is called "Beta."
- A is an optional parameter representing the lower bound of the random variable. A can be any real number. If A and B are not specified, the default value is $A = 0$.
- B is an optional parameter representing the upper bound of the random variable. B can be any real number greater than A. If A and B are not specified, the default value is $B = 1$.

Representation: Either $X \sim \beta(\alpha, \beta)$ or $X \sim \beta(\alpha, \beta, A, B)$

Support: $[A, \ B]$

Relationships to Other Distributions:

- To convert between two-parameter and four-parameter beta random variables, use these formulas: $\frac{\beta(\alpha, \beta, A, B) - A}{B - A} \sim \beta(\alpha, \beta)$ or $A + [B - A]\beta(\alpha, \beta) \sim \beta(\alpha, \beta, A, B)$
- A beta random variable with $\alpha = \beta = 1$ is also a uniform distribution. $\beta(1, 1, A, B) \sim Unif(A, B)$

- A beta random variable with $\alpha = 1$ and $\beta = 2$ is a triangular distribution with the most likely value equal to the lower bound. $\beta(1, 2, A, B) \sim Tri(A, A, B)$. By symmetry, a beta random variable with $\alpha = 2$ and $\beta = 1$ is a triangular distribution with the most likely value equal to the upper bound. $\beta(2, 1, A, B) \sim Tri(A, B, B)$.

- If X and Y are independent χ^2 random variables with ν_1 and ν_2 degrees of freedom, then $\frac{X}{X + Y} \sim \beta\left(\frac{\nu_1}{2}, \frac{\nu_2}{2}\right)$.

- If X and Y are independent gamma random variables with shape parameters α_1 and α_2, and the same scale parameter β, then $\frac{X}{X + Y} \sim \beta(\alpha_1, \alpha_2)$.

- Beta left-tail probabilities are related to F right-tail probabilities. If $X \sim \beta(\alpha, \beta)$ and $Y \sim F(2\beta, 2\alpha)$, then $P\left[X \leq \frac{\alpha}{\alpha + y\beta}\right] = P[Y > y]$. Also, $P[X \leq x] = P\left[Y > \frac{\alpha(1 - x)}{x\beta}\right]$.

Normalizing Transformation: Because of the enormous variation in the shapes of beta distributions, no simple normalizing transformation is available. In practical applications, automated transformation algorithms, such as Box-Cox or Johnson, may successfully find a normalizing transformation.

Process Control Tools: When the parameters α and β are known, or when they have been estimated, an individual X chart for beta data may be constructed using the formulas below. These formulas require a false alarm rate ε, which is 0.0027 for parity with normal-based Shewhart charts. Typically, the false alarm rate is symbolized by α, but since α is a parameter of the beta distribution, it is ε here.

$$UCL = F^{-1}_{\beta(\alpha, \beta, A, B)}\left(1 - \frac{\varepsilon}{2}\right)$$

$$CL = F^{-1}_{\beta(\alpha, \beta, A, B)}\left(\frac{1}{2}\right)$$

$$LCL = F^{-1}_{\beta(\alpha, \beta, A, B)}\left(\frac{\varepsilon}{2}\right)$$

These formulas use the beta inverse CDF $F^{-1}_{\beta(\alpha, \beta, A, B)}(p)$. In Excel, this function can be evaluated using the =BETAINV(p,α,β,A,B) function.

Estimating Parameter Values: When the bounds A and B are known, the shape parameters α and β can be estimated by these relatively simple functions of the sample mean $\overline{X}$ and the sample standard deviation s:

$$\hat{\alpha} = \overline{X}\left(\frac{\overline{X}(1 - \overline{X})}{s^2} - 1\right)$$

$$\hat{\beta} = (1 - \overline{X})\left(\frac{\overline{X}(1 - \overline{X})}{s^2} - 1\right)$$

In general, it is better to use a major statistical program such as MINITAB or STATGRAPHICS to estimate parameters for the beta distribution, using maximum likelihood methods.

Capability Metrics: For a process producing data with a beta distribution, calculate equivalent capability metrics using the beta CDF. Here are the formulas:

$$\text{Equivalent } P^{\%}_{PU} = \frac{-\Phi^{-1}(R_{\beta(\alpha,\beta,A,B)}(UTL))}{3}$$

$$\text{Equivalent } P^{\%}_{PL} = \frac{-\Phi^{-1}(F_{\beta(\alpha,\beta,A,B)}(LTL))}{3}$$

$$\text{Equivalent } P^{\%}_{P} = \frac{\text{Equivalent } P^{\%}_{PU} + \text{Equivalent } P^{\%}_{PL}}{2}$$

$$\text{Equivalent } P^{\%}_{PK} = Min\{\text{Equivalent } P^{\%}_{PU}, \text{Equivalent } P^{\%}_{PL}\}$$

$\Phi^{-1}(p)$ is the standard normal inverse CDF, evaluated in Excel with the function =NORMSINV(p). $F_{\beta(\alpha,\beta,A,B)}(x)$ is the CDF of the beta distribution, evaluated in Excel with the function =BETADIST(x,α,β,A,B). $R_{\beta(\alpha,\beta,A,B)}(x) = 1 - F_{\beta(\alpha,\beta,A,B)}(x)$ is the survival function of the beta distribution.

Probability Density Function:

For the two-parameter beta family:

$$f_{\beta(\alpha,\beta)}(x) = \begin{cases} \dfrac{x^{\alpha-1}(1-x)^{\beta-1}}{B(\alpha,\beta)} & 0 \le x \le 1 \\ 0 & \text{otherwise} \end{cases}$$

For the four-parameter beta family:

$$f_{\beta(\alpha,\beta,A,B)}(x) = \begin{cases} \dfrac{(x-A)^{\alpha-1}(B-x)^{\beta-1}}{(B-A)^{\alpha+\beta-1}B(\alpha,\beta)} & A \le x \le B \\ 0 & \text{otherwise} \end{cases}$$

In these formulas, $B(\alpha,\beta)$ is called the *beta function*, and it is defined so that the total area under the PDF is always 1. Therefore, $B(\alpha,\beta) = \int_0^1 t^\alpha(1-t)^\beta dt$. The beta function is easy to calculate from the gamma function using this formula:

$$B(\alpha, \beta) = \frac{\Gamma(\alpha)\Gamma(\beta)}{\Gamma(\alpha + \beta)}.$$

To calculate $f_{\beta(\alpha,\beta)}(x)$ in an Excel worksheet, use the formula $= x^\wedge(\alpha-1)*((1-x)^\wedge(\beta-1))/\text{EXP(GAMMALN}(\alpha)+\text{GAMMALN}(\beta)-\text{GAMMALN}(\alpha+\beta))$.

To calculate $f_{\beta(\alpha,\beta,A,B)}(x)$ in an Excel worksheet, use the formula $= (x-A)^\wedge(\alpha-1)*((B-x)^\wedge(\beta-1))/\text{EXP(GAMMALN}(\alpha)+\text{GAMMALN}(\beta)-\text{GAMMALN}(\alpha+\beta))/((B-A)^\wedge(\alpha+\beta-1))$.

Cumulative Distribution Function:

For the two-parameter beta family:

$$F_{\beta(\alpha,\beta)}(x) = \begin{cases} 0 & x < 0 \\ \dfrac{1}{B(\alpha, \beta)} \displaystyle\int_0^x t^{\alpha-1}(1 - t)^{\beta-1} dt & 0 \le x \le 1 \\ 1 & x > 1 \end{cases}$$

For the four-parameter beta family:

$$F_{\beta(\alpha,\beta,A,B)}(x) = \begin{cases} 0 & x < A \\ \dfrac{1}{(B - A)^{\alpha+\beta-1}B(\alpha, \beta)} \displaystyle\int_A^x (t - A)^{\alpha-1}(B - t)^{\beta-1} dt & A \le x \le B \\ 1 & x > B \end{cases}$$

To calculate the beta PDF in an Excel worksheet, use the formula =BETADIST(x,α,β) or =BETADIST(x,α,β,A,B)

Inverse Cumulative Distribution Function: Because of the complexity of the formulas, the beta inverse CDF must be calculated iteratively. To calculate $F^{-1}_{\beta(\alpha,\beta,A,B)}(p)$ in an Excel worksheet, use the formula =BETAINV(p,α,β,A,B). A and B are optional parameters in the =BETAINV function.

Random Number Generation: To calculate beta random numbers in an Excel worksheet, use the formula =BETAINV(RAND()$,\alpha,\beta,A,B)$. A and B are optional parameters in the =BETAINV function.

To calculate beta random numbers with Crystal Ball and Excel software, use the formula =CB.Beta2(A,B,α,β). In Crystal Ball functions, α and β

are restricted to the range $0.3 \leq \alpha \leq 1000$ and $0.3 \leq \beta \leq 1000$. Crystal Ball Excel functions also include a three-parameter beta distribution with α, β, and B parameters, assuming that $A = 0$. To generate random numbers for this three-parameter beta distribution, use the formula =CB.Beta(α,β,B).

Survival Function: $R_{\beta(\alpha,\beta,A,B)}(x) = 1 - F_{\beta(\alpha,\beta,A,B)}(x)$

Hazard Function: $h_{\beta(\alpha,\beta,A,B)}(x) = \dfrac{f_{\beta(\alpha,\beta,A,B)}(x)}{1 - F_{\beta(\alpha,\beta,A,B)}(x)}$

Figure 7-2 shows the PDF, CDF, survival, and hazard functions for several symmetric two-parameter beta random variables.

Figure 7-3 shows the PDF, CDF, survival, and hazard functions for several right-skewed two-parameter beta random variables. For any of the random variables in Figure 7-3, swapping the shape parameters α and β gives a left-skewed beta random variable, with a PDF that is a mirror image of the right-skewed random variable.

Mean (Expected Value):

For the two-parameter beta family: $E[\beta(\alpha, \beta)] = \dfrac{\alpha}{\alpha + \beta}$

For the four-parameter beta family: $E[\beta(\alpha, \beta, A, B)] = A + \dfrac{(B - A)\alpha}{\alpha + \beta}$

Mode: If $\alpha > 1$ and $\beta > 1$, the mode of the four-parameter beta family is $A + \dfrac{(B - A)(\alpha - 1)}{\alpha + \beta - 2}$. If either or both shape parameters are less than 1, then there are one or two modes at the upper or lower bounds.

Standard Deviation:

For the two-parameter beta family: $SD[\beta(\alpha, \beta)] = \dfrac{1}{\alpha + \beta}\sqrt{\dfrac{\alpha\beta}{\alpha + \beta + 1}}$

For the four-parameter beta family: $SD[\beta(\alpha, \beta, A, B)] = \dfrac{B - A}{\alpha + \beta}\sqrt{\dfrac{\alpha\beta}{\alpha + \beta + 1}}$

Variance:

For the two-parameter beta family: $V[\beta(\alpha, \beta)] = \dfrac{\alpha\beta}{(\alpha + \beta)^2(\alpha + \beta + 1)}$

For the four-parameter beta family: $V[\beta(\alpha, \beta, A, B)] = \dfrac{(B - A)^2\alpha\beta}{(\alpha + \beta)^2(\alpha + \beta + 1)}$

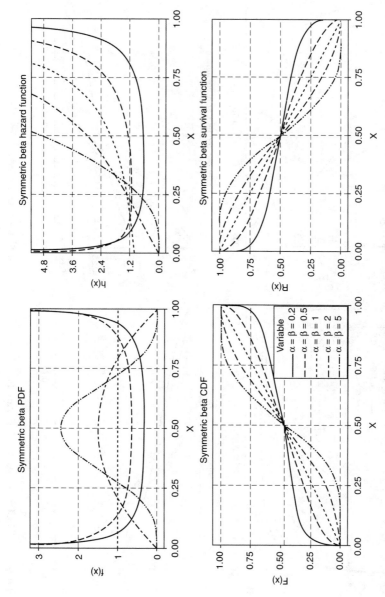

Figure 7-2 PDF, CDF, Survival, and Hazard Functions of Selected Symmetric Beta Random Variables.

165

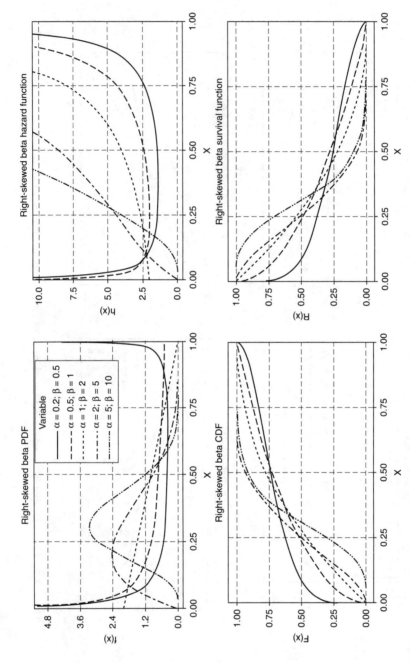

Figure 7-3 PDF, CDF, Survival, and Hazard Functions of Selected Right-Skewed Beta Random Variables.

Coefficient of Variation:

For the two-parameter beta family:

$$CV[\beta(\alpha, \beta)] = \sqrt{\frac{\beta}{\alpha(\alpha + \beta + 1)}}$$

For the four-parameter beta family:

$$CV[\beta(\alpha, \beta, A, B)] = \frac{B - A}{A\beta + B\alpha}\sqrt{\frac{\alpha\beta}{\alpha + \beta + 1}}$$

Coefficient of Skewness: $Skew[\beta(\alpha, \beta, A, B)] = \frac{2(\beta - \alpha)\sqrt{\alpha + \beta + 1}}{(\alpha + \beta + 2)\sqrt{\alpha\beta}}$

When $\beta > \alpha$, beta random variables are skewed to the right.

When $\beta < \alpha$, beta random variables are skewed to the left.

Coefficient of Kurtosis:

$$Kurt[\beta(\alpha, \beta, A, B)] = 6\frac{\alpha^3 - \alpha^2(2\beta - 1) + \beta^2(\beta + 1) - 2\alpha\beta(\beta + 2)}{\alpha\beta(\alpha + \beta + 2)(\alpha + \beta + 3)} + 3$$

Coefficient of Excess Kurtosis:

$$ExKurt[\beta(\alpha, \beta, A, B)] = 6\frac{\alpha^3 - \alpha^2(2\beta - 1) + \beta^2(\beta + 1) - 2\alpha\beta(\beta + 2)}{\alpha\beta(\alpha + \beta + 2)(\alpha + \beta + 3)}$$

Moment Generating Function:

For the two-parameter beta family:

$$m_{\beta(\alpha,\beta)}(t) = 1 + \sum_{k=1}^{\infty}\left(\prod_{r=0}^{k-1}\frac{\alpha + r}{\alpha + \beta + r}\right)\frac{t^k}{k!}$$

8

Binomial Distribution Family

A binomial random variable is a discrete random variable representing the count of defective items in a sample of n independent items, when the probability of any one item being defective is p. More generally, in any experiment consisting of n independent trials with two outcomes per trial, if the probability of outcome "A" is p on every trial, the observed count of outcome "A" is a binomial random variable.

A Bernoulli random variable, also known as Yes-No in Crystal Ball software, is a simple special case of the binomial random variable, with $n = 1$. A Bernoulli random variable has only two possible values, 0 and 1. The probability parameter p is the probability of observing the value 1. See Chapter 6 for a more complete description of Bernoulli (Yes-No) random variables.

An experiment consists of a series of Bernoulli trials if it meets these criteria:

- Every trial can have only two outcomes "A" and "B"
- The probability of outcome "A" is the same on every trial
- The trials are independent

In any series of n Bernoulli trials, the count of outcome "A" is a binomial random variable.

The name of the binomial distribution refers to the mathematical series known as the binomial expansion. In the expansion of $(p + q)^n$, where $q = 1 - p$, the xth term in the expansion is $f_X(x)$.

Example 8.1

Little Ben Junior likes to dump his piggy bank out onto the kitchen floor. After doing this, he counts the number of heads facing up, puts the coins back, and tries again. It is safe to assume that the coins are fair, and they fall independently of each other. The number of heads facing up after Ben dumps out his coins is a binomial random variable.

Example 8.2

Meanwhile, Big Ben Senior, a Six Sigma Green Belt, is planning an inspection process for a supplied product. Ben's company has suffered frequent field failures of field effect transistors (FETs). The failure mode appears to come from the manufacturer, but Ben must collect data to prove this. Unfortunately, the test for this failure mode is destructive, so Ben cannot test 100% of the incoming parts. He decides to test a sample of size n from each incoming lot. If each FET has probability p of being defective, the count of defective parts in each sample will be a binomial random variable.

To select a sample plan, Ben creates a Crystal Ball model, illustrated in Figure 8-1. The model contains an assumption in cell **B6** with a binomial distribution. Ben specifies the number of trials n and probability p by referring to cells in his Excel worksheet.

In cell **B8**, Ben enters the formula =B6>0, which is **TRUE** if the sample includes one or more defective units, or **FALSE** otherwise. In an Excel worksheet, the value of **TRUE** is 1, and the value of **FALSE** is 0. Ben designates cell **B8** as a forecast variable, so that Crystal Ball software remembers values of this cell for each trial in the simulation. The mean value of cell **B8** over many trials estimates the probability that the sample will include one or more defective units.

After running the simulation using the parameters shown in Figure 8-1, Ben finds that a sample size $n = 10$ has only a 0.4 probability of detecting a problem that affects 5% of the units. This is not good enough, so Ben tries a higher sample size.

Ben could add to this model the cost of inspection and the cost of defective parts accepted to support the decision to inspect or not.

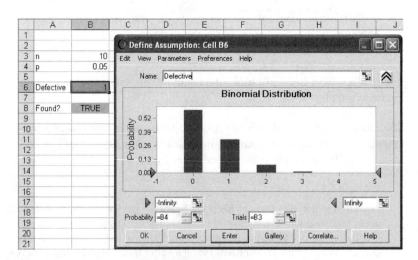

Figure 8-1   A Crystal Ball Model for Selecting a Sampling Plan.

Parameters: The binomial distribution family has two parameters:

- n represents the number of trials in the experiment, and must be a positive integer.
- p represents the probability of observing outcome "A" in any one trial, when X represents the count of outcomes "A." $0 \leq p \leq 1$

Note: In using the binomial distribution, be careful that the outcome with probability p is the same outcome counted by the random variable X. A common mistake is to define p as the probability of defective units, when X represents a count of nondefective units.

Representation: $X \sim Bin(n, p)$

Support: $\{0, 1, \ldots, n\}$

Relationships to Other Distributions

- A binomial random variable is the sum of n independent Bernoulli random variables with the same probability parameter p. If $X_i \overset{iid}{\sim} Bern(p)$ for $i = 1, \ldots, n$, then $\sum_{i=1}^{n} X_i \sim Bin(n, p)$
- A Bernoulli random variable (in Crystal Ball software, a Yes-No assumption) is a special case of a binomial random variable, with number of trials $n = 1$. If $X \sim Bin(1, p)$, then $X \sim Bern(p)$.
- As n gets large, and p is not too close to 0 or 1, a binomial random variable becomes approximately normally distributed. That is, when $X \sim Bin(n, p)$, then $\frac{X - np}{\sqrt{np(1 - p)}} \overset{D}{\longrightarrow} N(0, 1)$. When using this relationship to approximate binomial probabilities, use one of these formulas: $P[X \leq x] \cong \Phi\left[\frac{x + 0.5 - np}{\sqrt{np(1 - p)}}\right]$ or $P[X \geq x] \cong 1 - \Phi\left[\frac{x - 0.5 - np}{\sqrt{np(1 - p)}}\right]$. The 0.5 term in these formulas is a continuity correction, which improves the accuracy of the approximation. As a general rule of thumb, use this approximation only when $np(1 - p) \geq 10$.
- The Poisson random variable is a limiting form for the binomial random variable. This means that when p is small and np is large, a binomial random variable may be approximated by a Poisson random variable. That is, when $X \sim Bin(n, p)$, then $X \overset{D}{\longrightarrow} Pois(np)$. As a general rule of thumb, use this approximation only when $p < 0.1$ and $np > 20$.
- *Reproductive property.* The sum of independent binomial random variables with common probability parameter p is also a binomial

random variable. That is, if $X_i \sim Bin(n_i, p)$ and the $\{X_i\}$ are mutually independent, then $\sum_{i=1}^k X_i \sim Bin(\sum_{i=1}^k n_i, p)$

- The binomial random variable is a limiting form for the hypergeometric random variable. This means that as the lot size N gets large, and when the proportion of defective units $D/N = p$ remains constant, a hypergeometric random variable may be approximated by a binomial random variable. As a general rule, use this approximation only when $N/n \geq 10$.
- The multinomial random variable, not listed elsewhere in this book, is a generalization of the binomial random variable for situations where each trial has more than two possible outcomes. The multinomial random variable is multivariate, requiring k-1 counts to summarize the outcomes of n experiments with k outcomes in each experiment.

Process Control Tools: Two Shewhart control charts, the np-chart and the p-chart, are designed for processes producing binomial observations. When the subgroup size n is constant, the np-chart is simpler and easier to apply than the p-chart. When the subgroup size varies, use the p-chart, and note that the control limits change depending on the size of each subgroup.

Shore's general control charts for attributes (2000a) compensate for skewness in the distribution. When p is small, as it should be, skewness of binomial counts causes errors in the control limits for Shewhart charts. Applying Shore's method, the corrected control limits for an np-chart are:

$$UCL_{np} = \overline{np} + 3\sqrt{\overline{np}\left(1 - \frac{\overline{np}}{n}\right)} + 1.324\left(1 - 2\frac{\overline{np}}{n}\right) - 0.5$$

$$LCL_{np} = \overline{np} - 3\sqrt{\overline{np}\left(1 - \frac{\overline{np}}{n}\right)} + 1.324\left(1 - 2\frac{\overline{np}}{n}\right) + 0.5$$

Similarly, the corrected control limits for a p-chart are:

$$UCL_p = \overline{p} + 3\sqrt{\frac{\overline{p}(1 - \overline{p})}{n_i}} + \frac{1.324(1 - 2\overline{p}) - 0.5}{n_i}$$

$$LCL_p = \overline{p} - 3\sqrt{\frac{\overline{p}(1 - \overline{p})}{n_i}} + \frac{1.324(1 - 2\overline{p}) + 0.5}{n_i}$$

For processes where defective units occur very rarely, consider counting the number of conforming units between each defective unit. The count of conforming units between each defective unit is known to have a geometric distribution, X_0 version. To create a control chart for geometric counts, perform a double square root transformation $Y = X^{1/4}$ and plot the transformed data on an individual X chart, with control limits determined by the moving ranges. Only a single control chart is required. See the process control section of Chapter 15 for more information.

Normalizing Transformation: When applying regression models or hypothesis tests to binomial data, a transformation is useful to stabilize the variation between groups, and to normalize the residuals. One such transformation is arcsin $\sqrt{\frac{X}{n}}$, which is used in the two-sample proportions test listed in Sleeper (2006). Anscombe (1948) showed that arcsin $\sqrt{\dfrac{X + \frac{3}{8}}{n + \frac{3}{4}}}$ stabilizes variation better, and is asymptotically normal with mean arcsin $\sqrt{P}$ and standard deviation $\frac{1}{2\sqrt{n}}$.

Logistic regression is a type of regression that will fit a model $Y = f(X)$ when Y is pass-fail or 0-1 data. In Six Sigma projects and other problem-solving activity, logistic regression is extremely useful to determine which X variables have a significant effect on the probability of some event, like product failures. When logistic regression can be used, this method is preferable to using a normalizing transformation, followed by linear regression.

Estimating Parameter Values: The number of trials n is assumed to be known. A maximum likelihood, unbiased estimator for p is $\hat{p} = \dfrac{\overline{X}}{n} = \dfrac{1}{n^2}\sum_{i=1}^{n} X_i$.

Formulas for exact confidence intervals for p are not available, and these generally must be computed iteratively. MINITAB calculates exact confidence intervals as part of its Stat ⇨ Basic Statistics ⇨ 1 Proportion function.

Approximate confidence intervals for p can be calculated using the normal approximation. The lower bound of an approximate $100(1-\alpha)\%$ confidence interval for p is $L_p = \hat{p} - Z_{\alpha/2}\sqrt{\frac{\hat{p}(1 - \hat{p})}{n}}$. The corresponding upper bound is $U_p = \hat{p} + Z_{\alpha/2}\sqrt{\frac{\hat{p}(1 - \hat{p})}{n}}$. In these formulas, $Z_{\alpha/2}$ is the $\left(1 - \frac{\alpha}{2}\right)$ quantile of the standard normal random variable. To calculate $Z_{\alpha/2}$ in Excel, use =-NORMSINV(α/2).

When n trials produce zero failures, an upper $100(1-\alpha)\%$ confidence bound for p is $U_p = 1 - \alpha^{1/n}$. In the control phase of a Six Sigma project, this formula is particularly useful to analyze a verification test involving n trials and zero failures. Solving this formula for n gives a convenient sample size formula: $n = \frac{\ln \alpha}{\ln(1 - p)}$. Use this formula to calculate the minimum sample size for a verification test to prove that the population failure rate is less than p with $1 - \alpha$ confidence, assuming that zero failures occur.

Often, the terms *reliability* (R) and *confidence* (C) are applied to pass-fail verification tests. When these terms are used, $R = 1 - p$ and $C = 1 - \alpha$. R represents the proportion of the population which would pass the same test, and C represents the confidence that the population reliability exceeds R, based on the test. The minimum sample size required to demonstrate reliability R with confidence C is $n = \frac{\ln(1 - C)}{\ln R}$, when zero failures occur in the verification test.

Capability Metrics: When a binomial process with sample size n and probability p has tolerance limits LTL and UTL, calculate equivalent long-term capability metrics this way:

$$\text{Equivalent } P^\%_{PL} = \frac{-\Phi^{-1}(F_X(LTL - 1))}{3}$$

$$\text{Equivalent } P^\%_{PU} = \frac{-\Phi^{-1}(R_X(UTL))}{3}$$

$$\text{Equivalent } P^\%_{P} = (\text{Equivalent } P^\%_{PL} + \text{Equivalent } P^\%_{PU})/2$$

$$\text{Equivalent } P^\%_{PK} = Min\{\text{Equivalent } P^\%_{PL}, \text{Equivalent } P^\%_{PU}\}$$

If LTL is not an integer, replace ($LTL - 1$) with LTL. In these formulas, $F_X(x)$ is the CDF of a binomial random variable with parameters n and p, evaluated in Excel with the =BINOMDIST(x, n, p, TRUE) function. $R_X(x) = 1 - F_X(x)$ is the corresponding survival function. Also, $\Phi^{-1}(prob)$ is the standard normal inverse CDF, evaluated in Excel with the =NORMSINV($prob$) function. Table 8-1 lists selected values of equivalent $P^\%_{PL}$ and equivalent $P^\%_{PU}$ for binomial processes.

Probability Mass Function (PMF):

$$f_{Bin(n,p)}(x) = \begin{cases} \binom{n}{x} p^x (1 - p)^{n-x} & x \in \{0, 1, \ldots, n\}, \\ 0 & \text{otherwise} \end{cases}$$

where $\binom{n}{x} = \dfrac{n!}{x!(n - x)!}$

To calculate $f_{Bin(n,p)}(x)$ in Excel, use =BINOMDIST(x, n, p, FALSE)

Table 8-1 Selected Capability Values for Binomial Random Variables

\(n=100,\,p=0.01\)		\(n=100,\,p=0.1\)				\(n=1000,\,p=0.1\)			
UTL	Equiv. $P_{PU}^{\%}$	LTL	Equiv. $P_{PL}^{\%}$	UTL	Equiv. $P_{PU}^{\%}$	LTL	Equiv. $P_{PL}^{\%}$	UTL	Equiv. $P_{PU}^{\%}$
2	0.470	1	1.347	13	0.385	10	4.009	110	0.368
3	0.696	2	1.138	16	0.681	20	3.397	120	0.705
4	0.901	3	0.962	19	0.961	30	2.869	130	1.033
5	1.091	4	0.805	21	1.140	40	2.389	140	1.354
6	1.268	5	0.661	24	1.402	50	1.943	150	1.669
7	1.436	6	0.525	27	1.654	60	1.523	160	1.977
8	1.596	7	0.396	30	1.899	70	1.123	170	2.281
9	1.750	8	0.273	33	2.139	80	0.739	180	2.578
10	1.898	9	0.155	36	2.373	90	0.370	190	2.871
11	2.040								

Cumulative Distribution Function (CDF):

$$
F_{Bin(n,p)}(x) = \begin{cases} 0 & x < 0 \\ \sum_{i=0}^{x}\binom{n}{i}p^{i}(1-p)^{n-i} & i \le x < i+1,\, i \in \{0,1,\ldots,n-1\} \\ 1 & x \ge n \end{cases}
$$

Figure 8-2 shows the PMF and CDF of a $Bin(5, 1/6)$ random variable. If five fair six-sided dice are thrown, the count of dice with one dot on the top face has this distribution. Observe that this specific example has two modes, since $P[X = 0] = P[X = 1]$, and these outcomes are more likely than any other.

Figure 8-3 shows the PMF of a $Bin(20, 0.4)$ random variable. The smooth curve in this figure is a normal PDF that approximates this binomial distribution. The rule of thumb for using a normal approximation is $np(1 - p) \ge 10$, but in this case, $np(1 - p) = 4.8$. Even though the rule of thumb is not met, the binomial probabilities follow the normal curve closely in this case.

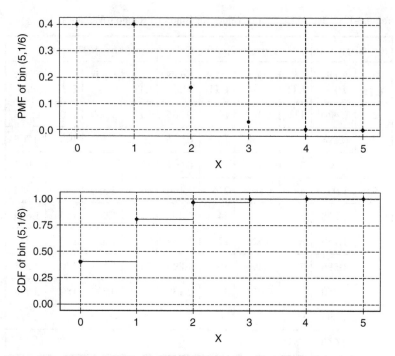

Figure 8-2   PMF and CDF of a *Bin*(5, 1/6) Random Variable.

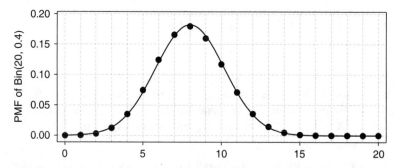

Figure 8-3   PMF of a *Bin*(20, 0.4) Random Variable. The Smooth Line Represents a Normal Density Function with $\mu = 8$ and $\sigma = \sqrt{4.8}$, which is an Approximation for *Bin*(20, 0.4).

To calculate $F_{Bin(n,p)}(x)$ in Excel, use =BINOMDIST(x, n, p, TRUE).

The Excel BINOMDIST algorithm changed significantly in Excel 2003. In versions earlier than 2003, BINOMDIST returned errors for valid x values in the middle of the distribution. In version 2003, BINOMDIST returns incorrect 0 values in the far left tail and correct values in the middle of the distribution. Although neither algorithm is perfect, the 2003 version is an improvement. See Chapter 2 or Knüsel (2005) for more information.

Inverse Cumulative Distribution Function:

$$F_{Bin(n,p)}^{-1}(prob) = x$$

if $\quad \sum_{i=0}^{x-1} \binom{n}{i} p^i (1-p)^{n-i} < prob \leq \sum_{i=0}^{x} \binom{n}{i} p^i (1-p)^{n-i}.$

In Excel, =CRITBINOM($n,p,prob$) returns $F_{Bin(n,p)}^{-1}(prob)$. Since the CRITBINOM function uses BINOMDIST function, its performance changed in Excel 2003, compared to earlier versions. In Excel 2003, for small values of $prob$, CRITBINOM may return an error when BINOMDIST returns 0 instead of a small positive number.

Random Number Generation: To generate $Bin(n, p)$ random numbers in Excel, use =CRITBINOM(n, p, RAND()). Very rarely, this method will return an error, for reasons explained above. A common alternate method is to generate n Bernoulli random numbers with =IF(RAND()<=1-p,0,1) and sum them.

To generate $Bin(n, p)$ random numbers in Excel with Crystal Ball, use =CB.Binomial(p,n)

Mean (Expected Value): $E[Bin(n, p)] = np$

Mode: The mode is any integer x satisfying $p(n + 1) - 1 \leq x \leq p(n + 1)$. When $p(n + 1)$ is an integer, there are two modes.

Standard Deviation: $SD[Bin(n, p)] = \sqrt{np(1-p)}$

Variance: $V[Bin(n, p)] = np(1-p)$

Coefficient of Variation: $CV[Bin(n, p)] = \sqrt{\frac{1-p}{np}}$

Coefficient of Skewness: $Skew[Bin(n, p)] = \dfrac{1 - 2p}{\sqrt{np(1-p)}}$

A binomial random variable is skewed to the right if $p < 0.5$ and to the left if $p > 0.5$

Coefficient of Kurtosis: $Kurt[Bin(n, p)] = \frac{1}{np(1-p)} - \frac{6}{n} + 3$

Coefficient of Excess Kurtosis: $ExKurt[Bin(n, p)] = \frac{1}{np(1-p)} - \frac{6}{n}$

A binomial random variable is platykurtic for mid-range p values, $\frac{1}{2} - \frac{1}{2\sqrt{3}} < p < \frac{1}{2} + \frac{1}{2\sqrt{3}}$, and leptokurtic otherwise.

Moment Generating Function: $M_{Bin(n,p)}(t) = (1 - p + pe^t)^n$

Chi-Squared Distribution Family

A chi-squared or χ^2 random variable is a continuous random variable representing the distribution of a sum of squares of independent standard normal random variables. Since the standard deviation σ of a normal random variable is estimated using a sum of squares, the χ^2 distribution is used in many testing procedures relating to the variation of normally distributed data. Because of the central limit theorem, sums and averages of a large number of random variables from virtually every distribution tend to be normally distributed. This fact leads to the successful solution of many statistical problems using the χ^2 distribution.

The χ^2 goodness-of-fit test, described in Chapter 1, is a familiar tool for many readers of this book. In a χ^2 goodness-of-fit test, the observed counts of data falling into a defined set of bins are compared to the expected counts of data falling into those same bins. The expected counts are calculated from the hypothesized distribution model. In each bin, the observed count of observations is a Poisson random variable. When the expected count is at least 5, the normal distribution is a reasonable approximation for the Poisson distribution, and the scaled sum of squares of the differences between observed and expected counts is approximately distributed as a χ^2 random variable.

The χ^2 family of distributions is a subset of the gamma family of distributions, described in Chapter 14. Specifically, a χ^2 random variable with v degrees of freedom is also a gamma random variable with shape parameter $\alpha = v/2$ and scale parameter $\beta = 2$. In symbols, $\chi^2(v) \sim \gamma\left(\frac{v}{2}, 2\right)$. Because of this relationship, tools intended for gamma distributions or χ^2 distributions work equally well for both families of distributions.

An exponential distribution with mean $\mu = 2$ is also a χ^2 random variable with 2 degrees of freedom. The sum of n mutually independent exponential random variables, with suitable scaling, is a χ^2 random variable with $2n$

degrees of freedom. Because of these facts, confidence intervals and inference procedures for exponentially distributed data can use the χ^2 distribution.

The square root of a χ^2 random variable, called a chi or χ random variable, has many important applications. The half-normal and Rayleigh distributions are scaled versions of χ random variables. Section 9.1 describes the χ family of random variables.

The sum of squared normal random variables with the same standard deviation and mean other than zero has a noncentral χ^2 distribution. The square root of this quantity has a noncentral χ distribution. Section 9.2 describes both of these families.

Since the χ^2 distribution is used in so many statistical tests and other tools, books that fully describe those tools are the best source of examples. Sleeper (2006) contains examples of one-sample standard deviation tests and tests of association for categorical data, two important tools using the χ^2 distribution.

Modelers will rarely select a χ^2 model to represent a process distribution. It is more common to select an equivalent gamma distribution. However, the relationship between χ^2 and gamma distributions is very convenient for modelers. Microsoft Office Excel functions can calculate the left-tail probability for the gamma distribution, but not the right-tail probability. However, other Excel functions can calculate the right-tail probability for the χ^2 distribution, which is also a gamma right-tail probability, when the shape parameter α is a multiple of $1/2$.

The following example uses the relationship between gamma and χ^2 distributions to solve a problem first described in Section 2.2.

Example 9.1

Scott is a Black Belt in a compressor manufacturing plant. A seal in a new product has critical surface texture. Surface texture has an upper limit of 45 μm. The machining process produces a surface texture in microns which has a gamma distribution with shape parameter $\alpha = 4$ and a scale parameter $\beta = 1$.

What is the capability index P_{PK} for this process? According to methods explained in Chapter 3, the most appropriate formula is based on Bothe's equivalent percentage method: Equivalent $P_{PK}^{\%} = \frac{-1}{3} \Phi(R_{\gamma(4,1)}(45))$.

The Excel **GAMMADIST** function, which calculates the left-tail probability $F_{\gamma(\alpha,\beta)}(x)$ reports that $F_{\gamma(4,1)}(45) = 1$. Therefore, $R_{\gamma(4,1)}(45) = 1 - F_{\gamma(4,1)}(45) = 0$, and equivalent $P_{PK}^{\%} = \infty$. Since the upper tail of the gamma distribution is unbounded, it is impossible to have infinite capability, regardless of where the upper limit is. Therefore, this calculation is incorrect.

The error occurred because numbers very close to 1 cannot be represented precisely in standard floating-point formats. To avoid this problem, Scott needs a way to calculate the right-tail probability for a gamma random variable. Excel functions do not include a gamma right-tail probability function, but the CHIDIST function calculates right-tail probabilities for a χ^2 random variable.

The surface texture X has a $\gamma(4, 1)$ distribution. Multiplying a γ random variable by a constant simply multiplies the scale parameter by the same constant. Therefore, $2X \sim \gamma(4, 2)$. Since $\chi^2(v) \sim \gamma(\frac{v}{2}, 2)$, this means that $2X \sim \gamma(4, 2) \sim \chi^2(8)$. Scott needs to know the probability that $X > 45$, which is also the probability that a $\chi^2(8) > 2 \times 45 = 90$. In an Excel worksheet, Scott enters the formula =CHIDIST(90,8), which returns the value 4.65×10^{-16}. The full formula for equivalent $P_{PK}^{\%}$ is =-NORMSINV (CHIDIST(90,8))/3, which returns the value 2.68.

Parameters: The χ^2 family of distributions has only one parameter:

- v, a positive integer, is the degrees of freedom parameter

Representation: $X \sim \chi^2(v)$

Support: $[0, \infty)$

Relationships to Other Distributions:

- A χ^2 random variable with v degrees of freedom is also a gamma random variable with shape parameter $\alpha = v/2$ and scale parameter $\beta = 2$. $\chi^2(v) \sim \gamma\left(\frac{v}{2}, 2\right)$
- A χ^2 random variable with 2 degrees of freedom is also an exponential random variable with mean $\mu = 2$. $\chi^2(2) \sim Exp(\mu = 2)$
- The sum of squares of v mutually independent standard normal random variables is a χ^2 random variable with v degrees of freedom. If $X_i \stackrel{iid}{\sim} N(0, 1)$, then $\sum_{i=1}^{v} X_i^2 \sim \chi^2(v)$
- If a sample consists of n mutually independent observations of a normally distributed process with standard deviation σ, then the sum of squared differences between the observations and the sample mean, divided by σ^2, is a χ^2 random variable with $n - 1$ degrees of freedom. $\frac{\sum(X_i - \overline{X})^2}{\sigma^2} = \frac{(n - 1)s^2}{\sigma^2} \sim \chi^2(n - 1)$ This fact is the basis for confidence intervals and hypothesis tests relating to the sample variance s^2 and sample standard deviation s.

- The square root of a χ^2 random variable with v degrees of freedom is a χ random variable with v degrees of freedom. $\sqrt{\chi^2(v)} \sim \chi(v)$
- If X and Y are independent χ^2 random variables with v_1 and v_2 degrees of freedom, then $\frac{X}{X+Y}$ is a beta random variable. Specifically, $\frac{X}{X+Y} \sim \beta\left(\frac{v_1}{2}, \frac{v_2}{2}\right)$
- The ratio of two independent χ^2 random variables, each divided by its respective degrees of freedom, is an F random variable. $\frac{\chi^2(v)/v}{\chi^2(\omega)/\omega} \sim F(v, \omega)$. This fact is the basis for F tests used to compare standard deviations and in all analysis of variance (ANOVA) procedures.
- A standard normal random variable, divided by the square root of an independent χ^2 random variable, divided by its degrees of freedom, is a Student's t random variable. $\frac{N(0,1)}{\sqrt{\chi^2(v)/v}} \sim t(v)$. This fact is the basis for many t-tests and confidence intervals for sample means.
- Right-tail probabilities of a χ^2 random variable are related to left-tail probabilities of a Poisson random variable by this relationship: $P[\chi^2(v) > x] = P\left[Pois\left(\frac{x}{2}\right) \le \left(\frac{v}{2}\right) - 1\right]$, or equivalently, $R_{\chi^2(v)}(x) = F_{Pois\left(\frac{x}{2}\right)}\left(\frac{v}{2} - 1\right)$
- *Reproductive property.* The sum of mutually independent χ^2 random variables is also a χ^2 random variable. If mutually independent $X_i \sim \chi^2(v_i)$, then $\sum X_i \sim \chi^2(\sum v_i)$

Normalizing Transformations: As v gets very large, a χ^2 random variable approaches a normal random variable in shape. Specifically, if $X \sim \chi^2(v)$, then $\frac{X - v}{\sqrt{2n}} \xrightarrow{D} N(0, 1)$. Practically, this is not very useful, since v must be very large for the approximation to be good.

The Wilson-Hilferty (1931) transformation is quite effective in transforming a χ^2 random variable into an approximate normal distribution:

$$\sqrt{\frac{9v}{2}}\left[\left(\frac{X}{v}\right)^{\frac{1}{3}} - 1 + \frac{2}{9v}\right] \approx N(0, 1).$$

The inverse of this formula can be used to approximate quantiles of the χ^2 distribution from $\Phi^{-1}(p)$, the standard normal inverse CDF:

$$F_{\chi^2(v)}^{-1}(p) \cong v\left(\sqrt{\frac{2}{9v}}\,\Phi^{-1}(p) + 1 - \frac{2}{9v}\right)^3$$

Process Control Tools: If a process is believed to be producing data with a $\chi^2(\nu)$ distribution, it can be normalized using the Wilson-Hilferty transformation above. The normalized data can be plotted on any normal-based Shewhart control chart.

Estimating Parameter Values: A relatively simple formula to estimate the degrees of freedom parameter ν is $\hat{\nu} = \frac{2\overline{X}^2}{\frac{n-1}{n}s^2}$, where $\overline{X}$ and s are the usual sample mean and standard deviation.

Bowman and Shenton (1982) studied this and other more complex estimators of the gamma shape parameter, when the scale and location parameters are fixed.

Capability Metrics: See chapter 14 on the gamma distribution.

In the following formulas, $\Gamma(x)$ represents the gamma function, which is calculated by the Excel function =EXP(GAMMALN(x)).

Probability Density Function (PDF):

$$
f_{\chi^2(\nu)}(x) = \begin{cases} \dfrac{e^{-x/2}x^{(\nu/2)-1}}{2^{\nu/2}\Gamma(\nu/2)} & x \geq 0 \\ 0 & x < 0 \end{cases}
$$

To calculate the χ^2 PDF in an Excel worksheet, use the formula =GAMMADIST(x,ν/2,2,FALSE).

Cumulative Distribution Function (CDF):

$$
F_{\chi^2(\nu)}(x) = \frac{1}{2^{\nu/2}\Gamma(\nu/2)} \int_0^x e^{-t/2}t^{(\nu/2)-1}\,dt
$$

To calculate the χ^2 CDF in an Excel worksheet, use the formula =1-CHIDIST(x,ν) or =GAMMADIST(x,ν/2,2,TRUE).

Inverse Cumulative Distribution Function: The χ^2 inverse CDF has no easy formula, and must be calculated iteratively.

To calculate the χ^2 inverse CDF $F_{\chi^2(\nu)}^{-1}(p)$ in Excel software, use =CHIINV(1-p,ν) or =GAMMAINV(p,ν/2,2).

Random Number Generation: A $\chi^2(\nu)$ random number is the sum of ν squared $N(0, 1)$ random numbers.

To generate χ^2 random numbers in Excel software, use =CHIINV(RAND(),ν)

To generate χ^2 random numbers with Crystal Ball and Excel software, use =CB.Gamma(0,2,ν/2).

Survival Function: $R_{\chi^2(\nu)}(x) = 1 - F_{\chi^2(\nu)}(x)$

Hazard Function: $h_{\chi^2(\nu)}(x) = \dfrac{f_{\chi^2(\nu)}(x)}{1 - F_{\chi^2(\nu)}(x)}$

χ^2 random variables have decreasing hazard function for $\nu = 1$, constant hazard function for $\nu = 2$, and increasing hazard function for $\nu > 2$.

Figure 9-1 illustrates the PDF, CDF, hazard, and survival functions for selected χ^2 random variables.

Mean (Expected Value): $E[\chi^2(\nu)] = \nu$

Median: $F^{-1}_{\chi^2(\nu)}(0.5)$, evaluated in Excel software by =CHIINV(0.5,ν)

Mode: The mode is 0 for $\nu \le 2$, and $\nu - 2$ for $\nu \ge 2$

Standard Deviation: $SD[\chi^2(\nu)] = \sqrt{2\nu}$

Variance: $V[\chi^2(\nu)] = 2\nu$

Coefficient of Variation: $CV[\chi^2(\nu)] = \sqrt{\frac{2}{\nu}}$

Coefficient of Skewness: $Skew[\chi^2(\nu)] = \sqrt{\frac{8}{\nu}}$

χ^2 random variables are always skewed to the right, but they approach an unskewed distribution as ν gets large.

Coefficient of Kurtosis: $Kurt[\chi^2(\nu)] = 3 + \frac{12}{\nu}$

Coefficient of Excess Kurtosis: $ExKurt(\chi^2(\nu)) = \frac{12}{\nu}$

χ^2 random variables are always leptokurtic, but they approach a mesokurtic distribution as ν gets large.

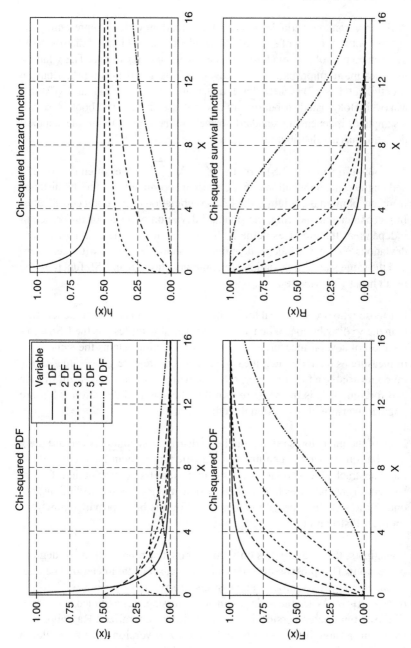

Figure 9-1 CDF, PDF, Hazard, and Survival Functions for Selected χ^2 Random Variables.

9.1   Chi Distribution Family

A chi, or χ, random variable is a continuous random variable representing the square root of the sum of squared independent standard normal distributions. The square root of a χ^2 random variable is a χ random variable. The χ family includes many important special cases. $\chi(1)$ is a half-normal distribution (Section 21.1), $\chi(2)$ is a Rayleigh distribution (Chapter 24), and $\chi(3)$ is a Maxwell-Boltzman distribution, not covered elsewhere in this book. Also, for a sample of independent standard normal observations, the sample standard deviation has a χ distribution.

The root-sum-square (RSS) formula, $Y = \sqrt{\sum X_i^2}$, arises often in scientific and engineering applications. In a Cartesian coordinate system, the distance from any point to the origin is the RSS combination of the point coordinates. In mechanical engineering, the standard deviation of a sum or difference of independent feature sizes is the RSS combination of the feature standard deviations. Whenever independent normal random variables with mean zero and the same standard deviation are combined through an RSS function, the result has a χ distribution.

For those who use statistical tools, the χ distribution is easier to understand than the χ^2 distribution. When a χ random variable represents the RSS combination of several measurements, the χ distribution retains the same units of measure as the individual measurements. This feature gives values of a χ random variable a familiar interpretation, while χ^2 values may seem strange and foreign. This is the same reason that the standard deviation is a more popular measure of dispersion than the variance.

Each of the many tools relying on the χ^2 distribution could instead use the χ distribution. However, centuries of statistical precedent have favored the more abstract χ^2 distribution over the more familiar χ distribution. Except for certain special cases like the half-normal and Rayleigh distributions, not one of the statistical programs mentioned in this book provides functions for the χ distribution.

Sometimes, the χ family has only one parameter, representing the degrees of freedom. This presentation adds an optional scale parameter σ. The one-parameter and two-parameter versions are related by $\frac{\chi(v,\,\sigma)}{\sigma} \sim \chi(v)$ and $\sigma\chi(v) \sim \chi(v,\,\sigma)$. When the scale parameter is missing, it is assumed that $\sigma = 1$. The two-parameter version is also called the generalized Rayleigh distribution, although there are other generalized versions of the Rayleigh family (see Chapter 24) that have been called the same thing.

Example 9.2

Suzette designs antennas to receive satellite TV signals. To maintain sufficient signal strength, the receiving element must be located no farther than 2.0 mm in any direction from the focal point of the antenna. This tolerance zone is a sphere with radius 2.0 mm, centered on a point in space. Considering positional tolerances, wind, and other environmental factors, each of the X, Y, and Z axis positions of the receiving element is normally distributed with mean $\mu = 0$ and standard deviation $\sigma = 0.5$ mm. What is the probability that the receiving element will fall outside the tolerance zone?

The distance D between the receiving element and the focal point is a function of the X, Y, and Z positions: $D = \sqrt{X^2 + Y^2 + Z^2}$. When X, Y, and Z are normally distributed with mean 0, D has a χ distribution with $\nu = 3$ degrees of freedom, and scale parameter $\sigma = 0.5$.

The probability that D is greater than 2.0 is $P(D > 2) = 1 - F_{\chi(3, 0.5)}(2)$. Using the formulas listed below, Suzette enters this formula into an Excel worksheet: =CHIDIST((2/0.5)^2,3) which returns the value 0.001154.

Example 9.3

The sample standard deviation s is a biased estimator of the population standard deviation σ. When the observations are normally distributed, how large is the bias? What is an unbiased estimator for σ?

The sample standard deviation of normally distributed data has a χ distribution,

specifically: $s = \sqrt{\frac{\Sigma(X_i - \bar{X})^2}{n - 1}} \sim \chi\left(n - 1, \frac{\sigma}{\sqrt{n - 1}}\right)$

Using the formulas listed below, the mean of a χ distribution is:

$$E[\chi(\nu, \sigma)] = \frac{\sigma\sqrt{2}\,\Gamma\left(\frac{\nu + 1}{2}\right)}{\Gamma\left(\frac{\nu}{2}\right)}.$$

Applying this formula to the sample standard deviation gives this result:

$$E[s] = \sigma\sqrt{\frac{2}{n - 1}}\,\frac{\Gamma\left(\frac{n}{2}\right)}{\Gamma\left(\frac{n - 1}{2}\right)}$$

In SPC literature, the symbol $c_4(n)$ is defined as $c_4(n) = \sqrt{\frac{2}{n - 1}}\Gamma\left(\frac{n}{2}\right)/\Gamma\left(\frac{n - 1}{2}\right)$. Based on the above derivation of $E[s]$, it is clear that $s/c_4(n)$ is an unbiased estimator of σ.

Parameters: The χ family of distributions has one or two parameters:

- ν, a positive integer, is the degrees of freedom parameter
- σ, a positive real number, is the optional scale parameter. When σ is missing, assume that $\sigma = 1$.

Representation: $X \sim \chi(\nu)$ or $X \sim \chi(\nu, \sigma)$

Support: $[0, \infty)$

Relationships to Other Distributions:

- The square of a $\chi(\nu)$ random variable is a $\chi^2(\nu)$ random variable. From this fact, the χ distribution family inherits all the relationships between the χ^2 family and other distribution families.
- A $\chi(1, \sigma)$ random variable is the absolute value of a normal random variable with mean $\mu = 0$. $|N(0, \sigma)| \sim \chi(1, \sigma)$
- A $\chi(1, \sigma)$ random variable is also a half-normal random variable. $\chi(1, \sigma) \sim HN(\sigma)$
- A Rayleigh random variable with threshold parameter $\tau = 0$ is also a χ random variable with 2 degrees of freedom. As used in this book, the scale parameters for the two families are related by $\beta = \sigma\sqrt{2}$. Therefore, $\chi(2, \sigma) \sim \text{Rayleigh}(\sigma\sqrt{2})$ and $\text{Rayleigh}(\beta) \sim \chi(2, \beta/\sqrt{2})$
- A $\chi(3, \sigma)$ random variable is also known as a Maxwell-Boltzmann random variable in physics literature. This distribution has been applied to model the velocity of a gas molecule in a closed system with no gas flow and equal pressure in all directions.
- The square root of the sum of squares (RSS) of ν mutually independent normal random variables with mean 0 and standard deviation σ_x is a $\chi(\nu, \sigma)$ random variable with $\sigma = \sigma_x\sqrt{n}$. If $X_i \overset{iid}{\sim} N(0, \sigma_x)$, then $\sqrt{\sum_{i=1}^{\nu} X_i^2} \sim \chi(\nu, \sigma_x\sqrt{n})$
- If a sample consists of n mutually independent observations of a normally distributed process with standard deviation σ, then the sample standard deviation s is a χ random variable with $n - 1$ degrees of freedom and scale parameter $\frac{\sigma}{\sqrt{n-1}}$. $s = \sqrt{\frac{\sum(X_i - \overline{X})^2}{n-1}} \sim \chi\left(n - 1, \frac{\sigma}{\sqrt{n-1}}\right)$. This fact is the basis for confidence intervals and hypothesis tests relating to the sample standard deviation.
- Right-tail probabilities of a χ random variable are related to left-tail probabilities of a Poisson random variable by this relationship: $P[\chi(\nu) > x] = P\left[Pois\left(\frac{\sqrt{x}}{2}\right) \leq \left(\frac{\nu}{2}\right) - 1\right]$, or equivalently, $R_{\chi(\nu)}(x) = F_{Pois\left(\frac{\sqrt{x}}{2}\right)}\left(\frac{\nu}{2} - 1\right)$
- The RSS combination of mutually independent χ random variables is also a χ random variable. If mutually independent $X_i \sim \chi(\nu_i)$, then $\sqrt{\sum X_i^2} \sim \chi(\sum \nu_i)$

Normalizing Transformations: The Wilson-Hilferty (1931) transformation for χ^2 random variables also works for the square of χ random variables. If $X \sim \chi(\nu, \sigma)$, then $\left(\frac{X}{\sigma}\right)^2 \sim \chi^2(\nu)$, and

$$\sqrt{\frac{9v}{2}}\left[\left(\frac{X^2}{\sigma^2 v}\right)^{\frac{1}{3}} - 1 + \frac{2}{9v}\right] \approx N(0, 1).$$

The inverse of this formula can be used to approximate quantiles of the χ distribution from $\Phi^{-1}(p)$, the standard normal inverse CDF:

$$F_{\chi^2(v)}^{-1}(p) \cong \sigma \sqrt{v\left(\sqrt{\frac{2}{9v}} \, \Phi^{-1}(p) + 1 - \frac{2}{9v}\right)^3}$$

Process Control Tools: If a process is believed to be producing data with a $\chi(v)$ distribution, it can be normalized using the Wilson-Hilferty transformation above. The normalized data can be plotted on any normal-based Shewhart control chart.

Estimating Parameter Values: When v, the degrees of freedom, is known, here is a maximum likelihood estimator for the scale parameter σ:

$$\hat{\sigma} = \sqrt{\frac{1}{nv} \sum_{i=1}^{n} X_i^2}$$

Capability Metrics: For a process producing data with a χ distribution, calculate equivalent capability metrics using the gamma CDF. Here are the formulas:

$$\text{Equivalent } P_{PU}^{\%} = \frac{-\Phi^{-1}(R_{\chi(v,\sigma)}(UTL))}{3}$$

$$\text{Equivalent } P_{PL}^{\%} = \frac{-\Phi^{-1}(F_{\chi(v,\sigma)}(LTL))}{3}$$

$$\text{Equivalent } P_{P}^{\%} = \frac{\text{Equivalent } P_{PU}^{\%} + \text{Equivalent } P_{PL}^{\%}}{2}$$

$$\text{Equivalent } P_{PK}^{\%} = Min\{\text{Equivalent } P_{PU}^{\%}, \text{Equivalent } P_{PL}^{\%}\}$$

$\Phi^{-1}(p)$ is the standard normal inverse CDF, evaluated in an Excel worksheet with the =NORMSINV(p) function. $F_{\chi(v,\sigma)}(x)$ is the CDF of the χ distribution, evaluated in an Excel worksheet with the function =GAMMADIST((x/σ)^2,v/2,2,TRUE). $R_{\chi(v,\sigma)}(x) = 1 - F_{\chi(v,\sigma)}(x)$ is the survival function of the χ distribution. When UTL is far above the mean, it is better to use this Excel function to calculate the survival function: =CHIDIST((x/σ)^2,v). Table 9-1 lists selected values of equivalent $P_{PL}^{\%}$ and $P_{PU}^{\%}$ for selected processes with a χ distribution.

Table 9-1 Selected Capability Values for χ Random Variables

ν = 1		ν = 2		ν = 3		ν = 10			
UTL	Equiv. $P^{\%}_{PU}$	UTL	Equiv. $P^{\%}_{PU}$	UTL	Equiv. $P^{\%}_{PU}$	LTL	Equiv. $P^{\%}_{PL}$	UTL	Equiv. $P^{\%}_{PU}$
2 σ	0.563	2 σ	0.367	2 σ	0.213	2 σ	0.540	2 σ	−0.540
3 σ	0.927	3 σ	0.762	3 σ	0.630	1.5 σ	0.838	3 σ	−0.027
4 σ	1.278	4 σ	1.134	4 σ	1.018	1 σ	1.193	4 σ	0.428
5 σ	1.622	5 σ	1.493	5 σ	1.389	0.8 σ	1.364	5 σ	0.851
6 σ	1.962	6 σ	1.846	6 σ	1.751	0.6 σ	1.564	6 σ	1.254
7 σ	2.301	7 σ	2.195	7 σ	2.107	0.4 σ	1.816	7 σ	1.644
8 σ	2.638	8 σ	2.540	8 σ	2.458	0.2 σ	2.188	8 σ	2.024

In the following formulas, $\Gamma(x)$ represents the gamma function, which is calculated by the Excel function =EXP(GAMMALN(x)).

Probability Density Function (PDF):

$$f_{\chi(\nu,\sigma)}(x) = \begin{cases} \dfrac{2e^{-x^2/(2\sigma^2)}x^{\nu-1}}{(2\sigma^2)^{\nu/2}\Gamma(\nu/2)} & x \geq 0 \\ 0 & x < 0 \end{cases}$$

The PDF of the χ distribution is related to the PDF of the χ^2 distribution this way: $f_{\chi(\nu,\sigma)}(x) = \frac{2x}{\sigma^2}f_{\chi^2(\nu)}\left(\frac{x^2}{\sigma^2}\right)$. Using this relationship, an Excel formula for the PDF of the χ distribution is =2*x/(σ^2)*GAMMADIST((x/σ)^2, ν/2,2,FALSE).

Cumulative Distribution Function (CDF): Using the relationship to the χ^2 distribution, $F_{\chi(\nu,\sigma)}(x) = F_{\chi^2(\nu)}\left(\frac{x^2}{\sigma^2}\right)$

To calculate the χ CDF in Excel software, use =1-CHIDIST((x/σ)^2,ν) or =GAMMADIST((x/σ)^2,ν/2,2,TRUE).

Inverse Cumulative Distribution Function: Using the relationship to the χ^2 distribution, $F_{\chi(\nu,\sigma)}^{-1}(p) = \sigma\sqrt{F_{\chi^2(\nu)}^{-1}(p)}$

To calculate the χ inverse CDF $F_{\chi(\nu,\sigma)}^{-1}(p)$ in Excel software, use =σ*SQRT(CHIINV(1-p,ν)) or =σ*SQRT(GAMMAINV(p,ν/2,2)).

Random Number Generation: A $\chi(\nu, \sigma)$ random number is σ times the square root of the sum of ν squared $N(0, 1)$ random numbers.

To generate $\chi(\nu, \sigma)$ random numbers in an Excel worksheet, use the formula =σ*SQRT(CHIINV(RAND(),ν)).

To generate $\chi(\nu, \sigma)$ random numbers with Crystal Ball and Excel software, use =σ*SQRT(CB.Gamma(0,2,ν/2)).

Survival Function: $R_{\chi(\nu,\sigma)}(x) = 1 - F_{\chi(\nu,\sigma)}(x)$

Hazard Function: $h_{\chi(\nu,\sigma)}(x) = \dfrac{f_{\chi(\nu,\sigma)}(x)}{1 - F_{\chi(\nu,\sigma)}(x)}$

Figure 9-2 illustrates the PDF, CDF, hazard, and survival functions for selected χ random variables.

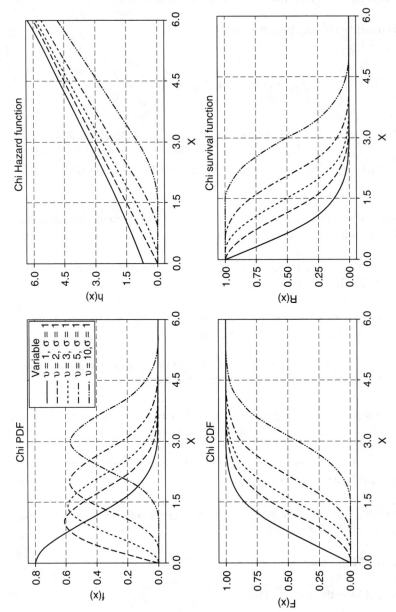

Figure 9-2 CDF, PDF, Hazard, and Survival Functions for Selected χ Random Variables.

Mean (Expected Value): $E[\chi(\nu, \sigma)] = \dfrac{\sigma\sqrt{2}\,\Gamma\left(\frac{\nu + 1}{2}\right)}{\Gamma\left(\frac{\nu}{2}\right)}$

Median: $F^{-1}_{\chi(\nu,\sigma)}(0.5)$, evaluated in an Excel worksheet with the formula
$$=\sigma*\text{SQRT(CHIINV}(0.5,\nu))$$

Mode: The mode is $\sigma\sqrt{\nu - 1}$

Standard Deviation:

$$SD[\chi(\nu, \sigma)] = \sigma\sqrt{\nu - 2\left(\frac{\Gamma\left(\frac{\nu + 1}{2}\right)}{\Gamma\left(\frac{\nu}{2}\right)}\right)^2}$$

Variance:

$$V[\chi(\nu, \sigma)] = \sigma^2\left(\nu - 2\left(\frac{\Gamma\left(\frac{\nu + 1}{2}\right)}{\Gamma\left(\frac{\nu}{2}\right)}\right)^2\right)$$

Coefficient of Variation:

$$CV[\chi(\nu, \sigma)] = \sqrt{\frac{\nu}{2}\left(\frac{\Gamma\left(\frac{\nu}{2}\right)}{\Gamma\left(\frac{\nu + 1}{2}\right)}\right)^2 - 1}$$

To calculate the skewness or kurtosis of X when $X \sim \chi(\nu, \sigma)$, first calculate these moments:

$$E[X] = \dfrac{\sigma\sqrt{2}\,\Gamma\left(\frac{\nu + 1}{2}\right)}{\Gamma\left(\frac{\nu}{2}\right)}$$

$$E[X^2] = \nu\sigma^2$$

$$E[X^3] = \dfrac{2^{3/2}\sigma^3\,\Gamma\left(\frac{\nu + 3}{2}\right)}{\Gamma\left(\frac{\nu}{2}\right)}$$

$$E[X^4] = \sigma^4(\nu^2 + 2\nu)$$

Coefficient of Skewness: $Skew[X] = \dfrac{E[X^3] - 3E[X^2]E[X] + 2(E[X])^3}{(V[X])^{3/2}}$

Coefficient of Kurtosis:

$$Kurt[X] = \dfrac{E[X^4] - 4E[X^3]E[X] + 6E[X^2](E[X])^2 - 3(E[X])^4}{(V[X])^2}$$

Coefficient of Excess Kurtosis:

$$ExKurt[X] = \frac{E[X^4] - 4E[X^3]E[X] + 6E[X^2](E[X])^2 - 3(E[X])^4}{(V[X])^2} - 3$$

9.2 Noncentral Chi-Squared and Chi Distribution Families

A noncentral chi-squared, or noncentral χ^2, random variable is a continuous random variable representing the square root of the sum of squared independent normal distributions with the same standard deviation and any mean value. The noncentral χ^2 distribution is a generalization of the (central) χ^2 distribution for cases where the component normal distributions have mean $\mu \neq 0$.

A noncentral chi, or noncentral χ, random variable is the square root of the corresponding noncentral χ^2 random variable.

The primary use of the noncentral χ^2 and χ distribution is in the calculation of power for inference tests using test statistics with a χ^2 distribution. As models for process behavior, these distributions may represent the sample variance and standard deviation of a process with unstable mean.

Parameters: The noncentral χ^2 and χ families of distributions have two parameters:

- ν, a positive integer, is the degrees of freedom parameter
- λ, a nonnegative real number, is the noncentrality parameter. When $\lambda = 0$, the noncentral χ^2 or χ distribution simplifies into the corresponding (central) χ^2 or χ distribution.

Representation: $X \sim NC\chi^2(\nu, \lambda)$ or $X \sim NC\chi(\nu, \lambda)$

Support: $[0, \infty)$

Relationships to Other Distributions:

- When ν mutually independent $X_i \sim N(\mu_i, \sigma_i)$, then:
  - $\sum_{i=1}^{\nu}\left(\frac{X_i}{\sigma_i}\right)^2 \sim NC\chi^2(\nu, \lambda)$, where the noncentrality parameter $\lambda = \sum_{i=1}^{\nu}\left(\frac{\mu_i}{\sigma_i}\right)^2$
  - $\sqrt{\sum_{i=1}^{\nu}\left(\frac{X_i}{\sigma_i}\right)^2} \sim NC\chi(\nu, \lambda)$, where the noncentrality parameter $\lambda = \sqrt{\sum_{i=1}^{\nu}\left(\frac{\mu_i}{\sigma_i}\right)^2}$

- Generally, it is more useful to assume that the component normal random variables have the same standard deviation. This results in the more familiar root-sum-square (RSS) combination of normal random variables, which has a noncentral χ distribution. When ν mutually independent $X_i \sim N(\mu_i, \sigma)$, then:

  - $\frac{1}{\sigma^2}\sum_{i=1}^{\nu} X_i^2 \sim NC\chi^2(\nu, \lambda)$, where the noncentrality parameter $\lambda = \frac{1}{\sigma^2}\sum_{i=1}^{\nu}\mu_i^2$
  - $\frac{1}{\sigma}\sqrt{\sum_{i=1}^{\nu} X_i^2} \sim NC\chi(\nu, \lambda)$, where the noncentrality parameter $\lambda = \frac{1}{\sigma}\sqrt{\sum_{i=1}^{\nu}\mu_i^2}$

- The square root of a $NC\chi^2(\nu, \lambda)$ random variable is a $NC\chi(\nu, \sqrt{\lambda})$ random variable.
- When $\lambda = 0$, the noncentral χ^2 and χ random variables simplify into their respective central random variables. $NC\chi^2(\nu, 0) \sim \chi^2(\nu)$ and $NC\chi(\nu, 0) \sim \chi(\nu)$
- *Reproductive Property.* The sum of mutually independent noncentral χ^2 random variables is also a noncentral χ^2 random variable. If independent $X_i \sim NC\chi^2(\nu_i, \lambda_i)$, then $\sum X_i \sim NC\chi^2(\sum \nu_i, \sum \lambda_i)$
- The noncentral χ^2 random variable is a mixture of (central) χ^2 random variables, with weights given by the Poisson probability mass function. The formulas given below for the CDF use this property.

Probability Density Function:

For the noncentral χ^2:

$$f_{NC\chi^2(\nu,\lambda)}(x) = \begin{cases} \sum_{j=0}^{\infty} \dfrac{e^{\frac{-(x+\lambda)}{2}} x^{\frac{\nu}{2}-1+j}\lambda^j}{j!\, 2^{\frac{\nu}{2}+2j}\Gamma(\frac{\nu}{2}+j)} & x \geq 0 \\[4mm] 0 & x < 0 \end{cases}$$

For the noncentral χ:

$$f_{NC\chi(\nu,\lambda)}(x) = \begin{cases} \sum_{j=0}^{\infty} \dfrac{e^{\frac{-(x^2+\lambda^2)}{2}} x^{\nu-1+2j}\lambda^{2j}}{j!\, 2^{\frac{\nu}{2}-1+2j}\,\Gamma(\frac{\nu}{2}+j)} & x \geq 0 \\[4mm] 0 & x < 0 \end{cases}$$

Of all the software products mentioned in this book, only JMP and STATGRAPHICS include functions for calculating the PDF of the noncentral χ^2 distribution. To convert these values into the PDF of the noncentral χ distribution, use this formula: $f_{NC\chi(\nu,\lambda)}(x) = 2x f_{NC\chi^2(\nu,\lambda^2)}(x^2)$

Cumulative Distribution Function: The easiest way to calculate the CDF is to use the fact that the noncentral χ^2 distribution is a Poisson-weighted sum of central χ^2 distribution.

For the noncentral χ^2 distribution:

$$F_{NC\chi^2(\nu,\lambda)}(x) = \sum_{j=0}^{\infty} f_{Pois(\lambda/2)}(j)\, F_{\chi^2(\nu+2j)}(x)$$

For the noncentral χ distribution:

$$F_{NC\chi(\nu,\lambda)}(x) = \sum_{j=0}^{\infty} f_{Pois(\lambda^2/2)}(j)\, F_{\chi^2(\nu+2j)}(x^2)$$

JMP, MINITAB, and STATGRAPHICS software contain functions to evaluate the CDF for the noncentral χ^2 distribution. To use these functions to calculate the CDF for the noncentral χ distribution, use this formula:
$$F_{NC\chi(\nu,\lambda)}(x) = F_{NC\chi^2(\nu,\lambda^2)}(x^2)$$

Inverse Cumulative Distribution Function: JMP, MINITAB, and STAT-GRAPHICS software contain functions to evaluate the inverse CDF for the noncentral χ^2 distribution. To use these functions to calculate the CDF for the noncentral χ distribution, use this formula:

$$F^{-1}_{NC\chi(\nu,\lambda)}(p) = \sqrt{F^{-1}_{NC\chi^2(\nu,\lambda^2)}(p)}$$

Random Number Generation: To generate $NC\,\chi^2(\nu,\lambda)$ random numbers, first generate ν independent normal random numbers: $X_i \sim N\!\left(\frac{\lambda}{\sqrt{\nu}}, 1\right)$. Then $\sum_{i=1}^{\nu} X_i^2 \sim NC\,\chi^2(\nu,\lambda)$.

To generate $NC\chi(\nu,\lambda)$ random numbers, first generate ν independent normal random numbers: $X_i \sim N\!\left(\frac{\lambda}{\nu}, 1\right)$. Then $\sqrt{\sum_{i=1}^{\nu} X_i^2} \sim NC\,\chi(\nu,\lambda)$.

Note: Many properties of the noncentral χ distribution are omitted here, because they are exceedingly complex. For many practical applications, it is easier to estimate moments and characteristics of complicated random variables like the noncentral χ by Monte Carlo simulation than by using formulas.

Mean (Expected Value):

For the noncentral χ^2: $E[NC\,\chi^2(\nu,\lambda)] = \nu + \lambda$

Standard Deviation:

For the noncentral χ^2: $SD[NC\,\chi^2(\nu,\lambda)] = \sqrt{2(\nu + 2\lambda)}$

Variance:

For the noncentral χ^2: $V[NC\chi^2(v, \lambda)] = 2(v + 2\lambda)$

Coefficient of Skewness:

For the noncentral χ^2: $Skew[NC\chi^2(v, \lambda)] = \dfrac{2^{3/2}(v + 3\lambda)}{(v + 2\lambda)^{3/2}}$

Coefficient of Kurtosis:

For the noncentral χ^2: $Kurt[NC\chi^2(v, \lambda)] = \dfrac{12(v + 4\lambda)}{(v + 2\lambda)^2}$

Coefficient of Excess Kurtosis:

For the noncentral χ^2: $ExKurt[NC\chi^2(v, \lambda)] = \dfrac{12(v + 4\lambda)}{(v + 2\lambda)^2} - 3$

Moment Generating Function:

For the noncentral χ^2: $m_{NC\chi^2(v,\lambda)}(t) = \dfrac{\exp\left(\frac{\lambda t}{1 - 2t}\right)}{(1 - 2t)^{v/2}}$

10

Discrete Uniform Distribution Family

A discrete uniform random variable is a discrete random variable equally likely to take any value from a discrete set of values. Most often, the values are adjacent integers. Discrete uniform random variables described in this chapter may have integer values between A and B.

Except for games such as dice, discrete uniform random variables have few applications as direct models for real processes. Most often, a discrete uniform distribution is indirectly applied to randomly select one of several alternatives. A good model for some processes is a mixture of two or more distributions, selected at random. In these models, a discrete uniform distribution provides an index or pointer, which is used to select values or distributions from a table.

A popular set of statistical tools known as resampling, or nonparametric bootstrapping, requires the random selection of observations from a sample. By resampling the sample data and analyzing the resamples, practitioners can calculate confidence intervals and perform statistical tests without assuming any particular distribution model. To implement resampling in a Crystal Ball simulation model, put the data in a table and use discrete uniform assumptions as indices into the table. To learn more about resampling methods, read Good and Good (2001) or Efron and Tibshirani (1994).

Example 10.1

In Example 1.12 from Chapter 1, Jill compiled expert opinions about the market share expected for a new product. To combine four distributions based on opinion into one composite distribution in her Crystal Ball model, Jill defined a discrete uniform assumption equally likely to have any of the values {1, 2, 3, 4}. Based on the value of the discrete uniform, Jill's model selects one of the four opinions at random. Figure 1-34 illustrates this process.

Example 10.2

In Example 1.10 from Chapter 1, Rhoda measured the voltage of 95 references and found the somewhat troubling distribution shown in Figure 1-31 None of the common distribution families will fit an apparently bimodal distribution like this. However, Rhoda needs to calculate a lower confidence limit on C_{PK}. She needs this as part of her company's approval process for supplied parts, but also as a way to illustrate the inferiority of this part.

Unfortunately, the usual ways of calculating C_{PK} with confidence intervals all assume normality. The equivalent capability metrics described elsewhere in this book require a selected distribution model, which is unavailable here.

So, Rhoda uses resampling to calculate a confidence interval for C_{PK}. In an Excel worksheet containing her 95 observations, she creates 95 Crystal Ball assumptions, each with a discrete uniform distribution from 1 to 95. Using the Excel **INDEX** function, Rhoda selects one of the observed values at random, 95 times. This set of 95 observations, sampled with replacement, forms a resample. Note that some observations may appear several times in the resample.

From the resample, Rhoda can calculate the mean, standard deviation, C_{PK}, or any other statistics of interest. These become the forecasts in Crystal Ball. Figure 10-1 shows the distribution of C_{PK}, based on 10,000 resamples of the observed data. This procedure assumes no distribution for the reference voltages, other than the empirical distribution provided by the sample of 95 units.

Based on resampling, Rhoda concludes that the 95% lower confidence bound on C_{PK} is 0.37, since 95% of the resamples had C_{PK} greater than 0.37.

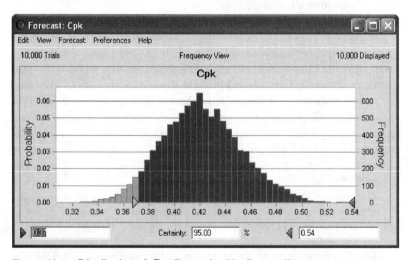

Figure 10-1   Distribution of C_{PK} Determined by Resampling.

Parameters: As used here, the discrete uniform distribution family has two parameters:

- A, an integer, is the minimum value.
- B, an integer, is the maximum value.

Representation: $X \sim DiscUnif(A, B)$

Support: $\{A, A + 1, \ldots, B\}$

Relationships to Other Distributions:

- $DiscUnif(0, 1)$ is a Bernoulli random variable with $p = 0.5$.
- $DiscUnif(A, B)$ is the same as a continuous uniform random variable, specifically $Unif(A - 0.5, B + 0.5)$, rounded to the nearest integer.

Probability Mass Function:

$$f_{DiscUnif(A,B)}(x) = \begin{cases} \dfrac{1}{B - A + 1} & A \leq x \leq B, x \text{ integer} \\ 0 & \text{otherwise} \end{cases}$$

Figure 10-2 illustrates the PMF of two discrete uniform random variables.

Cumulative Distribution Function:

$$F_{DiscUnif(A,B)}(x) = \begin{cases} 0 & x < A \\ \dfrac{\lfloor x \rfloor - A + 1}{B - A + 1} & A \leq x < B \\ 1 & x \geq B \end{cases}$$

where $\lfloor \; \rfloor$ means round down.

Inverse Cumulative Distribution Function:

$$F^{-1}_{DiscUnif(A,B)}(prob) = A + \lfloor prob \times (B - A + 1) \rfloor$$

Random Number Generation:

STATGRAPHICS software generates discrete uniform random numbers between any two integers. In MINITAB software, discrete uniform random numbers are called Integer random numbers. JMP also generates random

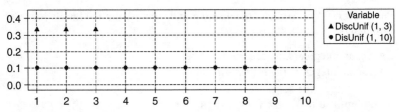

Figure 10-2   PMF of DiscUnif(1, 3) and DiscUnif(1, 10).

integer values between 1 and B. In an Excel worksheet, the formula =A+INT(RAND()*(B-A+1)) produces discrete uniform random numbers. With Crystal Ball and Excel, use =CB.DiscreteUniform(A,B) to generate discrete uniform random numbers.

Mean (Expected Value): $E[DiscUnif(A, B)] = \frac{A + B}{2}$

Standard Deviation: $SD[DiscUnif(A, B)] = \sqrt{\frac{(B - A + 1)^2 - 1}{12}}$

Variance: $V[DiscUnif(A, B)] = \frac{(B - A + 1)^2 - 1}{12}$

Coefficient of Variation:

$$CV[DiscUnif(A, B)] = \frac{1}{A + B}\sqrt{\frac{(B - A + 2)^2 - 1}{3}}$$

Coefficient of Skewness: $Skew[DiscUnif(A, B)] = 0$

Coefficient of Kurtosis:

$$Kurt[DiscUnif(A, B)] = \frac{9}{5}\left[\frac{(B - A + 1)^2 - 7/3}{(B - A + 1)^2 - 1}\right]$$

Coefficient of Excess Kurtosis:

$$ExKurt[DiscUnif(A, B)] = \frac{-6}{5}\left[\frac{(B - A + 1)^2 + 1}{(B - A + 1)^2 - 1}\right]$$

A discrete uniform random variable is always platykurtic.

Moment Generating Function: $M_{DiscUnif(A,B)}(t) = \dfrac{e^{tA} - e^{t(B+1)}}{(B - A + 1)(1 - e^t)}$

Exponential Distribution Family

An exponential random variable is a continuous random variable with a lower bound of zero. The exponential probability function always decreases, so the lower bound of zero is also the mode.

With only one parameter, the family of exponential distributions is one of the simplest of all distribution models. As a model for the time between events, the exponential distribution has a "lack of memory" property, which is unique among continuous distributions. This property means that if an exponential random variable is greater than any value, the distribution of future values has the same exponential distribution. In reliability applications, this property means that exponential random variables have a constant hazard rate. For this reason, the exponential model is often accepted as a default failure time model, when there is no information to support a more complex model.

The exponential distribution is a continuous analog to the geometric distribution, which is the only discrete distribution with the "lack of memory" property.

The family of exponential distributions has one parameter, which can be either a rate parameter λ, or a mean parameter $\mu = 1/\lambda$. The descriptions listed here are in terms of both parameters. In general, books, articles, and software may use either parameter. Be cautious to interpret parameters of this family correctly, as a rate or as a mean. The mean parameter is also called the scale parameter.

The family of exponential distributions is a special case of many other families, including gamma and Weibull. The chi-squared random variable with 2 degrees of freedom is also an exponential random variable with $\mu = 2$. There is an important relationship between a Poisson random variable and an exponential random variable. If the count of events per unit time is

Pois(λ), then the time between events is *Exp*(λ), with the same rate parameter λ.

In addition to the one-parameter exponential distribution, Section 11.1 describes a two-parameter exponential distribution, which includes a threshold or location parameter. Section 11.2 describes a truncated exponential distribution. The Laplace distribution family, described in Chapter 17, is a symmetric distribution formed by an exponential distribution on either side of a central point. For this reason, the Laplace distribution is often called "double exponential." However, "double exponential" may also refer to the largest extreme value or Gumbel distribution described in Chapter 12, because of the $e^{-e^{-x}}$ form of its probability function.

Early chapters of this book illustrated applications for the exponential distribution in examples 1.1, 1.3, 1.9, 3.4, and 3.5.

Example 11.1

Ty manages the IT department of a large manufacturing facility. One of Ty's customer satisfaction metrics is the time between unplanned server outages. After tracking this metric for some time, Ty wants to make a control chart to evaluate the ongoing stability of the process. Here are Ty's first 30 observations for time between server outages, in hours:

59	38	3	100	23	55	21	77	131	18
41	7	71	523	43	164	153	67	42	13
281	1	111	35	58	156	52	83	71	293

Before selecting a control chart, Ty must select a distribution model for this data. Since time values must be positive numbers, and some are close to the boundary of zero, a normal distribution is unlikely to work well. Ty prepares probability plots (not shown here) to test how well various distribution models fit this data. The normal probability plot shows a very bad fit, but any of the distribution families with zero as a lower bound (exponential, gamma, Weibull, or lognormal) fit the data well. Ty chooses the simplest of these models, the exponential distribution.

The sample mean of Ty's observations is $\overline{X} = 93$. Using formulas listed below, Ty calculates the center line and control limits for an exponential control chart. In these formulas, Ty uses the standard error rate of $\alpha = 0.0027$, which is consistent with normal Shewhart control charts.

$$UCL_X = -\overline{X}\ln(\alpha/2) = 614$$
$$CL_X = 0.6931\overline{X} = 64.4$$
$$LCL_X = -\overline{X}\ln(1 - \alpha/2) = 0.13$$

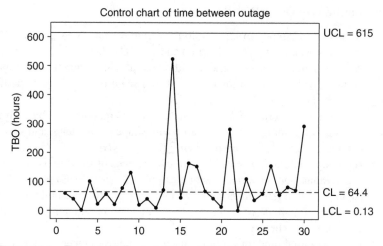

Figure 11-1   Exponential Control Chart of Time between Server Outages.

Figure 11-1 shows Ty's completed exponential control chart. This chart shows no signs of instability. In future, if Ty observes any point outside the control limits, or a series of seven consecutive points on either side of the center line, then Ty will have strong evidence of a special cause of variation affecting this process.

Parameters: The family of exponential random variables has one parameter, which can be either one of the following:

- λ is a rate parameter, representing the rate of occurrences. λ can be any positive number. $\lambda = 1/\mu$
- μ is a mean parameter, representing average time between occurrences. μ can be any positive number. $\mu = 1/\lambda$. The mean parameter is also called the scale parameter.

Representation:

In terms of the rate parameter: $X \sim Exp(\lambda)$
In terms of the mean parameter: $X \sim Exp(\mu)$

Since the notation $Exp()$ is ambiguous, the context must clearly specify whether the rate parameter or mean parameter is specified. One way to do this is to specify units of measurement for the parameter. If X is a measure of time in hours, then μ is also in hours, but the rate λ is in events per hour. Another way to specify clearly that the mean is 2 is $X \sim Exp(\mu = 2)$.

Support: $[0, \infty)$

Relationships to Other Distributions:

- A standard exponential random variable has $\lambda = \mu = 1$. If $X \sim Exp(1)$, then $\mu X \sim Exp(\mu)$ and $\frac{X}{\lambda} \sim Exp(\lambda)$.
- An exponential random variable with mean μ is also a gamma random variable with shape parameter $\alpha = 1$ and scale parameter $\beta = \mu$. If $X \sim Exp(\mu)$, then $X \sim \gamma(1, \mu)$
- The sum of n mutually independent exponential random variables with mean μ is a gamma random variable with shape parameter n and scale parameter μ. If $X_i \overset{iid}{\sim} Exp(\mu)$, then $\sum_{i=1}^{n} X_i \sim \gamma(n, \mu)$
- A chi-squared random variable with two degrees of freedom is also an exponential random variable with mean $\mu = 2$. $\chi^2(2) \sim Exp(\mu = 2)$
- An exponential random variable with mean μ is also a Weibull random variable with shape parameter $\alpha = 1$ and scale parameter $\beta = \mu$. If $X \sim Exp(\mu)$, then $X \sim Weibull(1, \mu)$
- An exponential random variable raised to a positive power is a Weibull random variable. If $X \sim Exp(1)$, then $\beta X^{1/\alpha} \sim Weibull(\alpha, \beta)$
- The minimum of a set of independent observations of an exponential random variable is also an exponential random variable. When $X_1, \ldots, X_n \overset{iid}{\sim} Exp(\mu)$, then $Min\{X_i\} \sim Exp(\mu/n)$, where μ represents the mean.
- The absolute value of a Laplace random variable minus its location parameter is an exponential random variable. If $X \sim Laplace(\beta, \tau)$, then $|X - \tau| \sim Exp(\beta)$, where β represents the mean of the exponential distribution.
- The difference between two independent exponential random variables is a Laplace random variable. If $X, Y \overset{iid}{\sim} Exp(\mu)$, then $X - Y \sim Laplace(\mu, 0)$.
- The natural log of an exponential random variable is a smallest extreme value random variable. If $X \sim Exp(\mu)$, then $\ln X \sim SEV(0, \mu)$. Also, the negative log of an exponential random variable is a largest extreme variable. $-\ln X \sim LEV(0, \mu)$.
- The square root of an exponential random variable is a Rayleigh random variable, with scale parameter $\beta = \sqrt{\mu}$. That is, if $X \sim Exp(\mu)$, then $\sqrt{X} \sim Rayleigh(\sqrt{\mu})$
- An exponential random variable raised to the 0.2654 power has a very close approximation to a normal distribution. If $X \sim Exp(\mu_X)$, and $Y = X^{0.2654}$, then $Y \approx N(\mu_Y = 0.9034\mu_X^{0.2654}, \sigma = 0.2675\mu_X^{0.2654})$.
- The negative natural log of a uniform random variable between 0 and 1 is an exponential random variable. If $U \sim Unif(0, 1)$, then $-\mu\ln U \sim Exp(\mu)$
- Any continuous random variable X with CDF $F_X(x)$ can be transformed into a standard exponential random variable by the function $-\ln(1 - F_X(X)) \sim Exp(1)$.

Normalizing Transformation: Yang and Xie (2000) derived a power transformation that converts an exponential distribution into a nearly normal distribution. If $X \sim Exp(\mu_X)$, and $Y = X^{0.2654}$, then $Y \approx N(\mu_Y = 0.9034\mu_X^{0.2654}, \sigma = 0.2675\mu_X^{0.2654})$.

Process Control Tools: There are two choices for control charts for individual observations of an exponentially distributed process:

- Plot the exponential data directly onto an individual X chart, and use these formulas for control limits and the central line:

$$UCL_X = -\overline{X}\ln(\alpha/2)$$
$$CL_X = 0.6931\overline{X}$$
$$LCL_X = -\overline{X}\ln(1 - \alpha/2)$$

- Transform the data into a normal distribution using $Y = X^{0.2654}$ and plot the data on an individual X chart, using moving ranges to calculate control limits.

Example 3.5 in Chapter 3 illustrates both these options. Unlike a normal process, an exponential process requires only a single control chart. Since the exponential distribution has only a single parameter, one chart is sufficient to detect changes in that parameter.

Estimating Parameter Values: The simplest situation for estimation is with a complete or uncensored dataset, in which every unit has an observed value. If a life test is allowed to run until every item has died, the observed lifetimes form a complete dataset. In this case, the unbiased, maximum likelihood estimator for the mean μ is $\hat{\mu} = \overline{X} = \frac{1}{n}\sum_{i=1}^{n}X_i$.

In reliability applications, the total test time for all units $T = \sum X_i$ and the number of failures experienced during that time, k, are sufficient to estimate the population parameter λ or μ. With this information, an unbiased estimator of the failure rate λ is $\hat{\lambda} = \frac{k}{T}$.

A $100(1-\alpha)\%$ confidence interval for λ has a lower limit of $L_\lambda = \frac{\chi^2_{1-\alpha/2,2k}}{2T}$ and an upper limit of $U_\lambda = \frac{\chi^2_{\alpha/2,2(k+1)}}{2T}$, where $\chi^2_{p,v}$ is the $1-p$ quantile of the chi-squared distribution with v degrees of freedom. In an Excel worksheet, $\chi^2_{p,v}$ is calculated by the =CHIINV(p,v) function.

A very common complication in estimating failure rates is censoring of the data. If some units in the test failed, but some did not, the dataset is censored. There are several types of censoring, including these:

- Type I censoring occurs when the test ends at a predetermined time, regardless of how many units have failed at that time.
- Type II censoring occurs when the test ends as soon as a predetermined number of failures have been observed.
- Type III censoring combines Types I and II. A test of this type ends after a certain time or after a certain number of failures, whichever happens first.
- Type IV censoring occurs when units start at different times. This is typical of warranty data analysis, when customers begin to use products at different times. At the time of analysis, some units are nearly new, while others are old.

The analysis of censored life data is too complex to be presented in any useful detail here. As a rough approximation, $\hat{\lambda} = \frac{k}{T}$ is a conservative estimate for failure rate in most cases, with the same approximate confidence interval given above. In many specific cases, better estimators are available. For most Six Sigma applications, it is better to entrust the analysis of life test data to software such as MINITAB or STATGRAPHICS, which include appropriate estimation methods.

An important special case occurs when zero failures occur after total test time T. After a reliability problem is fixed, one hopes for such a result in verification testing. With zero failures, the point estimate of failure rate λ is zero, but a $100(1 - \alpha)\%$ upper confidence limit on the rate λ is $U_\lambda = \frac{-\ln\alpha}{T}$. Similarly, a $100(1 - \alpha)\%$ lower confidence limit on the mean μ is $L_\lambda = \frac{T}{-\ln\alpha}$. With this one-sided confidence limit, one can decide when the total verification test time is sufficient.

For more information on the analysis of life data, see Nelson (2004) or Kececioglu (1991).

Capability Metrics: The mode of an exponential process is zero, so generally there will only be an upper tolerance limit UTL. Calculate equivalent long-term capability metrics this way:

$$\text{Equivalent } P_P^\% = \text{Equivalent } P_{PK}^\% = \frac{-\Phi^{-1}(\exp[-\lambda UTL])}{3} = \frac{-\Phi^{-1}(\exp[-UTL/\mu])}{3}$$

$\Phi^{-1}(p)$ is the standard normal inverse CDF, evaluated in Excel with the =NORMSINV(p) function. Table 11-1 lists selected values of equivalent $P_{PK}^\%$ for exponential processes.

Note that the standard deviation for an exponential random variable is equal to the mean. The bottom entry in Table 11-1 represents UTL at 20 standard

Table 11-1   Selected Capability Values for Exponential Random Variables

UTL	Equivalent $P_{PK}^{\%}$
3 μ	0.549
5 μ	0.824
7 μ	1.039
9 μ	1.222
11 μ	1.383
13 μ	1.529
15 μ	1.662
17 μ	1.787
19 μ	1.904
21 μ	2.014

deviations above the mean, at which point the defects are equivalent to a normal distribution with UTL at six standard deviations above the mean. Using conventional formulas for process capability, which assume a normal distribution, would seriously underestimate the defects.

Since an exponential distribution can be normalized with a simple transformation, applying normal-based formulas to the transformed data is another valid approach to capability metrics. Using this approach, both short-term and long-term metrics are available.

Probability Density Function:

In terms of the rate parameter: $f_{Exp(\lambda)}(x) = \begin{cases} \lambda\exp(-\lambda x) & x \geq 0 \\ 0 & x < 0 \end{cases}$

In terms of the mean parameter: $f_{Exp(\mu)}(x) = \begin{cases} \dfrac{1}{\mu}\exp\left(\dfrac{-x}{\mu}\right) & x \geq 0 \\ 0 & x < 0 \end{cases}$

In an Excel worksheet, calculate $f_{Exp(\lambda)}(x)$ with the formula =GAMMADIST(x,1,1/λ,FALSE). Calculate $f_{Exp(\mu)}(x)$ with the formula =GAMMADIST(x,1,μ,FALSE).

Cumulative Distribution Function:

In terms of the rate parameter: $F_{Exp(\lambda)}(x) = \begin{cases} 1 - \exp(-\lambda x) & x \geq 0 \\ 0 & x < 0 \end{cases}$

In terms of the mean parameter: $F_{Exp(\mu)}(x) = \begin{cases} 1 - \exp\left(\dfrac{-x}{\mu}\right) & x \geq 0 \\ 0 & x < 0 \end{cases}$

In an Excel worksheet, calculate $F_{Exp(\lambda)}(x)$ with the formula =GAMMADIST(x,1,1/λ,TRUE). Calculate $F_{Exp(\mu)}(x)$ with the formula =GAMMADIST(x,1,λ,TRUE).

Inverse Cumulative Distribution Function:

In terms of the rate parameter: $F_{Exp(\lambda)}^{-1}(prob) = \dfrac{-\ln(1 - prob)}{\lambda}$

In terms of the mean parameter: $F_{Exp(\mu)}^{-1}(prob) = -\mu \ln(1 - prob)$

In an Excel worksheet, calculate $F_{Exp(\lambda)}^{-1}(prob)$ with the formula =GAMMAINV($prob$,1,1/λ). Calculate $F_{Exp(\mu)}^{-1}(prob)$ with the formula =GAMMAINV($prob$,1,μ).

Random Number Generation: To generate exponential random numbers in an Excel worksheet, use either =-μ*LN(RAND()) or =-LN(RAND())/λ. With Crystal Ball and Excel software, use =CB.Exponential(λ)

Survival Function:

In terms of the rate parameter: $R_{Exp(\lambda)}(x) = \begin{cases} \exp(-\lambda x) & x \geq 0 \\ 1 & x < 0 \end{cases}$

In terms of the mean parameter: $R_{Exp(\mu)}(x) = \begin{cases} \exp\left(\dfrac{-x}{\mu}\right) & x \geq 0 \\ 1 & x < 0 \end{cases}$

Hazard Function:

In terms of the rate parameter: $h_{Exp(\lambda)}(x) = \lambda$

In terms of the mean parameter: $h_{Exp(\mu)}(x) = \dfrac{1}{\mu}$

Figure 11-2 illustrates the PDF, CDF, survival, and hazard functions for exponential random variables with means of 0.5, 1.0 and 2.0.

Mean (Expected Value): $E[Exp(\mu)] = \mu$ or $E[Exp(\lambda)] = \frac{1}{\lambda}$

Median: $\mu \ln 0.5 \cong 0.6931 \mu$

Mode: 0

Standard Deviation: $SD[Exp(\mu)] = \mu$ or $SD[Exp(\lambda)] = \frac{1}{\lambda}$

Variance: $V[Exp(\mu)] = \mu^2$ or $V[Exp(\lambda)] = \frac{1}{\lambda^2}$

Coefficient of Variation: $CV[Exp(\mu)] = CV[Exp(\lambda)] = 1$

Coefficient of Skewness: $Skew[Exp(\mu)] = Skew[Exp(\lambda)] = 2$

The exponential random variable is skewed to the right.

Coefficient of Kurtosis: $Kurt[Exp(\mu)] = Kurt[Exp(\lambda)] = 9$

Coefficient of Excess Kurtosis:

$ExKurt[Exp(\mu)] = ExKurt[Exp(\lambda)] = 6$

The exponential random variable is leptokurtic.

Moment Generating Function:

$$M_{Exp(\mu)}(t) = \frac{1}{1 - \mu t} \quad \text{or} \quad M_{Exp(\lambda)}(t) = \frac{\lambda}{\lambda - t}$$

11.1   Two-Parameter Exponential Distribution Family

A two-parameter exponential random variable is a one-parameter exponential random variable plus a constant. The added constant, τ, represents the minimum possible value of the random variable and is the second

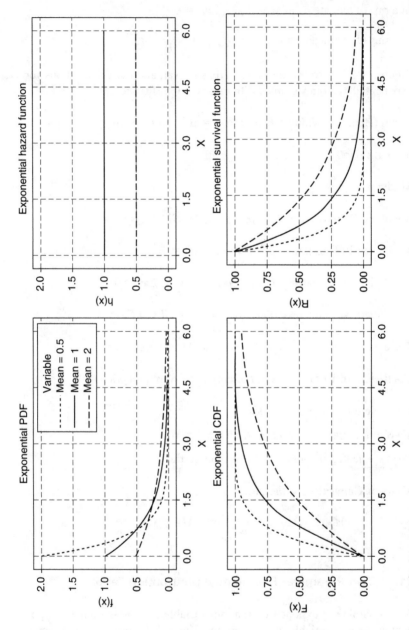

Figure 11-2 PDF, CDF, Hazard, and Survival Functions of Exponential Random Variables.

parameter of the distribution family. τ is called the threshold parameter or location parameter. The first parameter is usually called a scale parameter, β, corresponding to the mean, μ, of a one-parameter exponential distribution. Since the mean of a two-parameter exponential random variable is $\tau + \beta$, it is inappropriate to call β a mean parameter.

Because a two-parameter exponential random variable is a shifted version of the one-parameter exponential, the curves plotted in Figure 11-2 are the same for both families, except that the horizontal scale starts at τ instead of 0.

In Chapter 1, Example 1.1 and 1.3 illustrated one application of a two-parameter exponential distribution.

Parameters: The two-parameter exponential distribution family has two parameters:

- τ is a threshold parameter or location parameter, representing the minimum possible value. τ can be any real number.
- β is a scale parameter representing the average value of $X - \tau$. Also, $\beta + \tau$ is the value with $e^{-1} \cong 0.3679$ probability of occurrence greater than $\beta + \tau$ and $1 - e^{-1} \cong 0.6321$ probability of occurrence less than $\beta + \tau$.

Representation: $X \sim Exp(\beta, \tau)$

Support: $[\tau, \infty)$

Relationships to Other Distributions:

- A one-parameter exponential random variable has $\tau = 0$. $Exp(\beta, 0) \sim Exp(\beta)$, where β represents the mean or scale parameter.
- A one-parameter exponential random variable, plus a constant τ, is a two-parameter exponential random variable. $\tau + Exp(\beta) \sim Exp(\beta, \tau)$, where β represents the mean or scale parameter.
- A standard exponential random variable has $\tau = 0$ and $\lambda = \mu = 1$. If $X \sim Exp(1)$, then $\tau + \beta X \sim Exp(\beta, \tau)$.
- A two-parameter exponential random variable is also a three-parameter gamma random variable with shape parameter $\alpha = 1$. If $X \sim Exp(\beta, \tau)$, then $X \sim \gamma(1, \beta, \tau)$
- The sum of n mutually independent two-parameter exponential random variables with common scale parameter β and threshold parameters τ_i is a three-parameter gamma random variable with shape parameter n, scale parameter β, and threshold parameter $\tau = \sum \tau_i$ If $X_i \sim Exp(\beta, \tau_i)$, and the X_i are mutually independent, then $\sum_{i=1}^{n} X_i \sim \gamma(n, \beta, \sum_{i=1}^{n} \tau_i)$

- A two-parameter exponential random variable is also a three-parameter Weibull random variable with shape parameter $\alpha = 1$. If $X \sim Exp(\beta, \tau)$, then $X \sim Weibull(1, \beta, \tau)$
- A two-parameter exponential random variable minus τ and raised to the 0.2654 power has an approximately normal distribution. If $X \sim Exp(\beta, \tau)$, and $Y = (X - \tau)^{0.2654}$, then $Y \sim N(\mu = 0.9034\beta^{0.2654}, \sigma = 0.2675\beta^{0.2654})$.
- The negative natural log of a uniform random variable between 0 and 1, plus τ, is a two-parameter exponential random variable. If $U \sim Unif(0, 1)$, then $\tau - \beta \ln U \sim Exp(\beta, \tau)$

Normalizing Transformation: Yang and Xie (2000) derived a power transformation that converts an exponential distribution into an approximately normal distribution. If $X \sim Exp(\beta, \tau)$, and $Y = (X - \tau)^{0.2654}$, then $Y \approx N(\mu = 0.9034\beta^{0.2654}, \sigma = 0.2675\beta^{0.2654})$.

Process Control Tools: In most cases, the threshold parameter τ is known, because it corresponds to a physical boundary in the process. When τ is known, two-parameter exponential data may be charted in the same two ways used for one-parameter exponential data:

- Plot the exponential data directly onto an individual X chart, and use these formulas for control limits and the central line:

$$UCL_X = \tau - \overline{X}\ln(\alpha/2)$$

$$CL_X = \tau + 0.6931\overline{X}$$

$$LCL_X = \tau - \overline{X}\ln(1 - \alpha/2)$$

- Transform the data into an approximately normal distribution using $Y = (X - \tau)^{0.2654}$. Plot the data on an individual X chart, using moving ranges to calculate control limits. Only one chart is needed for one unknown parameter, so there is no need to plot the moving range control chart.

If the threshold parameter is estimated from the data, use this approach:

- Estimate τ with $\hat{\tau} = Min\{X_i\}$. Then normalize the data using the transformation $Y = (X - \hat{\tau})^{0.2654}$. Plot the data on an individual X, moving range (IX, MR) control chart. With two unknown parameters, two control charts are required. When using this approach, watch out for observed values less than $\hat{\tau}$. These values would produce a calculation error in the formula $Y = (X - \hat{\tau})^{0.2654}$, and should automatically be considered outside of the control limits.

Estimating Parameter Values: When the threshold parameter τ is known, the maximum likelihood estimator for the scale parameter β is $\hat{\beta} = \overline{X} - \tau$. Confidence intervals and other information about this estimator are analogous to the one-parameter exponential case, with corrections for the threshold parameter.

When both parameters are unknown, these are maximum likelihood estimators:

$$\hat{\tau} = Min\{X_i\}$$
$$\hat{\beta} = \overline{X} - \hat{\tau}$$

For more information on this estimation problem, see section 19.7 of Johnson *et al* (1994).

Capability Metrics: The mode of a two-parameter exponential process is τ, so generally there will only be an upper tolerance limit UTL. Calculate equivalent long-term capability metrics this way:

$$\text{Equivalent } P_P^\% = \text{Equivalent } P_{PK}^\% = \frac{-\Phi^{-1}(\exp[-(UTL - \tau)/\beta])}{3}$$

$\Phi^{-1}(p)$ is the standard normal inverse CDF, evaluated in Excel with the =NORMSINV(p) function.

Probability Density Function:

$$f_{Exp(\beta,\tau)}(x) = \begin{cases} \dfrac{1}{\beta}\exp\left(\dfrac{-(x - \tau)}{\beta}\right) & x \geq \tau \\ 0 & x < \tau \end{cases}$$

In an Excel worksheet, calculate $f_{Exp(\beta,\tau)}(x)$ with the formula =GAMMADIST(x-τ,1,β,FALSE).

Cumulative Distribution Function:

$$F_{Exp(\beta,\tau)}(x) = \begin{cases} 1 - \exp\left(\dfrac{-(x - \tau)}{\beta}\right) & x \geq \tau \\ 0 & x < \tau \end{cases}$$

In an Excel worksheet, calculate $F_{Exp(\beta,\tau)}(x)$ with the formula =GAMMADIST(x-τ,1,β,TRUE).

Inverse Cumulative Distribution Function:

$$F^{-1}_{Exp(\beta,\tau)}(prob) = \tau - \beta \ln(1 - prob)$$

In an Excel worksheet, calculate $F^{-1}_{Exp(\beta,\tau)}(x)$ with the formula =τ+GAMMAINV($prob$,1,β).

Random Number Generation: To generate exponential random numbers in an Excel worksheet, use the formula =τ-β*LN(RAND()). With Crystal Ball and Excel software, use =CB.Gamma(τ,β,1)

Survival Function:

$$R_{Exp(\beta,\tau)}(x) = \begin{cases} \exp\left(\dfrac{-(x - \tau)}{\beta}\right) & x \geq \tau \\ 1 & x < \tau \end{cases}$$

Hazard Function:

$$h_{Exp(\beta,\tau)}(x) = \frac{1}{\beta}$$

Mean (Expected Value): $E[Exp(\beta, \tau)] = \tau + \beta$

Median: $\tau + \beta \ln 0.5 \cong \tau + 0.6931\beta$

Mode: τ

Standard Deviation: $SD[Exp(\beta, \tau)] = \beta$

Variance: $V[Exp(\beta, \tau)] = \beta^2$

Coefficient of Variation: $CV[Exp(\beta, \tau)] = \dfrac{\beta}{\tau + \beta}$

Coefficient of Skewness: $Skew[Exp(\beta, \tau)] = 2$

The two-parameter exponential random variable is skewed to the right.

Coefficient of Kurtosis: $Kurt[Exp(\beta, \tau)] = 9$

Coefficient of Excess Kurtosis: $ExKurt[Exp(\beta, \tau)] = 6$

The two-parameter exponential random variable is leptokurtic.

Moment Generating Function: $M_{Exp(\beta,\tau)}(t) = \dfrac{e^{\tau t}}{1 - \beta t}$

11.2   Truncated Exponential Distribution Family

Occasional applications arise for truncated exponential distributions, especially when they are truncated from above. If an exponential distribution is truncated from below, the truncated distribution is identical to the original distribution, except that the minimum value is shifted to the point of truncation. This is a consequence of the "lack of memory" property. Therefore, the properties of a two-parameter exponential distribution also apply to left-truncated exponential distributions.

It is important to distinguish between truncation and censoring. Truncation applies to the distribution model describing the population; censoring is a property of the sample collected to estimate the population model. When an exponential distribution is truncated on the right at a value B, this has several consequences, including these:

- 100% of the values of X occur between 0 and B.
- The mean and standard deviation of X are now less than the scale parameter β, which is μ for the untruncated case.

A sample of observed values for X is censored if values of X are not observed for some of the items in the sample. In reliability applications, censoring is more common than truncation. The end of a warranty period represents the end of customers reporting failures to the manufacturer, but failures still happen after that period. If a product, like a missile, has a finite lifetime, failures for components are censored if they have not failed at the end of product life. Earlier, this chapter described censored data with references to other books for more information on estimation methods.

Example 11.2

Paula organizes surveys for a major political party. When a partisan organization performs surveys, the objective is to energize and motivate people to be more involved in the political process, and not to find out what they think. For a survey to accurately sample public opinion, respondents are selected at random and it is important to have a response from as many respondents possible from those selected. However, in a partisan survey, whoever responds before the end of the survey period is the sample of interest. These people will surely receive requests for contributions.

Paula measures the time to respond to a survey, sent by e-mail, and she finds that this time clearly has an exponential distribution. After a certain time, B, the survey period ends, and Paula compiles the survey results. Therefore, the time to respond is a truncated exponential random variable. For the purposes of the survey, people who did not respond by time B simply do not exist.

Parameters: For generality, a two-parameter exponential random variable may be truncated on the right at B, resulting in three-parameters:

- τ is a threshold parameter or location parameter, representing the minimum possible value. τ can be any real number.
- β is a scale parameter representing the mean value of $X - \tau$, before truncation.
- B is the right truncation parameter, which can be any number greater than τ.

Representation: $X \sim TExp(\beta, \tau, B)$

Support: $[\tau, B)$

Estimating Parameter Values: Estimating the parameters of a truncated exponential distribution, or more generally a truncated gamma distribution, can be complex and imprecise. See Chapman (1956) or section 17.8.1 of Johnson *et al* (1994) for more information.

Probability Density Function:

$$
f_{TExp(\beta,\tau,B)}(x) = \begin{cases} \dfrac{\exp(-(x - \tau)/\beta)}{\beta[1 - \exp(-(B - \tau)/\beta)]} & \tau \le x \le B \\ 0 & \text{otherwise} \end{cases}
$$

In Excel, calculate $f_{TExp(\beta,\tau,B)}(x)$ with =GAMMADIST(x-τ,1,τ,FALSE)/ GAMMADIST(B-τ,1,β,TRUE).

Cumulative Distribution Function:

$$
F_{TExp(\beta,\tau,B)}(x) = \begin{cases} 0 & x < \tau \\ \dfrac{1 - \exp(-(x - \tau)/\beta)}{1 - \exp(-(B - \tau)/\beta)} & \tau \le x \le B \\ 1 & x > B \end{cases}
$$

In Excel, calculate $F_{TExp(\beta,\tau,B)}(x)$ with =GAMMADIST(x-τ,1,β,TRUE)/ GAMMADIST(B-τ,1,τ,TRUE).

Inverse Cumulative Distribution Function:

$$
F_{TExp(\beta,\tau,B)}^{-1}(prob) = \tau - \beta \ln\left(1 - prob \times \left[1 - \exp\left(\frac{-(B - \tau)}{\beta}\right)\right]\right)
$$

In Excel, calculate $F_{TExp(\beta,\tau,B)}^{-1}(x)$ with =τ+GAMMAINV(*prob** GAMMADIST(*B*-τ,1,β,TRUE),1,β).

Random Number Generation: To generate exponential random numbers in an Excel worksheet, use =τ+GAMMAINV(RAND()*GAMMADIST (*B*-τ,1,β,TRUE),1,β). With Crystal Ball and Excel software, use =CB.Gamma(τ,β,1,,*B*)

Survival Function:

$$
R_{TExp(\beta,\tau,B)}(x) = \begin{cases} 1 & x < \tau \\ \dfrac{\exp(-(x - \tau)/\beta) - \exp(-(B - \tau)/\beta)}{1 - \exp(-(B - \tau)/\beta)} & \tau \le x \le B \\ 0 & x > B \end{cases}
$$

Hazard Function:

$$
h_{TExp(\beta,\tau,B)}(x) = \frac{\exp(-(x - \tau)/\beta)}{\beta[\exp(-(x - \tau)/\beta) - \exp(-(B - \tau)/\beta)]}
$$

Figure 11-3 illustrates the PDF, CDF, hazard, and survival functions of a standard exponential random variable untruncated and truncated at 4, 2, and 1.

The mean, standard deviation, skewness, and kurtosis of a truncated exponential distribution may be calculated in terms of the CDF of the gamma distribution, using the following formulas:

$$
\mu'_1 = \frac{\beta F_{\gamma(2,\beta)}\left(\frac{B - \tau}{\beta}\right)}{1 - \exp\left(\frac{-(B - \tau)}{\beta}\right)}
$$

$$
\mu'_2 = \frac{2\beta^2 F_{\gamma(3,\beta)}\left(\frac{B - \tau}{\beta}\right)}{1 - \exp\left(\frac{-(B - \tau)}{\beta}\right)}
$$

$$
\mu'_3 = \frac{6\beta^3 F_{\gamma(4,\beta)}\left(\frac{B - \tau}{\beta}\right)}{1 - \exp\left(\frac{-(B - \tau)}{\beta}\right)}
$$

$$
\mu'_4 = \frac{24\beta^4 F_{\gamma(5,\beta)}\left(\frac{B - \tau}{\beta}\right)}{1 - \exp\left(\frac{-(B - \tau)}{\beta}\right)}
$$

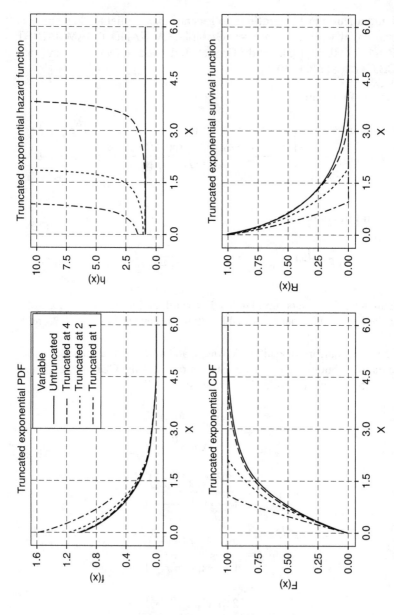

Figure 11-3 PDF, CDF, Hazard, and Survival Functions of a Standard Exponential Random Variable Untruncated, and Truncated at 4, 2, and 1.

In Excel, the CDF of the gamma distribution $F_{\gamma(\alpha,\beta)}(x)$ is calculated by the function =GAMMADIST(x,α,β,TRUE).

Mean (Expected Value): $E[\,TExp(\beta, \tau, B)] = \tau + \mu'_1$

Standard Deviation: $SD[\,TExp(\beta, \tau, B)] = \sqrt{\mu'_2 - (\mu'_1)^2}$

Variance: $V[\,TExp(\beta, \tau, B)] = \mu'_2 - (\mu'_1)^2$

Coefficient of Skewness:

$$Skew[\,TExp(\beta, \tau, B)] = \frac{\mu'_3 - 3\mu'_2\mu'_1 + 2(\mu'_1)^3}{(SD[\,TExp(\beta, \tau, B)])^3}$$

The truncated exponential random variable is skewed to the right, but as $B \to \tau$, the distribution approaches a symmetric, uniform distribution.

Coefficient of Kurtosis:

$$Kurt[\,TExp(\beta, \tau, B)] = \frac{\mu'_4 - 4\mu'_3\mu'_1 + 6\mu'_2(\mu'_1)^2 - 3(\mu'_1)^4}{(SD[\,TExp(\beta, \tau, B)])^4}$$

Coefficient of Excess Kurtosis:

$$ExKurt[\,TExp(\beta, \tau, B)] = \frac{\mu'_4 - 4\mu'_3\mu'_1 + 6\mu'_2(\mu'_1)^2 - 3(\mu'_1)^4}{(SD[\,TExp(\beta, \tau, B)])^4} - 3$$

The exponential random variable could be leptokurtic or platykurtic, depending on where it is truncated.

Extreme Value (Gumbel) Distribution Family

An extreme value, or Gumbel, random variable is a continuous random variable representing the distribution of a maximum or minimum of a large number of random variables, for many common situations. The extreme value distribution family includes two varieties, commonly known as the largest extreme value and the smallest extreme value distribution.

The theory of how extreme values are distributed is extremely important. An early application for extreme value theory with significant impact on government policy is in the prediction of floods. In 1950, the report of the President's Water Resources Policy Commission stated, "However big floods get, there will always be a bigger one coming; so says one theory of extremes, and experience suggests it is true." (U.S., 1950, volume 1, page 141). In the years since then, that somewhat fatalistic view has been replaced by quantitative predictions. With the benefit of a distribution model, policy makers can predict the probability that future floods will exceed a certain size, or they can predict the expected time between floods of a certain size. Common vernacular in flood-prone areas now includes terms such as "100-year flood" and "500-year flood," which originated in extreme value theory.

Reliability engineering is another major field of application for extreme value theory. Material strength, fatigue life, and other important parameters are often related to extremes. In Chapter 3, Example 3.3 describes an application for extreme value distributions as models for material strength.

It is important to realize that not all extreme values follow the extreme value distribution. Extreme value theory is a branch of statistics devoted to the study of maximums and minimums. Following the description of the extreme value distributions, Section 12.1 provides a high-level overview of extreme value theory for modelers and Six Sigma practitioners. The results presented here are relatively simple, but they have tremendous practical importance.

The largest and smallest extreme value distributions are often called Gumbel distributions, in honor of Emil Julius Gumbel (1891–1966). Besides his significant contributions to extreme value theory, Gumbel was a noted anti-Nazi writer and activist. His forced exile to France before the war very likely saved his life.

Example 12.1

Elias is studying a stainless steel sealing ring used in a cryogenic fuel pump. Elias needs a distribution model for the thickness of the sealing ring to predict leakage and other critical parameters. To measure thickness functionally, Elias places a ring between two very flat plates, and measures the distance between the plates. Figure 12-1 illustrates how these flat plates rest on top of whatever surface irregularities may exist on the sealing ring. Figure 12-1 exaggerates the surface irregularities of the part to show how the flat plates rest on top of the peaks on the surface.

The stem-and-leaf diagram below lists measurements of thickness for 100 sealing rings, in microns. In this diagram, the stems represent microns, and the leaves represent tenths of microns.

96	3
97	
98	345
99	159
100	0233466889
101	13699
102	04
103	33
104	
105	4678

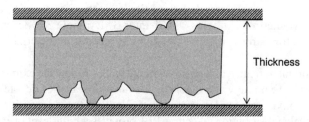

Figure 12-1   Measuring the Thickness of a Sealing Ring Functionally. Surface Irregularities are Exaggerated for Illustration.

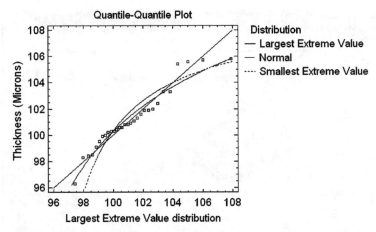

Figure 12-2   Quantile-Quantile Plot of Measured Thickness Data Showing the Fit to Three Possible Distribution Models.

To select a distribution for this data, Elias suspects that the distribution might have some sort of extreme value distribution, but he is not sure which one. In STATGRAPHICS software, Elias fits this data to smallest extreme value, largest extreme value, and normal distributions. Figure 12-2 is a quantile-quantile plot illustrating the fit of this data to these three models. Graphically, the largest extreme value distribution fits best, and the goodness-of-fit test statistics confirm this conclusion.

Therefore, Elias decides to use a largest extreme value model for ring thickness, with mode η = 100.13 and scale parameter β = 2.05, as estimated by STATGRAPHICS software. In Crystal Ball software, Elias defines an assumption with this distribution, illustrated in Figure 12-3. In Crystal Ball software, the largest extreme value distribution is called the maximum extreme distribution. Elias can now combine this distribution with formulas to predict the distribution of system parameters that depend on ring flatness.

Parameters: The smallest extreme value and largest extreme value distribution families each have two parameters:

- η (eta) is the mode of the distribution, and can be any real number.
- β (beta) is the scale parameter, and can be any positive number.

Representation:

For the smallest extreme value: $X \sim SEV(\eta, \beta)$

For the largest extreme value: $X \sim LEV(\eta, \beta)$

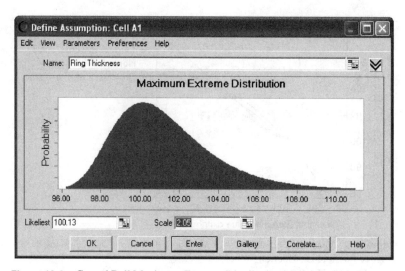

Figure 12-3   Crystal Ball Maximum Extreme Distribution Model for Ring Flatness.

Support: $(-\infty, \infty)$

Relationships to Other Distributions:

- For the extreme value families, η is a location parameter and β is a scale parameter. This means that if $X \sim SEV(\eta, \beta)$, then $\frac{X - \eta}{\beta} \sim SEV(0, 1)$. Also, if $X \sim LEV(\eta, \beta)$, then $\frac{X - \eta}{\beta} \sim LEV(0, 1)$.
- Smallest and largest extreme value families are negatives of each other. $-SEV(\eta, \beta) \sim LEV(-\eta, \beta)$.
- The natural antilog of a smallest extreme value random variable with mode $\eta = 0$ is an exponential random variable. If $X \sim SEV(\eta, \beta)$, then $e^{(X-\eta)} \sim Exp(\beta)$. Also, if $X \sim LEV(\eta, \beta)$, then $e^{-(X-\eta)} \sim Exp(\beta)$.
- The natural log of a Weibull random variable is a smallest extreme value random variable. If $X \sim Weibull(\alpha, \beta)$, then $\alpha \ln\left(\frac{X}{\beta}\right) \sim SEV(0, 1)$. Also, $-\alpha \ln\left(\frac{X}{\beta}\right) \sim LEV(0, 1)$.
- The difference between two independent extreme value random variables is a logistic random variable. If $X_1, X_2 \overset{iid}{\sim} SEV(\eta, \beta)$ or if $X_1, X_2 \overset{iid}{\sim} LEV(\eta, \beta)$ then $X_1 - X_2 \sim Logistic(0, \beta)$. It also follows that the sum of a smallest extreme value and an independent largest extreme value, with the same parameter values, is logistic. $SEV(\eta_1, \beta) + LEV(\eta_2, \beta) \sim Logistic(\eta_1 + \eta_2, \beta)$

Normalizing Transformations: If the mode and scale parameters are known, an extreme value random variable can be transformed into an approximately normal random variable with one of these transformations:

- $e^{0.2654(SEV(\eta,\beta)-\eta)} \approx N(\mu = 0.9034\beta^{0.2654}, \sigma = 0.2675\beta^{0.2654})$
- $e^{-0.2654(LEV(\eta,\beta)-\eta)} \approx N(\mu = 0.9034\beta^{0.2654}, \sigma = 0.2675\beta^{0.2654})$

This transformation combines the relationship between extreme value and exponential random variables with the work of Yang and Xie (2000).

In practice, the mode η is often estimated from sample data. Error in the estimation of η may cause this transformation to fail. If transformed data do not pass normality tests, an alternative approach is to use the Johnson transformation, with the selection algorithm built into MINITAB software.

Process Control Tools: For processes producing extreme value data, first estimate parameter values using software such as MINITAB or STATGRAPHICS. Then, transform the data into a normal distribution using this formula for smallest extreme value $Y = e^{0.2654(X-\hat{\eta})}$ or this formula for largest extreme value $Y = e^{0.2654(X-\hat{\eta})}$. Because of errors in estimating η, it is a good idea to create a normal probability plot and perform normality tests on the transformed data. If the transformed data is not sufficiently normal, then use a Johnson transformation.

Once the data is successfully normalized, plot the transformed data on a Shewhart control chart such as $\overline{X}$, s or IX, MR.

Estimating Parameter Values: No effective and simple formulas are available to estimate the mode or scale parameters of extreme value random variables. This is generally a job for software such as MINITAB or STATGRAPHICS.

For the largest extreme value distribution family, maximum likelihood estimates for η and β are the simultaneous solutions to these equations:

$$\hat{\beta} = \overline{X} - \frac{\sum_{i=1}^{n} X_i \exp\left(\frac{-X_i}{\hat{\beta}}\right)}{\sum_{i=1}^{n} \exp\left(\frac{-X_i}{\hat{\beta}}\right)}$$

$$\hat{\eta} = -\hat{\beta} \ln\left[\frac{1}{n}\sum_{i=1}^{n} \exp\left(\frac{-X_i}{\hat{\beta}}\right)\right]$$

To estimate the parameters of the smallest extreme value family, negate the data and then find solutions to the above equations using the negated data. Following estimation, the sign of the mode η must be reversed.

Capability Metrics: For a process producing the smallest extreme value distribution, calculate equivalent capability metrics this way:

$$\text{Equivalent } P_{PU}^{\%} = \frac{-\Phi^{-1}\left(\exp\left[-e^{\frac{(UTL-\eta)}{\beta}}\right]\right)}{3}$$

$$\text{Equivalent } P_{PL}^{\%} = \frac{-\Phi^{-1}\left(1 - \exp\left[-e^{\frac{(LTL-\eta)}{\beta}}\right]\right)}{3}$$

$$\text{Equivalent } P_{P}^{\%} = \frac{\text{Equivalent } P_{PU}^{\%} + \text{Equivalent } P_{PL}^{\%}}{2}$$

$$\text{Equivalent } P_{PK}^{\%} = Min\{\text{Equivalent } P_{PU}^{\%}, \text{Equivalent } P_{PL}^{\%}\}$$

For a process producing the largest extreme value distribution, calculate equivalent capability metrics this way:

$$\text{Equivalent } P_{PU}^{\%} = \frac{-\Phi^{-1}\left(1 - \exp\left[-e^{\frac{(UTL-\eta)}{\beta}}\right]\right)}{3}$$

$$\text{Equivalent } P_{PL}^{\%} = \frac{-\Phi^{-1}\left(\exp\left[-e^{\frac{(LTL-\eta)}{\beta}}\right]\right)}{3}$$

$$\text{Equivalent } P_{P}^{\%} = \frac{\text{Equivalent } P_{PU}^{\%} + \text{Equivalent } P_{PL}^{\%}}{2}$$

$$\text{Equivalent } P_{PK}^{\%} = Min\{\text{Equivalent } P_{PU}^{\%}, \text{Equivalent } P_{PL}^{\%}\}$$

$\Phi^{-1}(p)$ is the standard normal inverse CDF, evaluated in Excel with the =NORMSINV(p) function. Table 12-1 lists selected values of equivalent $P_{PL}^{\%}$ and $P_{PU}^{\%}$ for smallest and largest extreme value random variables.

Table 12-1 Selected Capability Values for Extreme Value Random Variables

Smallest Extreme Value				Largest Extreme Value			
LTL	Equiv. $P_{PL}^\%$	UTL	Equiv. $P_{PU}^\%$	LTL	Equiv. $P_{PL}^\%$	UTL	Equiv. $P_{PU}^\%$
$\eta - 2\beta$	0.381	$\eta + 0.5\beta$	0.290	$\eta - 0.5\beta$	0.290	$\eta + 2\beta$	0.381
$\eta - 4\beta$	0.698	$\eta + 1\beta$	0.502	$\eta - 1\beta$	0.502	$\eta + 4\beta$	0.698
$\eta - 6\beta$	0.937	$\eta + 1.5\beta$	0.760	$\eta - 1.5\beta$	0.760	$\eta + 6\beta$	0.937
$\eta - 8\beta$	1.134	$\eta + 2\beta$	1.077	$\eta - 2\beta$	1.077	$\eta + 8\beta$	1.134
$\eta - 10\beta$	1.305	$\eta + 2.5\beta$	1.471	$\eta - 2.5\beta$	1.471	$\eta + 10\beta$	1.305
$\eta - 13\beta$	1.529	$\eta + 3\beta$	1.964	$\eta - 3\beta$	1.964	$\eta + 13\beta$	1.529
$\eta - 17\beta$	1.787	$\eta + 3.5\beta$	2.588	$\eta - 3.5\beta$	2.588	$\eta + 17\beta$	1.787
$\eta - 21\beta$	2.014	$\eta + 4\beta$	3.378	$\eta - 4\beta$	3.378	$\eta + 21\beta$	2.014

Probability Density Function:

For the smallest extreme value:

$$f_{SEV(\eta,\beta)}(x) = \frac{1}{\beta}\, e^{\left(\frac{x-\eta}{\beta}\right)}\exp[-e^{\left(\frac{x-\eta}{\beta}\right)}]$$

For the largest extreme value:

$$f_{LEV(\eta,\beta)}(x) = \frac{1}{\beta}\, e^{-\left(\frac{x-\eta}{\beta}\right)}\exp[-e^{-\left(\frac{x-\eta}{\beta}\right)}]$$

Cumulative Distribution Function:

For the smallest extreme value:

$$F_{SEV(\eta,\beta)}(x) = 1 - \exp\left[-e^{\frac{(x-\eta)}{\beta}}\right]$$

For the largest extreme value:

$$F_{LEV(\eta,\beta)}(x) = \exp\left[-e^{\frac{-(x-\eta)}{\beta}}\right]$$

Inverse Cumulative Distribution Function:

For the smallest extreme value:

$$F^{-1}_{SEV(\eta,\beta)}(p) = \eta + \beta\ln\left[\ln\left(\frac{1}{1-p}\right)\right]$$

For the largest extreme value:

$$F^{-1}_{LEV(\eta,\beta)}(p) = \eta - \beta\ln\left[\ln\left(\frac{1}{p}\right)\right]$$

Random Number Generation:

To generate $SEV(\eta, \beta)$ random numbers, calculate $\eta + \beta\ln\left[\ln\left(\frac{1}{U}\right)\right]$, where $U \sim Unif(0, 1)$. In an Excel worksheet, the formula is =η+β*LN(LN(1/RAND())). To generate $LEV(\eta, \beta)$ random numbers, calculate $\eta - \beta\ln\left[\ln\left(\frac{1}{U}\right)\right]$. In an Excel worksheet, the formula is =η-β*LN(LN(1/RAND())).

With Excel and Crystal Ball software, calculate $SEV(\eta, \beta)$ random numbers with =CB.MinExtreme(η,β) and $LEV(\eta, \beta)$ random numbers with =CB.MaxExtreme(η,β).

Survival Function:

For the smallest extreme value:

$$R_{SEV(\eta,\beta)}(x) = \exp\left[-e^{\frac{(x-\eta)}{\beta}}\right]$$

For the largest extreme value:

$$R_{LEV(\eta,\beta)}(x) = 1 - \exp\left[-e^{\frac{-(x-\eta)}{\beta}}\right]$$

Hazard Function:

For the smallest extreme value:

$$h_{SEV(\eta,\beta)}(x) = \frac{1}{\beta}\, e^{\left(\frac{x-\eta}{\beta}\right)}$$

For the largest extreme value:

$$h_{LEV(\eta,\beta)}(x) = \frac{e^{-\left(\frac{x-\eta}{\beta}\right)}}{\beta\left(\exp\left[e^{-\left(\frac{x-\eta}{\beta}\right)}\right] - 1\right)}$$

Both the smallest and the largest extreme value distributions have an increasing hazard function, but the shapes of these functions are quite different. Figure 12-4 illustrates the PDF, CDF, survival and hazard functions for the smallest and the largest extreme value random variables, both with mode $\eta = 0$ and scale parameter $\beta = 1$.

Mean (Expected Value):

For the smallest extreme value: $E[SEV(\eta, \beta)] = \eta - \beta\gamma$

For the largest extreme value: $E[LEV(\eta, \beta)] = \eta + \beta\gamma$

In the above formulas for the mean, γ represents Euler's constant, which is approximately 0.5772156649. One of several integrals which defines γ is $\gamma = -\int_0^\infty e^{-x}\ln x\,dx$. Although γ has been studied extensively, it is unknown whether γ is a rational or irrational number.

Median:

For the smallest extreme value: $\eta + \beta\ln(\ln 2)$

For the largest extreme value: $\eta - \beta\ln(\ln 2)$

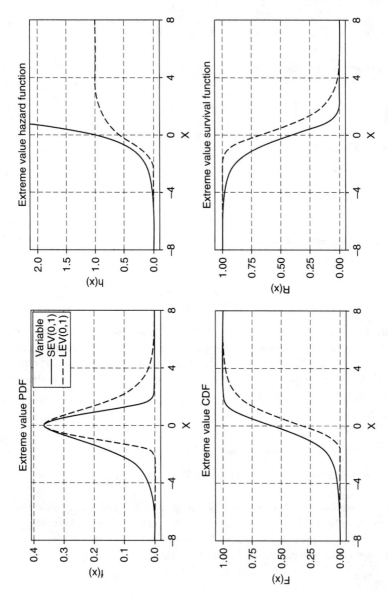

Figure 12-4 PDF, CDF, Survival, and Hazard Functions of Smallest and Largest Extreme Value Random Variables, with η = 0 and β = 1.

Mode: η

Standard Deviation: $SD[SEV(\eta, \beta)] = SD[LEV(\eta, \beta)] = \frac{\beta\pi}{\sqrt{6}}$

Variance: $V[SEV(\eta, \beta)] = V[LEV(\eta, \beta)] = \frac{\beta^2\pi^2}{6}$

Coefficient of Skewness: $Skew[SEV(\eta, \beta)] \cong -1.139547$ and $Skew[LEV(\eta, \beta)] \cong 1.139547$

Smallest extreme value random variables are skewed to the left. Largest extreme value random variables are skewed to the right.

Coefficient of Kurtosis: $Kurt[SEV(\eta, \beta)] = Kurt[LEV(\eta, \beta)] = 5.4$

Coefficient of Excess Kurtosis:

$ExKurt[SEV(\eta, \beta)] = ExKurt[LEV(\eta, \beta)] = 2.4$

All extreme value random variables are leptokurtic.

12.1 Overview of Extreme Value Theory

Extreme value theory is the study of the distribution of the maximum or minimum values in a dataset. Some of the earliest statisticians researched this topic, and most of today's results were derived in the first half of the twentieth century. Important references for extreme value theory include Fréchet (1927), Fisher and Tippett (1928), and Gnedenko (1943). These works and many more are summarized in Gumbel (1958).

The theorems and results of extreme value theory involve advanced mathematics. Even so, extreme values occur in many routine modeling situations and Six Sigma projects. To help practitioners successfully meet these challenges, this section briefly presents some simple results that have broad practical application.

The results presented here are limited to cases where the dataset consists of independent and identically distributed observations. In other words, the process must be a stable process, in which each observation does not influence any other observations.

For a sample consisting of n observations, $X_1, \ldots, X_n$, where the X_i are independent and identically distributed observations of a random variable with PDF $f(x)$, CDF $F(x)$, and survival function $R(x)$, define the minimum observed value $X_{(1)} = Min\{X_i\}$, and the maximum observed value $X_{(n)} = Max\{X_i\}$. This notation is consistent with the common statistical notation that $X_{(i)}$ is the ith *order statistic* of the sample.

The first important result is that the distributions of $X_{(1)}$ and $X_{(n)}$ are completely determined by the distribution of the individual observations. Some families of distributions have a similar, but mirrored distribution for both $X_{(1)}$ and $X_{(n)}$, while other families of distributions have different types of distributions for $X_{(1)}$ and $X_{(n)}$.

The second important result is that when n is known and reasonably small, it is relatively easy to calculate the exact probabilities that $X_{(1)}$ or $X_{(n)}$ are less than or greater than a specific number. Because the observations in the dataset are assumed to be mutually independent, joint probabilities are the product of individual probabilities. This leads to the following formulas:

$$R_{X_{(1)}}(x) = P[X_{(1)} > x] = P[X_i > x \text{ for all } i] = (P[X_i > x])^n = [R(x)]^n$$

$$F_{X_{(1)}}(x) = P[X_{(1)} \le x] = 1 - R_{X_{(1)}}(x) = 1 - [R(x)]^n$$

$$F_{X_{(n)}}(x) = P[X_{(n)} \le x] = P[X_i \le x \text{ for all } i] = (P[X_i \le x])^n = [F(x)]^n$$

$$R_{X_{(n)}}(x) = P[X_{(n)} > x] = 1 - F_{X_{(n)}}(x) = 1 - [F(x)]^n$$

Example 12.2

Ron is a reliability engineer on a program designing missiles that will intercept and destroy other missiles. The lifetime of each missile is 10 minutes after launch. Each missile contains three redundant navigation and guidance systems (NGS). Because each NGS is fast, complex, and delicate, it has numerous failure modes. The three NGS will test each other and cover for most failures, but if all three NGS fail before the missile reaches its target, the missile will fail.

Based on life testing of prototypes, Ron estimates that each NGS has a time from launch to failure which is a Weibull distribution with shape parameter $\alpha = 0.8$ and characteristic life $\beta = 60$ minutes. So if X is the time for one NGS to fail, the distribution of X is described by these formulas, for $x > 0$:

$$f_X(x) = \frac{\alpha x^{\alpha-1}}{\beta^\alpha} \exp\left(-\left[\frac{x}{\beta}\right]^\alpha\right) = \frac{0.8 x^{-0.2}}{60^{0.8}} \exp\left(-\left[\frac{x}{60}\right]^{0.8}\right)$$

$$F_X(x) = 1 - \exp\left(-\left[\frac{x}{\beta}\right]^\alpha\right) = 1 - \exp\left(-\left[\frac{x}{60}\right]^{0.8}\right)$$

$$R_X(x) = \exp\left(-\left[\frac{x}{\beta}\right]^\alpha\right) = \exp\left(-\left[\frac{x}{60}\right]^{0.8}\right)$$

Ron needs to calculate the probability that all three NGS will fail before 10 minutes, which is $P[X_{(3)} \leq 10] = F_{X_{(3)}}(10)$. To make this calculation, Ron assumes that each of the three NGS fails independently of each other. This may not be strictly true, since some failures of one NGS might induce failures in another NGS. However, the assumption of independence allows Ron to use the simple formulas listed above. If dependent failures are unlikely, then the formulas that assume independence will be nearly accurate.

The probability that all three NGS will fail before 10 minutes is:

$$F_{X_{(3)}}(10) = [F_X(10)]^3 = \left[1 - \exp\left(-\left[\frac{10}{60}\right]^{0.8}\right)\right]^3 = 0.009553$$

Ron also needs to predict the probability that none of the NGS will fail by the end of the flight. This calculation is necessary to justify the added cost and weight of the redundant system. Here is the probability that none of the NGS will fail before 10 minutes:

$$P[X_{(1)} > 10] = R_{X_{(1)}}(10) = [R_X(10)]^3 = \left[\exp\left(-\left[\frac{10}{60}\right]^{0.8}\right)\right]^3 = 0.488955$$

Because of this last calculation, Ron concludes that the probability of one or more NGS failing during the 10 minute flight is $1 - 0.489 = 0.511$

In general, the minimum of a set of observations of a Weibull random variable is also a Weibull random variable. When $X_1, \ldots, X_n \overset{iid}{\sim} Weibull(\alpha, \beta)$, then $X_{(1)} \sim Weibull(\alpha, \beta/n)$. This convenient fact also applies to exponential distributions. The exact distribution of the maximum $X_{(n)}$ does not belong to an easily recognized distribution family.

In many situations, the relatively simple calculations above are not possible, because the distribution or number of the individual random variables is unknown. Consider the problem of predicting the time (or cycles) before a piece of material experiences a fatigue failure. The physics of fatigue suggests that as time accumulates, the weakest molecular link eventually fails. After the weakest link has failed, stress increases on the remainder of the material and the entire piece will soon fail. Therefore, the time to fatigue failure is related to the time when the weakest link fails.

Clearly, characteristics of the weakest link are an extreme value problem. The number of such links in a sample of material is certainly very large, but the exact number is unknown. Also, the distribution of failure time for an individual link cannot be measured without considerable difficulty, if at all. Even the concept of a "link" is an abstraction, describing an extremely complex network of interatomic relationships. For all the above reasons, a calculation such as $F_{X_{(n)}} = [F_X]^n$ is impossible for fatigue life and for many other real applications.

To fit a model to a dataset that represents an extreme value, here are some rules of thumb:

- If the data probably represents a minimum, two candidate distribution families are:
  o Smallest extreme value family
  o Three-parameter Weibull family
- If the data probably represents a maximum, calculate the negative of the data. Then, attempt to fit either the smallest extreme value or a three-parameter Weibull distribution to $-X$. An alternative is to fit the largest extreme value distribution to X, but common software tools do not have a three-parameter negative Weibull distribution. Practically, it is easier to negate the data, to compare both options.
- If it is unclear whether the data represents a maximum or minimum, try all four possibilities.

Besides the extreme value and Weibull distributions, there is a third possible distribution, but it rarely occurs in practice, and is not included in any leading statistical software.

The rest of this section explains the theory behind these rules of thumb. To help more readers understand the general concepts of extreme values, this discussion omits much theoretical detail. Those who need more detail are referred to Gumbel (1958) or other texts on the subject.

When the distribution or number of the individual X is unknown, selecting a distribution model for extreme values requires an asymptotic distribution. Asymptotic distributions are limiting distributions, which a statistic would follow if the sample size n were infinitely large. For example, the central limit theorem states that within specific limitations, the sample mean has an asymptotic normal distribution, regardless of the distribution of the individual X. In symbols, $\dfrac{\overline{X} - \mu}{s/\sqrt{n}} \xrightarrow{D} N(0, 1)$. In the symbol $\xrightarrow{D}$, the arrow denotes a limit or asymptote, and the D denotes distribution.

The most significant result in extreme value theory is this: if the individual X follows any continuous distribution, there are only three possible families of asymptotic distributions for the minimum $X_{(1)}$ and maximum $X_{(n)}$. The nature of the tails of the X distribution determines which of the three types of distributions $X_{(1)}$ and $X_{(n)}$ follows. For some families of X distributions, $X_{(1)}$ and

$X_{(n)}$ have the same asymptotic distribution family, and for others, they are different. Using the terminology of Gumbel (1958), the three types of X tail behaviors are called *exponential type*, *Cauchy type*, and *bounded type*. If one can identify the type of the upper (or lower) tail of the X distribution, then one knows the asymptotic distribution of the sample maximum (or minimum).

- *Exponential type tails.* A distribution with a tail of exponential type has unlimited values, but the density function goes to zero quickly, in a way similar to an exponential distribution. This is true if the hazard function $h_X(x) = \frac{f_X(x)}{1 - F_X(x)}$ is asymptotically equal to $\frac{-d \ln f_X(x)}{dx}$ for values approach $+\infty$ or $-\infty$. Many important distributions have one or both tails of exponential type, including normal, lognormal, gamma, logistic, and exponential. Extreme values of a tail of exponential type have an asymptotic Gumbel or extreme value distribution. The asymptotic CDF of the minimum is of the form $F_{X_{(1)}}(x) = 1 - \exp\left[-e^{\frac{(x-\eta)}{\beta}}\right]$, which is the smallest extreme value distribution. The asymptotic CDF of the maximum is of the form $F_{X_{(n)}}(x) = \exp\left[-e^{\frac{-(x-\eta)}{\beta}}\right]$, which is the largest extreme value distribution.

- *Cauchy type tails.* A distribution with a tail of Cauchy (also called Pareto) type has unlimited values, but the density function goes to zero not as quickly as a tail of exponential type. Families of distributions with Cauchy type tails include Student's t, Pareto, and Cauchy. The PDF of the Cauchy distribution has the form $f_X(x) = \left(\pi\beta\left[1 + \left(\frac{x-\tau}{\beta}\right)^2\right]\right)^{-1}$. Because of space limitations, this book does not discuss the Cauchy family further. Chapter 7 of Evans *et al* (2000) provides more information about the Cauchy family. Tails of Cauchy type can be so heavy that distributions with these tails may not have values for moments such as variance, skewness, and kurtosis. Some, such as $t(1)$, do not have a mean, although the median and mode of $t(1)$ are both 0. For a lower tail of Cauchy type, the asymptotic CDF of the minimum is of the form $F_{X_{(1)}}(x) = 1 - \exp\left[-\left(\frac{\beta}{\tau - x}\right)^\alpha\right]$ for $x \le \tau$. For an upper tail of Cauchy type, the asymptotic CDF of the maximum is of the form $F_{X_{(n)}}(x) = \exp\left[-\left(\frac{\beta}{x - \tau}\right)^\alpha\right]$ for $x \ge \tau$. These asymptotic distributions are called Fréchet distributions. It is rare in practice for observed data to follow a distribution with Cauchy type tails. For this reason, these distributions for extremes are also rare, and no leading statistical software includes functions for calculating or fitting these distributions.

- *Bounded type tails.* A distribution with a tail of bounded type has a limited range of possible values. The lower tails of Weibull, gamma, and lognormal are bounded, as are both tails of uniform, triangular, and beta. For a lower tail of bounded type, the asymptotic CDF of the minimum has the form $F_{X_{(1)}}(x) = 1 - \exp\left[-\left(\frac{x - \tau}{\beta}\right)^\alpha\right]$ for $x \ge \tau$,

Table 12-2   Asymptotic Extreme Value Distributions

Type of Tail for Individual Observations	Asymptotic CDF of Smallest Extreme Value	Asymptotic CDF of Largest Extreme Value
Exponential type	SEV (Gumbel) $F_{X_{(1)}}(x) = 1 - \exp\left[-e^{\frac{(x-\eta)}{\beta}}\right]$	LEV (Gumbel) $F_{X_{(n)}}(x) = \exp\left[-e^{\frac{-(x-\eta)}{\beta}}\right]$
Cauchy type	Fréchet $F_{X_{(1)}}(x) = 1 - \exp\left[-\left(\dfrac{\beta}{\tau - x}\right)^{\alpha}\right]$ for $x \le \tau$	Fréchet $F_{X_{(n)}}(x) = \exp\left[-\left(\dfrac{\beta}{x - \tau}\right)^{\alpha}\right]$ for $x \ge \tau$
Bounded type	Three-parameter Weibull $F_{X_{(1)}}(x) = 1 - \exp\left[-\left(\dfrac{x - \tau}{\beta}\right)^{\alpha}\right]$ for $x \ge \tau$	(Three-parameter Weibull) $F_{X_{(n)}}(x) = \exp\left[-\left(\dfrac{\tau - x}{\beta}\right)^{\alpha}\right]$ for $x \le \tau$

which is a three-parameter Weibull distribution. For an upper tail of bounded type, the asymptotic CDF of the maximum has the form $F_{X_{(n)}}(x) = \exp\left[-\left(\frac{\tau - x}{\beta}\right)^{\alpha}\right]$ for $x \le \tau$. Therefore, $-X_{(n)}$ has a three-parameter Weibull distribution.

Table 12-2 summarizes the asymptotic distributions for smallest and largest extreme values for the three types of tails which continuous distributions might have.

13

F Distribution Family

An F random variable is a continuous random variable representing the ratio of two independent χ^2 random variables, each divided by its respective degrees of freedom. Using symbols, if $X \sim \chi^2(\nu_1)$, $Y \sim \chi^2(\nu_2)$, and X and Y are independent, then $\frac{X/\nu_1}{Y/\nu_2} \sim F(\nu_1, \nu_2)$. The two parameters are both degrees of freedom, and the order of the two parameters is important.

The F distribution is used almost exclusively in statistical testing. The ratio of sample variances from two populations with the same variance has an F distribution. Also, in an analysis of variance (ANOVA), ratios of mean squares have an F distribution if the null hypothesis is true.

F random variables are skewed to the right and have a lower bound of zero. While many real processes are also skewed to the right with a lower bound of zero, the F family is not a good family for modeling process behavior. The limitation of the F family for modelers is that the two parameters must be integers. This means that the F family includes a relatively limited variety of shapes. Better models for right-skewed data with a lower bound of zero are gamma, Weibull, lognormal, or loglogistic families. Because the F random variable has few applications for modelers, this chapter does not include many of the features included in other chapters for modelers.

The F distribution is also called Snedecor's F distribution, in honor of mathematician and statistician George Waddel Snedecor (1881–1974). Another name for the F distribution is the Fisher-Snedecor distribution, in honor of Sir Ronald Aylmer Fisher (1890–1962). As a statistician, biologist, and geneticist, Fisher made voluminous and revolutionary contributions to modern statistical methods and to the design of experiments. Fisher was the first to develop the F distribution, in the form $\ln F/2$. (Fisher, 1924)

After a listing of the major properties of the F distribution, Section 13.1 describes the noncentral F distribution, a generalized version of the F distribution used for power and sample size calculations.

Parameters: The F family of distributions has two parameters:

- v_1 is the first degrees of freedom, or the numerator degrees of freedom. v_1 can be any positive integer.
- v_2 is the second degrees of freedom, or the denominator degrees of freedom. v_2 can be any positive integer.

Representation: $X \sim F(v_1, v_2)$

Support: $[0, \infty)$

Relationships to Other Distributions:

- The reciprocal of an F random variable is also an F random variable with the parameters swapped. $\frac{1}{F(v_1, v_2)} \sim F(v_2, v_1)$. Since this operation swaps the tails, this relationship can be used to calculate left-tail F probabilities using a program that has a function for calculating right-tail F probabilities. $P[F(v_1, v_2) < x] = P[F(v_2, v_1) > \frac{1}{x}]$
- The ratio of two χ^2 random variables, each divided by its degrees of freedom, is an F random variable. If $X \sim \chi^2(v_1)$, $Y \sim \chi^2(v_2)$, and X and Y are independent, then $\frac{X/v_1}{Y/v_2} \sim F(v_1, v_2)$.
- If s_1 is the sample standard deviation of a random sample of n_1 observations from a normally distributed population with standard deviation σ_1, and s_2 is the sample standard deviation of a second independent random sample of n_2 observations from a normally distributed population with standard deviation σ_2, then the ratio $\left(\frac{s_1/\sigma_1}{s_2/\sigma_2}\right)^2 \sim F(n_1 - 1, n_2 - 1)$. This fact is the basis for all statistical tests in ANOVA.
- Beta left-tail probabilities are related to F right-tail probabilities. If $X \sim \beta(\alpha, \beta)$ and $Y \sim F(2\beta, 2\alpha)$, then $P\left[X \le \frac{\alpha}{\alpha + y\beta}\right] = P[Y > y]$. Also, $P[X \le x] = P\left[Y > \frac{\alpha(1 - x)}{x\beta}\right]$.

Probability Density Function:

$$f_{F(v_1, v_2)}(x) = \frac{1}{x B\left(\frac{v_1}{2}, \frac{v_2}{2}\right)} \sqrt{\frac{(v_1 x)^{v_1} v_2^{v_2}}{(v_1 x + v_2)^{v_1 + v_2}}}, \text{ for } x > 0.$$

In these formulas, $B(\alpha, \beta)$ is called the "beta function," and it is defined as $B(\alpha, \beta) = \int_0^1 t^\alpha (1 - t)^\beta dt$. The beta function is easy to calculate from the gamma function using this formula: $B(\alpha, \beta) = \frac{\Gamma(\alpha)\Gamma(\beta)}{\Gamma(\alpha + \beta)}$. To calculate $B(\alpha, \beta)$ in an Excel worksheet, use the formula: =EXP(GAMMALN(α)+GAMMALN(β)-GAMMALN($\alpha+\beta$))

Figure 13-1 illustrates the PDF and CDF of selected *F* random variables.

Cumulative Distribution Function and Survival Function: Most testing applications require the right-tail probability, also known as the survival function $R_{F_{(\nu_1,\nu_2)}}(x) = P[F(\nu_1, \nu_2) > x]$. To calculate this probability in an Excel worksheet, use the formula =FDIST(x,ν_1,ν_2).

To calculate a left tail probability or CDF $F_{F_{(\nu_1,\nu_2)}}(x) = P[F(\nu_1, \nu_2) \leq x]$ in an Excel worksheet, use the formula =FDIST($1/x,\nu_2,\nu_1$).

Inverse Cumulative Distribution Function and Inverse Survival Function: To calculate a critical value for most *F* tests requires the inverse survival function $R_{F_{(\nu_1,\nu_2)}}^{-1}(p)$, which is the number x that satisfies the equation $P[F(\nu_1, \nu_2) > x] = p$. To calculate this value in an Excel worksheet, use the formula =FINV(p,ν_1,ν_2).

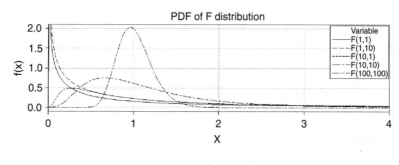

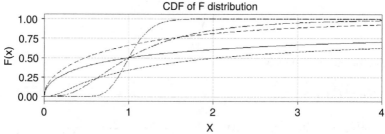

Figure 13-1   PDF and CDF of Selected *F* Random Variables.

The inverse CDF $F^{-1}_{F(v_1,v_2)}(p)$ is the number x that satisfies the equation $P[F(v_1, v_2) \leq x] = p$. To calculate this value in an Excel worksheet, use the formula =1/FINV(p,v_2,v_1).

Random Number Generation: To generate $F(v_1, v_2)$ random numbers in an Excel worksheet, use the formula =FINV(RAND(),v_1,v_2).

Note: For small values of v_2, the mean and other moments listed below do not exist, because the tail of the F distribution is so heavy.

Mean (Expected Value): $E[F(v_1, v_2)] = \dfrac{v_2}{v_2 - 2}$ for $v_2 > 2$.

Mode: When $v_1 \leq 2$, the mode is 0. When $v_2 > 2$, the mode is $\dfrac{(v_1 - 2)v_2}{v_1(v_2 + 2)}$

Standard Deviation:

$$SD[F(v_1, v_2)] = \frac{v_2}{v_2 - 2} \sqrt{\frac{2(v_1 + v_2 - 2)}{v_1(v_2 - 4)}} \text{ for } v_2 > 4.$$

Variance:

$$V[F(v_1, v_2)] = \frac{2v_2^2(v_1 + v_2 - 2)}{v_1(v_2 - 2)^2(v_2 - 4)} \text{ for } v_2 > 4.$$

Coefficient of Variation:

$$CV[F(v_1, v_2)] = \sqrt{\frac{2(v_1 + v_2 - 2)}{v_1(v_2 - 4)}} \text{ for } v_2 > 4.$$

Coefficient of Skewness:

$$Skew[F(v_1, v_2)] = \frac{(2v_1 + v_2 - 2)}{(v_2 - 6)} \sqrt{\frac{8(v_2 - 4)}{v_1(v_1 + v_2 - 2)}} \text{ for } v_2 > 6.$$

Coefficient of Kurtosis:

$$Kurt[F(v_1, v_2)] =$$

$$\frac{12[(v_2 - 2)^2(v_2 - 4) + v_1(v_1 + v_2 - 2)(5v_2 - 22)]}{v_1(v_2 - 6)(v_2 - 8)(v_1 + v_2 - 2)} + 3 \text{ for } v_2 > 8.$$

Coefficient of Excess Kurtosis:

$$ExKurt[F(v_1, v_2)] =$$

$$\frac{12\left[(v_2 - 2)^2(v_2 - 4) + v_1(v_1 + v_2 - 2)(5v_2 - 22)\right]}{v_1(v_2 - 6)(v_2 - 8)(v_1 + v_2 - 2)} \text{ for } v_2 > 8.$$

13.1 Noncentral *F* Distribution Family

A noncentral *F* random variable is a continuous random variable representing the ratio of a noncentral χ^2 random variable divided by a (central) χ^2 random variable, each divided by its respective degrees of freedom. Using symbols, if $X \sim NC\chi^2(v_1, \lambda)$, $Y \sim \chi^2(v_2)$, and X and Y are independent, then $\frac{X/v_1}{Y/v_2} \sim NCF(v_1, v_2, \lambda)$.

The noncentral *F* distribution is a generalization of the (central) *F* distribution used for power and sample size calculations for tests using *F* statistics. *F* statistics usually represent ratios like s_1^2/s_2^2, where s_1 and s_2 are estimates of σ_1 and σ_2. If in fact $\sigma_1 = \sigma_2$ (the null hypothesis is true), then s_1^2/s_2^2 has a (central) *F* distribution. If in fact $\sigma_1 > \sigma_2$ (the alternative hypothesis is true), then s_1^2/s_2^2 has a noncentral *F* distribution.

Parameters: The noncentral *F* family of distributions has three parameters:

- v_1 is the first degrees of freedom, or the numerator degrees of freedom. v_1 can be any positive integer.
- v_2 is the second degrees of freedom, or the denominator degrees of freedom. v_2 can be any positive integer.
- λ is the noncentrality parameter, which can be any nonnegative number. If $\lambda = 0$, the noncentral *F* distribution simplifies into the (central) *F* distribution.

Representation: $X \sim NCF(v_1, v_2, \lambda)$

Support: $[0, \infty)$

Relationships to Other Distributions:

- When the noncentrality parameter $\lambda = 0$, the noncentral *F* distribution simplifies into the (central) *F* distribution. $NCF(v_1, v_2, 0) \sim F(v_1, v_2)$
- The ratio of a noncentral χ^2 random variable divided by a (central) χ^2 random variable, each divided by its respective degrees of freedom is a noncentral *F* random variable. If $X \sim NC\chi^2(v_1, \lambda)$, $Y \sim \chi^2(v_2)$, and X and Y are independent, then $\frac{X/v_1}{Y/v_2} \sim NCF(v_1, v_2, \lambda)$.

Probability Density Function:

$$f_{NCF(v_1, v_2, \lambda)}(x) = \frac{1}{x B\left(\frac{v_1}{2}, \frac{v_2}{2}\right)} \sqrt{\frac{e^{-v_1}(v_1 x)^{v_1} v_2^{v_2}}{(v_1 x + v_2)^{v_1 + v_2}}}$$

$$\times \sum_{j=0}^{\infty} \left[\frac{1}{j!} \left(\frac{\lambda v_1 x}{2(v_1 x + v_2)} \right)^j \prod_{k=0}^{j-1} \frac{v_1 + v_2 + 2k}{v_1 + 2k} \right]$$

In this formula, $B(\alpha, \beta)$ is called the "beta function," which can be calculated as: $B(\alpha, \beta) = \frac{\Gamma(\alpha)\Gamma(\beta)}{\Gamma(\alpha + \beta)}$. To calculate $B(\alpha, \beta)$ in an Excel worksheet, use the formula: =EXP(GAMMALN(α)+GAMMALN(β)-GAMMALN(α+β))

Cumulative Distribution Function and Survival Function: To calculate left-tail probabilities (CDF) or right-tail probabilities (survival function) for noncentral F random variables, use a major statistical package such as JMP, MINITAB, or STATGRAPHICS. Of these, only STATGRAPHICS software provides the survival function.

Inverse Cumulative Distribution Function and Inverse Survival Function: To calculate p-quantiles (inverse CDF) for noncentral F random variables, use a major statistical package such as JMP, MINITAB, or STATGRAPHICS.

Random Number Generation: Random numbers with a noncentral F distribution can be generated from their component parts. To generate $X \sim NC\chi^2(v_1, \lambda)$ random numbers, first generate v_1 independent normal random numbers: $X_i \sim N\left(\frac{\lambda}{\sqrt{v_1}}, 1\right)$. Then $X = \sum_{i=1}^{v_1} X_i^2 \sim NC\chi^2(v_1, \lambda)$. To generate $Y \sim \chi^2(v_2)$ random numbers either use the same process with $\lambda = 0$ or use a built-in χ^2 inverse CDF function as in this Excel formula: =CHIINV(RAND(),v_2). Then the ratio $\frac{X/v_1}{Y/v_2} \sim NCF(v_1, v_2, \lambda)$

Mean (Expected Value):

$$E[NCF(v_1, v_2, \lambda)] = \frac{v_2(v_1 + \lambda)}{v_1(v_2 - 2)} \text{ for } v_2 > 2.$$

Standard Deviation:

$$SD[NCF(v_1, v_2, \lambda)] =$$

$$\frac{v_2}{v_1(v_2 - 2)} \sqrt{\frac{2[(v_1 + \lambda) + (v_1 + 2\lambda)(v_2 - 2)]}{(v_2 - 4)}} \text{ for } v_2 > 4.$$

14

Gamma Distribution Family

A gamma random variable is a continuous random variable with a lower bound of zero. Distributions of gamma random variables have a variety of shapes including the exponential distribution as a special case, when the shape parameter $\alpha = 1$. As the shape parameter α increases, gamma random variables become very similar in shape to normal random variables.

The gamma family of distributions may be the most popular distribution models that people have never heard of. The reason for this is the many important special cases of the gamma distribution that are known by other names. In addition to the family of exponential distributions, the chi-squared (χ^2) family is a subset of the gamma family. The chi-squared distribution has so many important uses that it is the subject of Chapter 9. When the shape parameter α is an integer, gamma random variables are also called Erlang random variables. The sum of α mutually independent exponential random variables is an Erlang (gamma) random variable with shape parameter α. Because of this relationship to the exponential distribution, the gamma family is useful in queuing theory, reliability analysis and other applications of the exponential distribution.

The Erlang distribution family is named in honor of engineer and statistician Agner Krarup Erlang (1878–1929). While working at the Copenhagen Telephone Company, Erlang developed probability models to determine how many circuits and how many operators were required to provide an acceptable level of telephone service. From his pioneering work, new fields of queuing theory and traffic engineering have solved similar problems in many fields.

Both gamma and Weibull distribution families include the exponential distribution as a special case, but the exponential is the only distribution shared by the gamma and Weibull families. A generalized gamma distribution is a four-parameter version of the gamma distribution, which includes Weibull

Figure 14-1   Agner Krarup Erlang.

as a special case. The generalized gamma has several important limiting cases, including lognormal and Pareto families. Because the generalized gamma family has limited practical importance to modelers, this book does not discuss it further.

The gamma distribution family described here has two parameters, a shape parameter α, and a scale parameter β. This chapter provides formulas in two forms, one for the Erlang distribution with integer α, and one for the gamma distribution with any positive α. Section 14.1 describes a three-parameter gamma distribution, which adds a threshold parameter τ, representing the minimum possible value.

Example 14.1

Ric is evaluating constraints in a printer production line. The final inspection station is automated, and when a printer fails any part of the test, the tester diverts the printer into a repair queue. However, the queue only has enough space for three printers. If another printer fails when the repair queue is full, the line shuts down. The production line produces 125 printers per hour, and the probability that any printer will fail the final inspection station is 0.01.

If the repair technician empties the repair queue once per hour, what is the probability that the queue will overfill with four printers in less than one hour?

Since each printer has the same probability of failing the test, the time between failures has an approximately exponential distribution with rate parameter $\lambda = 0.01\frac{\text{failures}}{\text{printer}} \times 125\frac{\text{printers}}{\text{hour}} = 1.25\frac{\text{failures}}{\text{hour}}$. Ric needs to know the probability that the sum of four $Exp(\lambda = 1.25)$ random variables is less than one hour.

This question is easy to answer using Crystal Ball software. In an Excel worksheet, Ric defines four Crystal Ball assumptions, each with an exponential distribution and a rate parameter of 1.25. Then, Ric enters a formula to sum the four exponential values and defines this sum to be a forecast, with a lower specification limit of 1.0 hour. Then, Ric runs a Monte Carlo simulation.

Figure 14-2 shows the forecast distribution of the time until the fourth printer fails. This Crystal Ball forecast window reports that in 96.11% of the trials, the time to the fourth failure is greater than 1.0 hour. Therefore, Ric concludes that there is a 3.89% probability that four or more printers will fail in less than 1.0 hour. In this same window, Crystal Ball reports that the gamma distribution model fits this simulated data better than other distribution models.

Ric's Crystal Ball simulation is easy and quick, but it is not the only way to solve this problem. If Ric knows that the sum of independent exponential random variables is a gamma random variable, then there is a direct way to calculate the same answer.

Since each exponential random variable has a rate of 1.25 failures per hour, it also has a mean of $\mu = 1/\lambda = 0.80$ hours per failure. The sum of four mutually independent exponential random variables with the same mean μ is a gamma

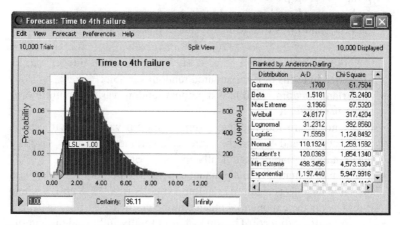

Figure 14-2 Crystal Ball Forecast Window Showing the Distribution of the Time until the Fourth Printer Fails.

random variable with shape parameter $\alpha = 4$ and scale parameter $\beta = \mu$. That is, the time until the fourth printer fails is a $\gamma(4, 0.80)$ random variable.

The probability that four printers will fail in less than one hour is $P[\gamma(4, 0.80) < 1] = F_{\gamma(4,0.80)}(1)$. In Excel, Ric calculates the gamma CDF with this formula =GAMMADIST(1,4,0.80,TRUE), which returns the value 0.0383. Using this calculation method, there is a 3.83% chance that the repair queue will overflow in less than one hour.

For a different approach to this same problem, see Example 20.1 in the chapter on the negative binomial family of distributions.

Parameters: The gamma distribution family has two parameters:

- α is a shape parameter, which may be any positive number. When $\alpha = 1$, gamma random variables are also exponential random variables. When α is an integer, the gamma random variable is also called an Erlang random variable.
- β is a scale parameter, which may be any positive number.

Representation: $X \sim \gamma(\alpha, \beta)$

Support: $[0, \infty)$

Relationships to Other Distributions:

- A gamma random variable with shape parameter $\alpha = 1$ is an exponential random variable with mean equal to the scale parameter β. $\gamma(1, \beta) \sim Exp(\beta)$.
- A chi-squared random variable with ν degrees of freedom is also a gamma random variable with shape parameter $\alpha = \frac{\nu}{2}$ and scale parameter $\beta = 2$. $\chi^2(\nu) \sim \gamma\left(\frac{\nu}{2}, 2\right)$. Conversely, any gamma random variable with shape parameter α such that 2α is an integer can be expressed in terms of a chi-squared random variable: $\frac{2\gamma(\alpha, \beta)}{\beta} \sim \chi^2(2\alpha)$
- An Erlang random variable is a gamma random variable with integer shape parameter α.
- If X and Y are independent gamma random variables with shape parameters α_1 and α_2, and the same scale parameter β, then $\frac{X}{X + Y}$ is a beta random variable. Specifically, $\frac{X}{X + Y} \sim \beta(\alpha_1, \alpha_2)$.
- The sum of α mutually independent exponential random variables with the same mean β is a gamma random variable. If $X_i \overset{iid}{\sim} Exp(\beta)$, where β represents the exponential mean parameter, then $\sum_{i=1}^{\alpha} X_i \sim \gamma(\alpha, \beta)$.
- *Reproductive property.* The sum of n mutually independent gamma random variables with possibly different shape parameters α_i and the same scale parameter β is a gamma random variable with shape

parameter equal to the sum of the component shape parameters. If $X_i \sim \gamma(\alpha_i, \beta)$ and the X_i are mutually independent, then $\sum_{i=1}^{n} X_i \sim \gamma(\sum_{i=1}^{n} \alpha_i, \beta)$.

Normalizing Transformations: Because of the wide variety of shapes of gamma random variables, there may be no single normalizing transformation. In the special case of exponential distributions ($\alpha = 1$), Yang and Xie (2000) determined that $X^{0.2654}$ is approximately normal. For large α, the natural log of X approaches a normal distribution faster than X itself. Olshen (1937) found that $Skew[\ln X] \cong \dfrac{-1}{\sqrt{\alpha - \frac{1}{2}}}$ and $Kurt[\ln X] \cong 3 + \dfrac{2}{\alpha - \frac{1}{2}}$.

Since both log and power transformations are both Box-Cox transformations, practitioners may have good results by applying Box-Cox transformations to gamma process data.

Process Control Tools: When the parameters α and β are known, or when they have been estimated, an individual X chart for gamma data may be constructed using the formulas below. These formulas require a false alarm rate ε, which is 0.0027 for parity with normal-based Shewhart charts. Typically, the false alarm rate is symbolized by α, but since α is a parameter of the gamma distribution, it is ε here.

$$UCL = F^{-1}_{\gamma(\alpha,\beta)}\left(1 - \frac{\varepsilon}{2}\right)$$

$$CL = F^{-1}_{\gamma(\alpha,\beta)}\left(\frac{1}{2}\right)$$

$$LCL = F^{-1}_{\gamma(\alpha,\beta)}\left(\frac{\varepsilon}{2}\right)$$

These formulas use the gamma inverse CDF $F^{-1}_{\gamma(\alpha,\beta)}(p)$. In an Excel worksheet, this function can be evaluated using the =GAMMAINV(p,α,β) function.

When α is large, a simpler approach is to plot the natural log of the gamma data, which is approximately normally distributed, on a Shewhart control chart such as the IX, MR chart for individual data or the $\overline{X}$, s chart for subgrouped data.

Since the sum or average of independent gamma or exponential random variables is a gamma random variable, any of these methods may be used to create an $\overline{X}$ chart for gamma or exponential process data. When the process is stable, $\overline{X} \sim \gamma(n\alpha, \frac{\beta}{n})$

Estimating Parameter Values: When the sample size n is large, the following formulas provide estimates of α and β:

$$\hat{\alpha} = \left(\frac{\overline{X}}{s_n}\right)^2$$

$$\hat{\beta} = \frac{s_n^2}{\overline{X}}$$

In these formulas, $\overline{X} = \frac{1}{n}\sum X_i$ and $s_n = \sqrt{\frac{1}{n}\sum(X_i - \overline{X})^2}$. The Excel function for $\overline{X}$ is =AVERAGE, for s_n is =STDEVP. These estimates are easy to calculate, but they are biased and not very precise. Maximum likelihood estimates are more precise, but these require iterative solutions of formulas including the "digamma" function, which is not readily available in Excel software.

For modelers and Six Sigma practitioners, it is best to use a major statistical package to estimate gamma parameters. Chapter 13, Section 7 of Johnson *et al* (1994) summarizes many methods of gamma estimation.

Capability Metrics: For a process producing data with a gamma distribution, calculate equivalent capability metrics using the gamma CDF. Here are the formulas:

$$\text{Equivalent } P_{PU}^\% = \frac{-\Phi^{-1}(R_{\gamma(\alpha,\beta)}(UTL))}{3}$$

$$\text{Equivalent } P_{PL}^\% = \frac{-\Phi^{-1}(F_{\gamma(\alpha,\beta)}(LTL))}{3}$$

$$\text{Equivalent } P_P^\% = \frac{\text{Equivalent } P_{PU}^\% + \text{Equivalent } P_{PL}^\%}{2}$$

$$\text{Equivalent } P_{PK}^\% = Min\{\text{Equivalent } P_{PU}^\%, \text{Equivalent } P_{PL}^\%\}$$

$\Phi^{-1}(p)$ is the standard normal inverse CDF, evaluated in Excel with the =NORMSINV(p) function. $F_{\gamma(\alpha,\beta)}(x)$ is the CDF of the gamma distribution, evaluated in Excel with the =GAMMADIST(x,α,β,TRUE) function. $R_{\gamma(\alpha,\beta)}(x) = 1 - F_{\gamma(\alpha,\beta)}(x)$ is the survival function of the gamma distribution. Table 14-1 lists selected values of equivalent $P_{PL}^\%$ and $P_{PU}^\%$ for selected gamma processes.

Each tolerance limit in Table 14-1 is listed as a multiple of the scale parameter β. The values of the scale parameter α represented in the table, 0.5, 1, 2, and 3.6, are the same values used for Table 28-1 on the Weibull distribution. Comparing these tables provides some insight into differences between gamma and Weibull families. One characteristic of the gamma family is that the mean value is $\alpha\beta$, the product of both parameters. In the Weibull family, β is the "characteristic life" or 63.2% probability point, regardless of the value of α.

Probability Density Function:

For the Erlang distribution:

$$f_{\gamma(\alpha,\beta)}(x) = \begin{cases} \dfrac{x^{\alpha-1}e^{-x/\beta}}{\beta^{\alpha}(\alpha-1)!} & x \geq 0 \\ 0 & x < 0 \end{cases}$$

For the gamma distribution:

$$f_{\gamma(\alpha,\beta)}(x) = \begin{cases} \dfrac{x^{\alpha-1}e^{-x/\beta}}{\beta^{\alpha}\Gamma(\alpha)} & x \geq 0 \\ 0 & x < 0 \end{cases}$$

In this formula and several formulas below, $\Gamma(\alpha)$ refers to the gamma function, which can be evaluated in an Excel worksheet with the formula =EXP(GAMMALN(α)). The gamma or Erlang PDF may be calculated in an Excel worksheet with the formula =GAMMADIST(x,α,β,FALSE).

Cumulative Distribution Function:

For the Erlang distribution:

$$F_{\gamma(\alpha,\beta)}(x) = \begin{cases} 1 - e^{-x/\beta}\sum_{i=0}^{\alpha-1}\dfrac{(x/\beta)^{i}}{i!} & x \geq 0 \\ 0 & x < 0 \end{cases}$$

For the gamma distribution:

$$F_{\gamma(\alpha,\beta)}(x) = \begin{cases} \dfrac{1}{\beta^{\alpha}\Gamma(\alpha)}\int_{0}^{x}u^{\alpha-1}e^{-u/\beta}du & x \geq 0 \\ 0 & x < 0 \end{cases}$$

Table 14-1 Selected Capability Values for Gamma Random Variables

Shape α = 0.5		Shape α = 1 (Exponential)		Shape α = 2				Shape α = 3.6			
UTL	Equiv. $P^\%_{PU}$	UTL	Equiv. $P^\%_{PU}$	UTL	Equiv. $P^\%_{PU}$	LTL	Equiv. $P^\%_{PL}$	UTL	Equiv. $P^\%_{PU}$	LTL	Equiv. $P^\%_{PL}$
2β	0.563	3β	0.549	6β	0.704	1β	0.210	6β	0.409	2β	0.277
4β	0.866	5β	0.824	8β	0.915	0.5β	0.446	8β	0.637	1.5β	0.421
6β	1.091	7β	1.039	10β	1.097	0.3β	0.596	10β	0.831	1β	0.605
8β	1.278	9β	1.222	12β	1.258	0.25β	0.645	12β	1.002	0.5β	0.879
10β	1.441	11β	1.383	14β	1.405	0.2β	0.703	14β	1.157	0.3β	1.055
12β	1.587	13β	1.529	16β	1.540	0.1β	0.866	17β	1.367	0.1β	1.380
14β	1.721	15β	1.662	16β	1.666	0.01β	1.297	20β	1.556	0.05β	1.557
16β	1.846	17β	1.787	20β	1.784	0.001β	1.631	23β	1.729	0.02β	1.768
18β	1.962	19β	1.904	22β	1.896	0.0001β	1.910	26β	1.889	0.01β	1.914
20β	2.072	21β	2.014	24β	2.002	0.00001	2.156	29β	2.039	0.006β	2.015

There is no easy simplification for these formulas. Fortunately, Excel software and most statistical programs will calculate this function quickly. The gamma or Erlang CDF may be calculated in an Excel worksheet with the formula =GAMMADIST(x,α,β,TRUE).

Inverse Cumulative Distribution Function: There is no easy formula to invert the gamma or Erlang CDF. To calculate $F^{-1}_{\gamma(\alpha,\beta)}(prob)$ in Excel, use =GAMMAINV($prob,\alpha,\beta$).

Random number generation: Erlang random numbers are the sum of independent exponential random numbers. One way to generate Erlang random numbers is to generate α independent uniform random numbers U_i, convert them into exponential random numbers with $-\ln U_i$, sum them, and multiply by the scale parameter β.

Excel software provides an easier way to generate gamma or Erlang random numbers with the formula =GAMMAINV(RAND(),α,β).

With Crystal Ball and Excel, generate gamma or Erlang random numbers with =CB.Gamma(0,β,α).

Survival Function:

For the Erlang distribution:

$$R_{\gamma(\alpha,\beta)}(x) = \begin{cases} e^{-x/\beta}\sum_{i=0}^{\alpha-1}\dfrac{(x/\beta)^i}{i!} & x \geq 0 \\ 1 & x < 0 \end{cases}$$

For the gamma distribution:

$$R_{\gamma(\alpha,\beta)}(x) = \begin{cases} 1 - \dfrac{1}{\beta^\alpha\Gamma(\alpha)}\int_0^x u^{\alpha-1}e^{-u/\beta}du & x \geq 0 \\ 1 & x < 0 \end{cases}$$

Hazard Function:

For the Erlang distribution:

$$h_{\gamma(\alpha,\beta)}(x) = \frac{x^{\alpha-1}}{\beta^\alpha(\alpha-1)!\sum_{i=0}^{\alpha-1}\dfrac{x^i}{\beta^i i!}}$$

For the gamma distribution, use the relationship between the hazard function, PDF and CDF:

$$h_{\gamma(\alpha,\beta)}(x) = \frac{f_{\gamma(\alpha,\beta)}(x)}{1 - F_{\gamma(\alpha,\beta)}(x)}$$

Figure 14-3 illustrates the PDF, CDF, hazard, and survival functions for selected gamma random variables. All five examples illustrated have the same scale parameter $\beta = 1$. For comparison with the Weibull distribution (Chapter 28), this figure illustrates the same shape parameters used for Figure 28-4.

Mean (Expected Value): $E[\gamma(\alpha, \beta)] = \alpha\beta$

Median: $F^{-1}_{\gamma(\alpha,\beta)}(0.5)$, evaluated in an Excel worksheet with the formula =GAMMAINV(0.5,α,β).

Mode: The mode is 0 when $\alpha \leq 1$ and $\beta(\alpha - 1)$ when $\alpha \geq 1$

Standard Deviation: $SD[\gamma(\alpha, \beta)] = \beta\sqrt{\alpha}$

Variance: $V[\gamma(\alpha, \beta)] = \alpha\beta^2$

Coefficient of Variation: $CV[\gamma(\alpha, \beta)] = \frac{1}{\sqrt{\alpha}}$

Coefficient of Skewness: $Skew[\gamma(\alpha, \beta)] = \frac{2}{\sqrt{\alpha}}$

Gamma and Erlang random variables are always skewed to the right, but they approach an unskewed, normal distribution as α gets large.

Coefficient of Kurtosis: $Kurt[\gamma(\alpha, \beta)] = 3 + \frac{6}{\alpha}$

Coefficient of Excess Kurtosis: $ExKurt[\gamma(\alpha, \beta)] = \frac{6}{\alpha}$

Gamma and Erlang random variables are always leptokurtic, but they approach a mesokurtic, normal distribution as α gets large.

Moment Generating Function:

$$M_{\gamma(\alpha,\beta)}(t) = \frac{1}{(1 - \beta t)^\alpha}$$

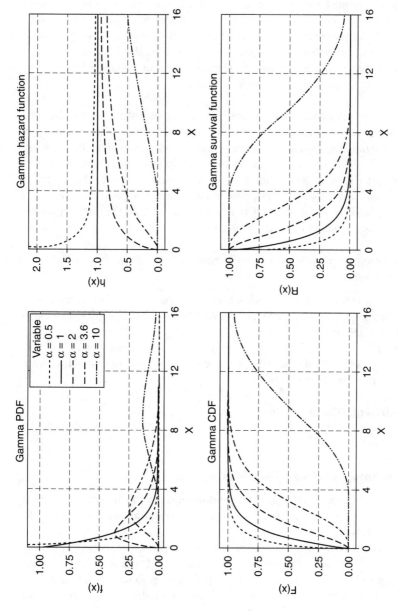

Figure 14-3 PDF, CDF, Hazard and Survival Functions for Selected Gamma Random Variables, All with Scale Parameter $\beta = 1$.

14.1 Three-Parameter Gamma Distribution Family

A three-parameter gamma random variable is a continuous random variable with a lower bound τ. Probability functions of three-parameter gamma random variables have a variety of shapes including the two-parameter exponential distribution as a special case, when the shape parameter $\alpha = 1$. As the shape parameter α increases, gamma random variables become very similar in shape to normal random variables.

The three-parameter gamma distribution is a useful model for the time between events, when there is a minimum possible time τ. It has been widely applied in reliability analysis, queueing theory, and many other fields. When selecting a distribution because of its fit to observed data, the three-parameter gamma distribution is a very popular. It is always skewed to the right, and by changing the shape parameter α, the skewness may be adjusted to any positive value. When data is skewed to the left, the three-parameter gamma distribution might fit the negative of the data.

Parameters: The three-parameter gamma distribution family has these parameters:

- α is a shape parameter, which may be any positive number. When $\alpha = 1$, three-parameter gamma random variables are also two-parameter exponential random variables.
- β is a scale parameter, which may be any positive number.
- τ is a threshold parameter, representing the minimum possible value. τ may be any real number.

Representation: $X \sim \gamma(\alpha, \beta, \tau)$

Support: $[\tau, \infty)$

Relationships to Other Distributions:

- A three-parameter gamma random variable with shape parameter $\alpha = 1$ is a two-parameter exponential random variable. $\gamma(1, \beta, \tau) \sim Exp(\beta, \tau)$.
- The sum of α mutually independent two-parameter exponential random variables with the same scale parameter β and possibly different threshold parameters τ_i is a three-parameter gamma random variable. If $X_i \sim Exp(\beta, \tau_i)$ and the X_i are mutually independent, then $\sum_{i=1}^{\alpha} X_i \sim \gamma(\alpha, \beta, \sum_{i=1}^{\alpha} \tau_i)$.

- *Reproductive property.* The sum of n mutually independent three-parameter gamma random variables with possibly different shape parameters α_i and threshold parameters τ_i and the same scale parameter β is a three-parameter gamma random variable with shape and threshold parameters equal to the sum of the corresponding component parameters. If $X_i \sim \gamma(\alpha_i, \beta, \tau_i)$ and the X_i are mutually independent, then

$$\sum_{i=1}^{n} X_i \sim \gamma\left(\sum_{i=1}^{n} \alpha_i, \beta, \sum_{i=1}^{n} \tau_i\right)$$

Normalizing Transformations: If the threshold parameter τ is known, then $X - \tau$ is a two-parameter gamma random variable. The methods discussed earlier for normalizing a two-parameter gamma random variable may be effective on then $X - \tau$. If τ is unknown, then either Box-Cox or Johnson transformations may be successful, or they may not. Because of the wide variety of shapes of gamma random variables, there may be no single normalizing transformation.

Process Control Tools: When the parameters α, β and τ are known, or when they have been estimated, an individual X chart for three-parameter gamma data may be constructed using the formulas below. These formulas require a false alarm rate ε, which is 0.0027 for parity with normal-based Shewhart charts. Typically, the false alarm rate is symbolized by α, but since α is a parameter of the gamma distribution, it is ε here.

$$UCL = F^{-1}_{\gamma(\alpha,\beta,\tau)}\left(1 - \tfrac{\varepsilon}{2}\right)$$
$$CL = F^{-1}_{\gamma(\alpha,\beta,\tau)}\left(\tfrac{1}{2}\right)$$
$$LCL = F^{-1}_{\gamma(\alpha,\beta,\tau)}\left(\tfrac{\varepsilon}{2}\right)$$

These formulas use the three-parameter gamma inverse CDF $F^{-1}_{\gamma(\alpha,\beta,\tau)}(p)$. In Excel, this function can be evaluated using the $=\tau+\text{GAMMAINV}(p,\alpha,\beta)$ function.

Since the sum or average of independent three-parameter gamma or two-parameter exponential random variables is a three-parameter gamma random variable, these methods may be used to create an $\overline{X}$ chart for gamma or exponential process data. When the process is stable, $\overline{X} \sim \gamma(n\alpha, \tfrac{\beta}{n}, \tau)$. Note that as the shape parameter gets large, a gamma random variable approaches the shape of a normal random variable. Therefore, the normal-based Shewhart $\overline{X}$ chart may work well on processes with a gamma distribution, if the shape parameter α is large.

Estimating Parameter Values: When the sample size n is large, the following formulas provide estimates of α, β and τ:

$$\hat{\alpha} = \frac{4 s_n^6}{m_3^2}$$

$$\hat{\beta} = \frac{m_3}{2 s_n^2}$$

$$\hat{\tau} = \overline{X} - \frac{2 s_n^4}{m_3}$$

In these formulas, $\overline{X} = \frac{1}{n}\sum X_i$, $s_n = \sqrt{\frac{1}{n}\sum(X_i - \overline{X})^2}$, and $m_3 = \frac{1}{n}\sum(X_i - \overline{X})^3$. The Excel function for $\overline{X}$ is =AVERAGE, for s_n is =STDEVP, and for m_3 is =SKEW($data$)*(STDEV($data$)^3)*(n-1)* (n-2)/(n^2). These estimates are easy to calculate, but they are biased and not very precise. Maximum likelihood estimates are more precise, but these require iterative solutions of formulas including the "digamma" function, which is not readily available in Excel.

For modelers and Six Sigma practitioners, it is best to use a major statistical package to estimate gamma parameters. Chapter 13, Section 7 of Johnson *et al* (1994) summarizes many methods of gamma estimation.

Capability Metrics: For a process producing data with a three-parameter gamma distribution, calculate equivalent capability metrics using the gamma CDF. Here are the formulas:

$$\text{Equivalent } P_{PU}^{\%} = \frac{-\Phi^{-1}(R_{\gamma(\alpha,\beta,\tau)}(UTL))}{3}$$

$$\text{Equivalent } P_{PL}^{\%} = \frac{-\Phi^{-1}(F_{\gamma(\alpha,\beta,\tau)}(LTL))}{3}$$

$$\text{Equivalent } P_{P}^{\%} = \frac{\text{Equivalent } P_{PU}^{\%} + \text{Equivalent } P_{PL}^{\%}}{2}$$

$$\text{Equivalent } P_{PK}^{\%} = Min\{\text{Equivalent } P_{PU}^{\%}, \text{Equivalent } P_{PL}^{\%}\}$$

$\Phi^{-1}(p)$ is the standard normal inverse CDF, evaluated in Excel with the =NORMSINV(p) function. $F_{\gamma(\alpha,\beta,\tau)}(x)$ is the CDF of the three-parameter gamma distribution, evaluated in Excel with the =GAMMADIST (x-τ,α,β,TRUE) function. $R_{\gamma(\alpha,\beta,\tau)}(x) = 1 - F_{\gamma(\alpha,\beta,\tau)}(x)$ is the survival function of the three-parameter gamma distribution.

Probability Density Function:

$$
f_{\gamma(\alpha,\beta,\tau)}(x) = \begin{cases} \dfrac{(x-\tau)^{\alpha-1}e^{-(x-\tau)/\beta}}{\beta^{\alpha}\Gamma(\alpha)} & x \geq \tau \\[2mm] 0 & x < \tau \end{cases}
$$

In this formula and several formulas below, $\Gamma(\alpha)$ refers to the gamma function, which can be evaluated in an Excel worksheet with =EXP(GAMMALN(α)). The three-parameter gamma PDF may be calculated in an Excel worksheet with =GAMMADIST(x-τ,α,β,FALSE).

Cumulative Distribution Function:

$$
F_{\gamma(\alpha,\beta,\tau)}(x) = \begin{cases} \dfrac{1}{\beta^{\alpha}\Gamma(\alpha)}\int_0^{x-\tau} u^{\alpha-1}e^{-u/\beta}\,du & x \geq \tau \\[2mm] 0 & x < \tau \end{cases}
$$

There is no easy simplification for these formulas. Fortunately, Excel and statistical programs will calculate this function quickly. The gamma CDF may be calculated in an Excel worksheet with the formula =GAMMADIST(x-τ,α,β,TRUE).

Inverse Cumulative Distribution Function: There is no easy formula to invert the three-parameter gamma CDF. To calculate $F_{\gamma(\alpha,\beta,\tau)}^{-1}(prob)$ in an Excel worksheet, use the formula =τ+GAMMAINV($prob$,α,β).

Random number generation: To generate three-parameter gamma random numbers in an Excel worksheet, use the formula =τ+GAMMAINV (RAND(),α,β).

With Crystal Ball and Excel software, generate three-parameter gamma random numbers with =CB.Gamma(τ,β,α).

Survival Function:

$$
R_{\gamma(\alpha,\beta,\tau)}(x) = \begin{cases} 1 - \dfrac{1}{\beta^{\alpha}\Gamma(\alpha)}\int_0^{x-\tau} u^{\alpha-1}e^{-u/\beta}\,dx & x \geq \tau \\[2mm] 1 & x < \tau \end{cases}
$$

Hazard Function: To calculate the hazard function, use the relationship between the hazard function, PDF and CDF:

$$h_{\gamma(\alpha,\beta,\tau)}(x) = \frac{f_{\gamma(\alpha,\beta,\tau)}(x)}{1 - F_{\gamma(\alpha,\beta,\tau)}(x)}$$

Mean (Expected Value): $E[\gamma(\alpha, \beta, \tau)] = \tau + \alpha\beta$

Median: $F^{-1}_{\gamma(\alpha,\beta,\tau)}(0.5)$, evaluated in Excel with =τ+GAMMAINV(0.5,α,β).

Mode: The mode is τ when $\alpha \leq 1$ and $\tau + \beta(\alpha - 1)$ when $\alpha \geq 1$

Standard Deviation: $SD[\gamma(\alpha, \beta, \tau)] = \beta\sqrt{\alpha}$

Variance: $V[\gamma(\alpha, \beta, \tau)] = \alpha\beta^2$

Coefficient of Variation: $CV[\gamma(\alpha, \beta, \tau)] = \frac{\beta\sqrt{\alpha}}{\tau + \alpha\beta}$

Coefficient of Skewness: $Skew[\gamma(\alpha, \beta, \tau)] = \frac{2}{\sqrt{\alpha}}$

Gamma random variables are always skewed to the right, but they approach an unskewed, normal distribution as α gets large.

Coefficient of Kurtosis: $Kurt[\gamma(\alpha, \beta, \tau)] = 3 + \frac{6}{\alpha}$

Coefficient of Excess Kurtosis: $ExKurt[\gamma(\alpha, \beta, \tau)] = \frac{6}{\alpha}$

Gamma random variables are always leptokurtic, but they approach a mesokurtic, normal distribution as α gets large.

Moment Generating Function:

$$M_{\gamma(\alpha,\beta,\tau)}(t) = \frac{e^{\tau t}}{(1 - \beta t)^{\alpha}}$$

15

Geometric Distribution Family

A geometric random variable is a discrete random variable representing the count of nondefective items before the first defective item in a series of independent items, each with p probability of being defective. More generally, in any experiment consisting of a series of independent trials with two possible outcomes, if the probability of outcome "A" is p on every trial, the count of "B" outcomes before the first "A" outcome is a geometric random variable.

An experiment consists of a series of Bernoulli trials if it meets these criteria:

- Every trial can have only two outcomes "A" and "B"
- The probability of outcome "A" is the same on every trial
- The trials are independent

In any series of Bernoulli trials, the count of outcome "B" before the first outcome "A" is a geometric random variable.

The geometric random variable has a "lack of memory" property, which no other discrete random variable has. Regardless of the history of "A" or "B" outcomes, the count of future "B" outcomes before the next "A" outcome always has the same distribution. The exponential distribution (Chapter 11) is the only continuous distribution with the same "lack of memory" property. Because of this shared property, the geometric distribution is a discrete analog to the exponential distribution.

Two versions of the geometric random variable are commonly used, and both are described here. X_0 represents the number of "B" outcomes before the first "A" outcome, and has a minimum value of 0. X_1 represents the total number of trials up to and including the first "A" outcome, and has a minimum value of 1. When referring to publications or when using software, be careful to use the correct version. The geometric functions in JMP, STATGRAPHICS, and Excel software use X_0. Crystal Ball software uses X_1. Starting in release 15, MINITAB has functions for both X_0 and X_1.

The geometric distribution family is an important special case of the negative binomial and Pascal distribution families with $k = 1$. The geometric distribution has also been called the Furry distribution (Furry, 1937).

Example 15.1

Nic plays the same numbers in each week's lottery drawing. Nic's numbers are $\{1, 2, 3, 4, 5, 6\}$ because he once found this astonishing sequence in a deck of new playing cards.

In this lottery game, the probability of winning the big prize is $p = \binom{42}{6}^{-1} = \frac{1}{5,245,786}$. The number of times Nic loses before his first win is a geometric random variable, with a minimum possible value of 0. Since the mean of a geometric random variable (X_0 version) is $\frac{1}{p} - 1$, Nic can expect to lose 5,245,785 times before his first win of the big prize.

Regardless of how many times Nic plays this game, whether he wins or loses, the expected number of losses before his next win is still 5,245,785.

Example 15.2

Nic's brother Ric is a Green Belt supporting a printer production line. Each time a printer fails any part of the automated final inspection process, the tester diverts the printer into a repair queue. On an average, the tester diverts 1 out of 100 printers into the repair queue.

Ric uses a Crystal Ball simulation model to evaluate flow and constraints in the production line. As part of this model, Ric assumes that the number of printers which pass the final test between failed units, and including the failed unit, is a geometric random variable with probability $p = 0.01$. Figure 15-1 shows the Crystal Ball window defining the distribution for this assumption. Note that Crystal Ball geometric assumptions use the X_1 version, with a minimum value of 1.

The same problem can be modeled and analyzed in different ways. For alternative approaches to the same problem, see examples in Chapter 14 on the gamma distribution and Chapter 20 on the negative binomial distribution.

Parameters: The geometric distribution family has one parameter:

- p represents the probability of observing outcome "A" in any one trial, when X_0 represents the count of outcomes "B" before the first occurrence of outcome "A" $0 \le p \le 1$.

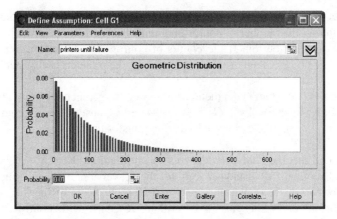

Figure 15-1   Crystal Ball Define Assumption Form with a $Geom_1(0.01)$ Assumption.

Representation: $X_0 \sim Geom_0(p)$ or $X_1 \sim Geom_1(p)$

Support:

The support of X_0 is $\{0, 1, 2 \ldots\}$

The support of X_1 is $\{1, 2, 3 \ldots\}$

Relationships to Other Distributions:

- In a sequence of independent Bernoulli (Yes-No) random variables, the count of 0 values before the first 1 value is a geometric random variable, 0 version. If $X_i \overset{iid}{\sim} Bern(p)$, then the count of 0 values before the first 1 value in the sequence is $Geom_0(p)$.
- A geometric random variable is a special case of a negative binomial random variable with $k = 1$. $NB_0(1, p) \sim Geom_0(p)$ and $NB_k(1, p) \sim Geom_1(p)$.
- The sum of k independent geometric random variables is a negative binomial random variable. If $X_i \overset{iid}{\sim} Geom_0(p)$, then $\sum_{i=1}^{k} X_i \sim NB_0(k, p)$. If $X_i \overset{iid}{\sim} Geom_1(p)$, then $\sum_{i=1}^{k} X_i \sim NB_k(k, p)$.
- An exponential random variable is a continuous version of a geometric random variable.

Process Control Tools: Control charts for geometric processes have been researched extensively, because of their application to attribute processes where defects occur rarely. Standard Shewhart control charts for binomial (p and np) and Poisson (c and u) processes are not practical to apply when defects occur very rarely. As an alternative, Calvin (1983) proposed counting and plotting the number of conforming units between defects, which has a geometric distribution. If the counts of conforming units form a stable geometric process, then the process producing the products is also stable.

Xie *et al* (2000) studied the methods proposed for geometric control charts, including Glushkovsky's *G*-chart (1994) and Quesenberry's geometric *Q*-chart (1995), plus various normalizing transformations.

Xie found that the *double square root* transformation, $Y = X^{1/4}$, tends to normalize geometric data very well, for the X_0 version. After applying this transformation, the transformed data may be plotted on any Shewhart chart for normal data, such as an individual X chart, with control limits determined from the moving range. Note that since the geometric distribution has only a single parameter p, only a single control chart is needed to detect changes in p.

Interestingly, Yang and Xie (2000) found that almost the same transformation $Y = X^{0.2654}$ normalizes exponentially distributed data. Because the exponential and geometric distributions share the "lack of memory" property, this coincidence is not surprising.

Normalizing Transformation: Because the geometric distribution is discrete, no exact normalizing transformation is possible. However, Xie *et al* (2000) found $Y = X^{1/4}$ works well to normalize geometric data.

Estimating Parameter Values: A maximum likelihood estimator for p is $\hat{p} = \frac{1}{1 + \overline{X}_0}$ when observed counts exclude the final trial in the series, and $\hat{p} = \frac{1}{\overline{X}_1}$ when observed counts include the final trial in the series.

Capability Metrics: Since the mode of a geometric random variable is always 0, assume there is only a single upper tolerance limit UTL and no lower tolerance limit.

$$\text{Equivalent } P^{\%}_{PK} = \text{Equivalent } P^{\%}_{PU} = \frac{-\Phi^{-1}(R_X(UTL))}{3}$$

In this formula, $R_X(x) = 1 - F_X(x)$ is the survival function of a geometric random variable with parameter p. Excel does not have a function which calculates $R_X(x)$ directly, but it is easy to calculate as $R_{X_0}(x) = (1 - p)^{1+x}$ for the X_0 version or $R_{X_1}(x) = (1 - p)^x$ for the X_1 version. Alternatively, statistical software such as STATGRAPHICS can calculate $R_{X_0}(x)$.

Table 15-1 Selected Capability Values for Geometric Random Variables, X_0 version.

$p = 0.5$		$p = 0.2$		$p = 0.01$	
UTL	$P_{PU}^{\%}$	*UTL*	$P_{PU}^{\%}$	*UTL*	$P_{PU}^{\%}$
3	0.511	10	0.455	200	0.371
6	0.806	20	0.786	400	0.701
9	1.032	30	1.031	600	0.941
12	1.223	40	1.234	800	1.138
15	1.390	50	1.412	1000	1.309
18	1.540	60	1.571	1200	1.463
21	1.678	70	1.716	1400	1.602
24	1.807	80	1.851	1600	1.731
27	1.927	90	1.977	1800	1.852
30	2.040	100	2.095	2000	1.966

$\Phi^{-1}(prob)$ is the standard normal inverse CDF, evaluated in Excel with the =NORMSINV($prob$) function. Table 15-1 lists selected values of equivalent $P_{PU}^{\%}$ for geometric processes.

In Table 15-1, the tolerance limits on the bottom line are approximately 20 standard deviations above the mean. Only at this point is equivalent $P_{PU}^{\%} = 2$, equivalent to a normal distribution with a tolerance limit six standard deviations above the mean. This illustrates how heavy the tail is for the geometric distribution.

Probability Mass Function:

For the X_0 version:

$$f_{Geom_0(p)}(x) = \begin{cases} p(1 - p)^x & x \in \{0, 1, \ldots\} \\ 0 & \text{otherwise} \end{cases}$$

For the X_1 version:

$$f_{Geom_1(p)}(x) = \begin{cases} p(1 - p)^{x-1} & x \in \{1, 2, \ldots\} \\ 0 & \text{otherwise} \end{cases}$$

To calculate $f_{Geom_0(p)}(x)$ in Excel, use =NEGBINOMDIST$(x, 1, p)$

Figure 15-2 illustrates the PMF for two geometric random variables, with $p = 0.1$ and $p = 0.5$. The PMFs for all geometric random variables have the same shape, with different scales.

Cumulative Distribution Function:

For the X_0 version: $F_{Geom_0(p)}(x) = \sum_{i=0}^{x} p(1 - p)^x = 1 - (1 - p)^{1+x}$

For the X_1 version: $F_{Geom_1(p)}(x) = \sum_{i=1}^{x} p(1 - p)^{x-1} = 1 - (1 - p)^x$

Inverse Cumulative Distribution Function:

$F^{-1}_{Geom_0(p)}(prob) = \lceil \frac{\ln(1 - prob)}{\ln(1 - p)} \rceil - 1$, where $\lceil \; \rceil$ means round up to the next higher integer. Excel contains several rounding functions with different behavior at integer values and on either side of zero. This formula is recommended to calculate $F^{-1}_{Geom_0(p)}(prob)$ for a $Geom_0(p)$ random variable: =MAX(0,ROUNDUP(LN(1-$prob$)/LN(1-p),0)-1). Add 1 to this formula for the $Geom_1(p)$ inverse CDF.

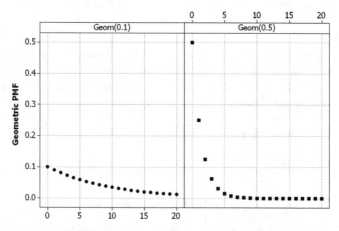

Figure 15-2   PMF of $Geom_0(0.1)$ and $Geom_0(0.5)$ Random Variables.

Random Number Generation: To calculate geometric random numbers in an Excel worksheet, use the inverse CDF formula given above with RAND() in place of *prob*.

To calculate geometric random numbers for the X_1 version with Excel and Crystal Ball software, use =CB.Geometric(p).

An alternate method to calculate geometric random numbers is to generate a sequence of uniform random numbers between 0 and 1, and count the number of uniforms required to produce a value less than p. This algorithm avoids any complexity associated with calculating logarithms, but it may take a long time for small values of p.

Survival Function:

$$R_{Geom_0(p)}(x) = (1 - p)^{1+x}$$
$$R_{Geom_1(p)}(x) = (1 - p)^{x}$$

Hazard Function:

$$h_{Geom_0(p)}(x) = h_{Geom_1(p)}(x) = \frac{p}{1 - p}$$

Note that the hazard function is a constant. This is a mathematical expression of the "lack of memory" property described above.

Mean (Expected Value):

For the X_0 version: $E[Geom_0(p)] = \dfrac{1 - p}{p}$

For the X_1 version: $E[Geom_1(p)] = \dfrac{1}{p}$

Mode: 0

Standard Deviation: $SD[Geom(p)] = \dfrac{\sqrt{1 - p}}{p}$

Variance: $V[Geom(p)] = \dfrac{1 - p}{p^2}$

Coefficient of Variation:

For the X_0 version: $CV[Geom_0(p)] = \dfrac{1}{\sqrt{1 - p}}$

For the X_1 version: $CV[Geom_1(p)] = \sqrt{1 - p}$

Coefficient of Skewness:

$$Skew[\,Geom(p)] = \frac{2 - p}{\sqrt{1 - p}}$$

A geometric random variable is always skewed to the right.

Coefficient of Kurtosis: $Kurt[\,Geom(p)] = \frac{p^2}{1 - p} + 9$

Coefficient of Excess Kurtosis: $ExKurt[\,Geom(p)] = \frac{p^2}{1 - p} + 6$

A geometric random variable is always leptokurtic.

Moment Generating Function:

For the X_0 version: $M_{Geom_0(p)}(t) = \dfrac{p}{1 - (1 - p)e^t}$

For the X_1 version: $M_{Geom_1(p)}(t) = \dfrac{pe^t}{1 - (1 - p)e^t}$

16

Hypergeometric
Distribution Family

A hypergeometric random variable is a discrete random variable representing a count of defective units in a random sample selected without replacement from a finite population. More generally, if a population of finite size contains items of type "A" or type "B," the number of type "A" items in a random sample selected without replacement from the population is a hypergeometric random variable.

The hypergeometric distribution is often applied to the selection of sampling plans, when the population is finite. In this application, the hypergeometric distribution is similar to a binomial distribution, except that the binomial distribution applies to populations of infinite size. The binomial distribution is based on a series of Bernoulli trials, each of which has the same probability of outcome "A." The hypergeometric distribution is not based on Bernoulli trials, because when items are selected from a finite population, the probabilities change for the next item selected.

Games of chance are another important application for the hypergeometric distribution. For instance, in card games, some cards would help one's hand, while the rest would not. The number of helpful cards dealt next from the deck has a hypergeometric distribution.

Example 16.1

Larry is playing draw poker, and he has the following hand: A♣, 7♦, 6♥, 4♦, 3♣. Larry must decide whether to discard the ace and hope to draw a 5 on one card, or to discard the four low cards and hope to draw another ace or something better.

The dealer has a standard deck of 52 cards. Larry knows where five cards are, so there are 47 other cards he might get when he draws. Actually, Larry's buddies have several of those other cards in their hands, but Larry cannot see those other cards. Larry has to figure his chances as if he has an equal probability of receiving each of the unseen 47 cards.

Suppose Larry discards the ace. Of the remaining 47 cards, there are 4 helpful cards (fives) and 43 unhelpful cards. The probability that Larry draws a five is $4/47 = 0.085$. The same probability can also be calculated using the hypergeometric distribution. The number of fives Larry draws on one card is a hypergeometric random variable X, with population size $N = 47$, sample size $n = 1$, and the number of helpful cards $D = 4$. To calculate $P[X = 1]$ in Excel on Larry's PDA, he could enter =HYPGEOMDIST(1,1,4,47), which returns the value 0.085.

Suppose Larry discards the four low cards, and hopes for aces in the next four cards he receives. The number of aces in the next four cards Larry receives is a hypergeometric random variable with population size $N = 47$, sample size $n = 4$, and the number of helpful cards $D = 3$, since there are three unseen aces. The Excel formula for the probability of zero aces is =HYPGEOMDIST(0,4,3,47), which returns a value of 0.761. Therefore, Larry has $1 - 0.761 = 0.239$ probability of drawing one or more aces in four cards.

Never draw to an inside straight, Larry.

Example 16.2

Cindy is finishing a Black Belt project, which corrected a casting porosity problem. Now in the control phase, Cindy must select a sample size n for the control plan. Castings are manufactured in lots of 150. Before shipping each lot, a random sample of n castings will be inspected for porosity. Cindy wants to select n to provide 95% confidence of detecting a problem in 10% of the castings.

Suppose 10% of the castings, 15 out of a lot of 150, have a porosity defect. If Cindy selects a random sample of size n, the number of defective castings in the sample is a hypergeometric random variable X, with lot size $N = 150$, the number of defects $D = 15$, and sample size n. To detect the problem means that the sample contains one or more defective castings. To have 95% confidence of detecting the problem, Cindy wants to choose n so that $P[X \geq 1] = 0.95$.

That is quite difficult to compute, because $P[X \geq 1]$ is the sum of $P[X = 1]$, $P[X = 2]$, all the way up to $P[X = n]$. However, there is an easier way. If Cindy wants 95% confidence of detecting the problem, another way of saying this is that Cindy wants 5% probability of not detecting the problem. That is, Cindy wants to choose n so that $P[X = 0] = 0.05$. This is easy to compute.

Cindy knows that $N = 150$ and $D = 15$, representing a 10% rate of defective castings. In an Excel spreadsheet, she sets up one cell for the sample size n, and enters this formula into another cell: =HYPGEOMDIST(0,n,15,150). This formula calculates $P[X = 0]$. By trying different values for n, she finds that with $n = 26$, $P[X = 0] = 0.04912$. With $n = 25$, $P[X = 0] = 0.05582$. Therefore, a sample size of 26 is the smallest sample size that provides 95% confidence of detecting a porosity problem affecting 10% of each lot of castings.

It is common to approximate the hypergeometric distribution with a binomial distribution, which assumes that the probability of selecting a defective unit does not change as the sample is selected. How would this assumption affect Cindy's sample plan?

To approximate a hypergeometric random variable by a binomial, calculate $p = D/N = 15/150 = 0.1$. In an Excel spreadsheet, enter =BINOMDIST(0,n, 0.1,TRUE) which calculates $P[X = 0]$. With $n = 29$, $P[X = 0] = 0.04701$. With $n = 28$, $P[X = 0] = 0.05234$. Therefore, a sample size of 29 is the smallest sample size that provides the required 95% confidence. Using the approximation results in a sample size 3 larger than required by the exact calculation.

In sampling applications, the binomial approximation leads to the same or a larger sample size than the exact sample size given by the hypergeometric model. Therefore, it is a conservative assumption.

Parameters: The hypergeometric distribution has three parameters:

- N is the population size. N is an integer > 0. In Excel HYPGEOMDIST function, N is the Number_pop parameter. In Crystal Ball software, N is the Population parameter.
- D is the count of units in the population with the attribute counted by the random variable X. In most sampling scenarios, D is the count of defective units in the population. D is an integer such that $0 \le D \le N$. In Excel HYPGEOMDIST function, D is the Population_s parameter. In Crystal Ball software, D is the Success parameter.
- n is the size of a sample drawn at random without replacement from the population. n is an integer such that $1 \le n \le N$. In Excel HYPGEOMDIST function, n is the Number_sample parameter. In Crystal Ball software, n is the Trials parameter.

Representation: $X \sim Hypergeom(N, D, n)$

Support: X may assume any integer value between $Max\{0, n - N + D\}$ and $Min\{D, n\}$. The lower limit of $n - (N - D)$ represents the situation where there are fewer than n nondefective items in the population; in this case, the sample must have at least $n - (N - D)$ defective items. The upper limit of D or n, whichever is less, is the maximum number of defective items the sample might have.

Relationships to Other Distributions:

- The hypergeometric distribution describes the probabilities of defects in the sample, when sampling without replacement from a finite population. Variations on this story lead to the binomial distribution. If items are sampled with replacement from a finite population, the number of defective items in the sample is a binomial random variable. Also, if items are sampled with or without replacement from an infinite population, the number of defective items in the sample is a binomial random variable. If the population size is finite but unknown, it is common to assume that it is infinite and to apply the binomial distribution.

- *Hypergeom*(N,D,n) may be approximated by a binomial random variable, specifically *Bin*$(n,D/N)$. As a rule of thumb, only use this approximation when $N \geq 10n$. Example 16.2 illustrates the error that might occur if this rule of thumb is not satisfied.

- Because of the above property, and further approximations of the binomial distribution, either the Poisson or normal distribution may approximate the hypergeometric distribution. When N, D, and n are large, and calculation of hypergeometric probabilities is impractical, use this normal approximation:

$$F_{Hypergeom(N,D,n)}(x) = P[X \leq x] \cong \Phi\left(\frac{X + 0.5 - \frac{nD}{N}}{\sqrt{n\left(\frac{D}{N}\right)\left(\frac{N-D}{N}\right)\left(\frac{N-n}{N-1}\right)}}\right)$$

Even when $N < 10n$, and the binomial rule of thumb is not met, this approximation by Ord (1968) uses the binomial PMF, and is quite accurate:

$$P[X = x] \approx \binom{n}{x}p^x(1-p)^{n-x}\left[1 + \frac{x(1-2p) + np^2 - (x-np)^2}{2Np(1-p)}\right]$$

- Numerous generalizations of the hypergeometric distribution are available to describe more complex sampling situations. Chapter 6 of Johnson *et al* (2005) is an excellent reference on these topics.

Normalizing Transformation: Because of the wide variety of shapes that the hypergeometric distribution may take, no general normalizing transformation is available. However, for cases where a binomial or Poisson approximation is appropriate, a direct approximation of the hypergeometric distribution by a normal distribution may be appropriate.

Process Control Tools: The usual process control method for hypergeometric processes is to apply control charts designed for binomial processes, such as the np-chart. Since the binomial distribution has a longer tail than the hypergeometric, applying an np-chart to hypergeometric data reduces the power of the chart to detect upward shifts in the rate of defects.

Shore's general control charts for attributes (2000a) have control limits that compensate for skewness in the distribution. For the hypergeometric case, here are control chart formulas, based on observed values of X, with N and n known.

$$\hat{D} = \left\lfloor \frac{\overline{X}(N + 1)}{n} \right\rfloor, \text{ where } \lfloor\,\rfloor \text{ means round down.}$$

$$\hat{\sigma} = \sqrt{n\left(\frac{\hat{D}}{N}\right)\left(1 - \frac{\hat{D}}{N}\right)\left(\frac{N - n}{N - 1}\right)}$$

$$\hat{Skew} = \frac{(N - 2\hat{D})(N - 2n)}{N - 2}\sqrt{\frac{N - 1}{n\hat{D}(N - \hat{D})(N - n)}}$$

$$UCL_{NB} = n\frac{\hat{D}}{N} + 3\hat{\sigma} + 1.324\hat{\sigma}\hat{Skew} - 0.5$$

$$CL_{NB} = n\frac{\hat{D}}{N}$$

$$LCL_{NB} = n\frac{\hat{D}}{N} - 3\hat{\sigma} + 1.324\hat{\sigma}\hat{Skew} + 0.5$$

Estimating Parameter Values:

A typical problem is to estimate D from an observed count of defective items x, if the population size N and sample size n are known. A maximum likelihood estimator for D is $\hat{D} = \lfloor \frac{x(N + 1)}{n} \rfloor$, where $\lfloor\,\rfloor$ denotes rounding down. If $\frac{x(N + 1)}{n}$ is an integer, then both $\frac{x(N + 1)}{n}$ and $\frac{x(N + 1)}{n} - 1$ are maximum likelihood estimators. For confidence intervals, see tables published by Chung and DeLury (1950) or Owen (1962).

Capability Metrics: When a hypergeometric process has tolerance limits LTL and UTL, calculate equivalent long-term capability metrics this way:

$$\text{Equivalent } P_{PL}^{\%} = \frac{-\Phi^{-1}(F_X(LTL - 1))}{3}$$

$$\text{Equivalent } P_{PU}^{\%} = \frac{-\Phi^{-1}(R_X(UTL))}{3}$$

$$\text{Equivalent } P_P^{\%} = (\text{Equivalent } P_{PL}^{\%} + \text{Equivalent } P_{PU}^{\%})/2$$

$$\text{Equivalent } P_{PK}^{\%} = Min\{\text{Equivalent } P_{PL}^{\%}, \text{Equivalent } P_{PU}^{\%}\}$$

If LTL is not an integer, replace $(LTL - 1)$ with LTL. In these formulas, $F_X(x)$ is the CDF of a hypergeometric random variable, and $R_X(x) = 1 - F_X(x)$ is the corresponding survival function. Excel software has no built-in function for the hypergeometric CDF, and summing individual terms can be tedious and confusing. Statistical programs such as JMP and MINITAB can compute $F_X(x)$. STATGRPAHICS computes both $F_X(x)$ and $R_X(x)$. Also, $\Phi^{-1}(prob)$ is the standard normal inverse CDF, evaluated in Excel with the =NORMSINV($prob$) function. Table 16-1 lists selected values of equivalent $P_{PU}^\%$ for selected hypergeometric processes, all with $D/N = 0.01$, and a binomial process with $p = 0.01$. This table illustrates how the hypergeometric distribution changes as the population size grows, with the proportion of defects held constant.

Probability Mass Function:

$$f_{Hypergeom(N,D,n)}(x) = \frac{\binom{D}{x}\binom{N-D}{n-x}}{\binom{N}{n}},$$

$$\text{for } Max\{0, n - N + X\} \le x \le Min\{D, n\}$$

To calculate $f_{Hypergeom(N,D,n)}(x)$ in Excel, use =HYPGEOMDIST(x,n,D,N)

Figure 16-1 illustrates the PMF of six hypergeometric random variables with the same sample size n.

Cumulative Distribution Function:

$$F_{Hypergeom(N,D,n)}(x) = \sum_{i=Max\{0,n-N+D\}}^{Min\{x,D,N\}} \frac{\binom{D}{i}\binom{N-D}{n-i}}{\binom{N}{n}}$$

There is no convenient way to calculate $F_{Hypergeom(N,D,n)}(x)$ in an Excel worksheet. In an Excel worksheet, calculate each term in the sum with the HYPGEOMDIST function and add them. MINITAB, JMP, and STATGRAPHICS software all have functions to calculate the PMF and CDF of the hypergeometric distribution.

Inverse Cumulative Distribution Function: There is no easy formula or convenient way to calculate $F_{Hypergeom(N,D,n)}^{-1}(prob)$ in Excel worksheet.

Table 16-1 Equivalent Capability Metrics for Selected Hypergeometric Random Variables and a Limiting Binomial Random Variable.

Equivalent $P_{PU}^\%$

UTL	Hypergeometric(N,D,n)					Binomial
	(300, 3, 100)	(600, 6, 100)	(1000, 10, 100)	(2000, 20, 100)	(5000, 50, 100)	$n = 100, p = 0.01$
2	0.598	0.515	0.494	0.481	0.474	0.470
3	–	0.798	0.749	0.720	0.705	0.696
4	–	1.077	0.988	0.940	0.916	0.901
5	–	1.374	1.217	1.146	1.111	1.091
6	–	–	1.441	1.341	1.295	1.268
7	–	–	1.665	1.528	1.470	1.436
8	–	–	1.895	1.710	1.637	1.596
9	–	–	2.142	1.886	1.798	1.750
10	–	–	–	2.059	1.954	1.898
11	–	–	–	2.229	2.105	2.040

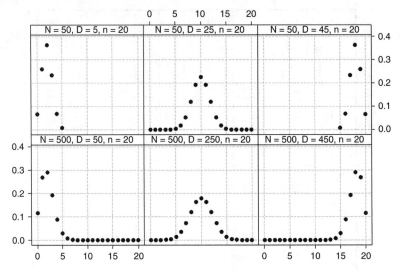

Figure 16-1   PMF of Six Hypergeometric Random Variables with $n = 20$.

Example 16.2 illustrates how to solve this problem by substituting in values for x until $F_{Hypergeom(N,D,n)}(x)$ satisfies the desired criteria. MINITAB or STATGRAPHICS software will calculate the hypergeometric inverse CDF.

Random Number Generation: Crystal Ball software provides two functions for generating hypergeometric random numbers. To use a probability parameter $p = D/N$, enter =CB.Hypergeometric(p,n,N). To use the parameters as described in this chapter, enter =CB.Hypergeometric2(D,n,N).

A good algorithm to generate hypergeometric random numbers is to generate n independent $Unif(0, 1)$ random numbers $u_1 \ldots u_n$. Let $p_1 = D/N$, and let $d_1 = 1$ if $u_1 < p_1$, or 0 otherwise. For u_2 through u_n, calculate $p_{i+1} = \frac{(N - i + 1)p_i - d_i}{N - i}$ and let $d_{i+1} = 1$ if $u_{i+1} < p_{i+1}$, or 0 otherwise. The sum of the d_i is a hypergeometric random number.

Mean (Expected Value): $E[Hypergeom(N, D, n)] = \frac{nD}{N}$

Standard Deviation:

$$SD[Hypergeom(N, D, n)] = \sqrt{n\left(\frac{D}{N}\right)\left(1 - \frac{D}{N}\right)\left(\frac{N - n}{N - 1}\right)}$$

If $p = D/N$, the first three terms under the square root correspond to $\sqrt{np(1 - p)}$ which is the standard deviation of a binomial distribution. The remaining term, $\frac{N - n}{N - 1}$ is a finite population correction factor, which goes to 1 as N goes to infinity.

Variance: $V[Hypergeom(N, D, n)] = n\left(\frac{D}{N}\right)\left(1 - \frac{D}{N}\right)\left(\frac{N - n}{N - 1}\right)$

Coefficient of Variation:

$$CV[Hypergeom(N, D, n)] = \sqrt{\frac{(N - D)(N - n)}{nD(N - 1)}}$$

Coefficient of Skewness:

$$Skew[Hypergeom(N, D, n)] = \frac{(N - 2n)(N - 2D)}{N - 2}$$

$$\times \sqrt{\frac{(N - 1)}{nD(N - D)(N - n)}}$$

Coefficient of Kurtosis:

$$Kurt[Hypergeom(N, D, n)] = \left[\frac{N^2(N - 1)}{n(N - 2)(N - 3)(N - n)}\right]$$

$$\times \left[\frac{N(N + 1) - 6n(N - n)}{D(N - D)}\right.$$

$$\left. + \frac{3n(N - n)(N + 6)}{N^2} - 6\right] + 3$$

Coefficient of Excess Kurtosis:

$$ExKurt[Hypergeom(N, D, n)] = \left[\frac{N^2(N - 1)}{n(N - 2)(N - 3)(N - n)}\right]$$

$$\times \left[\frac{N(N + 1) - 6n(N - n)}{D(N - D)}\right.$$

$$\left. + \frac{3n(N - n)(N + 6)}{N^2} - 6\right]$$

17

Laplace Distribution Family

A Laplace random variable is a symmetric, continuous random variable that appears to be a two-parameter exponential random variable, with its mirror image on either side of the location parameter τ. Other names for the Laplace distribution are the double exponential, bilateral exponential, and two-tailed exponential distribution. Note that the term *double exponential* may also refer to a very different distribution, the Gumbel or extreme value distribution (Chapter 12).

For modelers, the Laplace distribution is useful as a symmetric distribution with heavier tails and a sharper central peak than a normal distribution. Users of Crystal Ball software will not find Laplace in the distribution gallery. To generate a Laplace assumption in Crystal Ball, set up two assumptions: Y, a Yes-No (Bernoulli) distribution with $p = 0.5$, and X, an exponential distribution with rate parameter $\lambda = 1/\beta$. To calculate a $Laplace(\beta, \tau)$ random variable in a model, enter a formula for $\tau + (2Y - 1) \times X$.

The name of the Laplace distribution honors Pierre Simon, Marquis de Laplace (1749–1827) who made great contributions to mathematics, probability theory, astronomy, and other sciences. Besides publishing the Laplace distribution as the "first law of error" (1774), he also described the normal distribution as the "second law of error." On a grander scale, Laplace proved the dynamic stability of the solar system and postulated the existence of black holes.

To place probability theory in proper context, Laplace's expressed his philosophy of causal determinism this way in *Essai philosophique sur les probabilités*:

"We may regard the present state of the universe as the effect of its past and the cause of its future. An intellect which at a certain moment would know

Figure 17-1   Pierre Simon Laplace.

all forces that set nature in motion, and all positions of all items of which nature is composed, if this intellect were also vast enough to submit these data to analysis, it would embrace in a single formula the movements of the greatest bodies of the universe and those of the tiniest atom; for such an intellect nothing would be uncertain and the future just like the past would be present before its eyes."

According to Ball (1908), Laplace presented a copy of his masterwork *Méchanique Céleste* to Napoleon. During their meeting, Napoleon asked, "Mr. Laplace, they tell me you have written this large book on the system of the universe, and have never even mentioned its Creator."

Laplace replied, *"Je n'avais pas besoin de cette hypothèse-là."* (I have no need of that hypothesis.)

Example 17.1

Tanja is an ecologist building a model of climate change in the Northern Hemisphere. One of the parameters in her model is the thickness of permafrost in northern Lapland. Tanja's data consists of measured thicknesses of ice over many years. The differences from one year to the next have a sample skewness of -0.09, and a sample kurtosis of 3.02. The sample kurtosis is an estimate of excess kurtosis, so that a normal random variable produces a kurtosis of 0.

A skewness near to zero indicates a nearly symmetric distribution. However, a kurtosis of 3 indicates heavier tails than a normal distribution. Since a Laplace distribution is symmetric and has an excess kurtosis of 3, Tanja decides to model the differences from year to year as a Laplace random variable.

After selecting a Laplace model, Tanja's next step is to estimate the parameters τ and β. In an Excel worksheet, she calculates the difference in thickness from one year to the next. Using the formulas below, Tanja estimates τ with the median difference, which is 0.0515. The formula to estimate β is $\hat{\beta} = \frac{1}{n}\sum_{i=1}^{n}\left|X_i - \hat{\tau}\right|$. In a column of the worksheet, Tanja calculates $\left|X_i - \hat{\tau}\right|$ and averages those values. The result is $\hat{\beta} = 0.184$. Figure 17-2 illustrates these calculations. In column H, Tanja defines two assumptions with yes-no and exponential distributions, and she combines these in cell H8 to form a Laplace distribution. The formula bar shows the formula from cell H8.

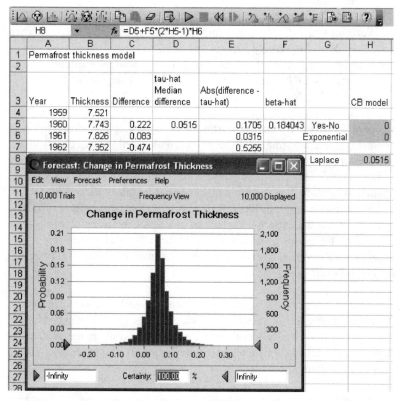

Figure 17-2   Excel Worksheet with Laplace Model of Changes in Permafrost Thickness.

Parameters: The Laplace distribution family has two parameters

- β is a scale parameter, which can be any positive number. β is the mean of the exponential distribution formed by $|X - \tau|$.
- τ is a location parameter, which is the mean, median, and mode of the Laplace distribution.

Representation: $X \sim Laplace(\beta, \tau)$

Support: $(-\infty, \infty)$

Relationships to Other Distributions:

- The absolute value of a Laplace random variable minus its location parameter is an exponential random variable. If $X \sim Laplace(\beta, \tau)$, then $|X - \tau| \sim Exp(\beta)$, where β represents the mean of the exponential distribution.
- The difference between two independent exponential random variables is a Laplace random variable. If $X, Y \overset{iid}{\sim} Exp(\mu)$, then $X - Y \sim Laplace(\mu, 0)$.
- The absolute value of the ratio of two independent Laplace random variables is an F random variable with 2 and 2 degrees of freedom. If $X, Y \overset{iid}{\sim} Laplace(\beta, 0)$ then $\left|\frac{X}{Y}\right| \sim F(2, 2)$.

Process Control Tools: Since the shape of a Laplace distribution is symmetric, like a normal distribution, normal-based control charts might work acceptably well. When these charts are applied to Laplace data, the rate of false alarms is higher than with normal data. A simulation of $\overline{X}$ charts applied to Laplace data showed an average run length (ARL) of 70–180 for the $\overline{X}$ chart with subgroup sizes 2–10. For normal data, the ARL is 370. The ARL measures the expected number of plot points between points outside the control limits, for a stable process. If the ARL is 100, then one would expect a stable process to exceed the control limits one out of 100 points on the control chart.

For either range or standard deviation charts, the ARL is approximately 25, which is unacceptably small. If Laplace data is plotted using standard Shewhart control charts, out of control indications on the range or standard deviation charts should be interpreted very cautiously, since many of these may be false alarms.

Another alternative is to plot individual observations on a control chart with control limits determined for the Laplace distribution. Here are the formulas, based on the median $\widetilde{X}$, and a false alarm rate α, which is typically 0.0027.

$$\hat{\beta} = \tfrac{1}{n}\sum_{i=1}^{n}\left|X_i - \widetilde{X}\right|$$
$$UCL = \widetilde{X} - \hat{\beta}\ln(\alpha)$$
$$CL = \widetilde{X}$$
$$LCL = \widetilde{X} + \hat{\beta}\ln(\alpha)$$

This individual chart for Laplace data has the desirable property that ARL = $1/\alpha$. Its disadvantage is that, as a single chart, it is unable to detect arbitrary changes to either parameter of the distribution. However, if interpreted with a variety of control chart rules, it is likely to perform well.

Estimating Parameter Values: When β and τ are both unknown, maximum likelihood estimators are these:

- $\hat{\tau} = \widetilde{X}$, where $\widetilde{X}$ is the sample median
- $\hat{\beta} = \tfrac{1}{n}\sum_{i=1}^{n}\left|X_i - \hat{\tau}\right|$

When τ is known, then $\hat{\beta} = \tfrac{1}{n}\sum_{i=1}^{n}\left|X_i - \tau\right|$

A confidence interval for β is available, when τ is known. The limits of a $100(1-\alpha)\%$ confidence interval for β are $L_\beta = \dfrac{2n\hat{\beta}}{\chi^2_{\alpha/2,2n}}$ and $U_\beta = \dfrac{2n\hat{\beta}}{\chi^2_{1-\alpha/2,2n}}$, where $\chi^2_{\alpha,\nu}$ is the quantile of the chi-squared distribution with ν degrees of freedom, and α probability to the right. Evaluate $\chi^2_{\alpha,\nu}$ in Excel worksheet with =CHIINV(α,ν).

Capability Metrics: For a process producing a Laplace distribution, calculate equivalent capability metrics using the Laplace CDF. Here are the formulas:

$$\text{Equivalent } P^{\%}_{PU} = \frac{-\Phi^{-1}\!\left(\tfrac{1}{2}\exp\!\left[\tfrac{-(UTL - \tau)}{\beta}\right]\right)}{3}$$

$$\text{Equivalent } P^{\%}_{PL} = \frac{-\Phi^{-1}\!\left(\tfrac{1}{2}\exp\!\left[\tfrac{-(\tau - LTL)}{\beta}\right]\right)}{3}$$

$$\text{Equivalent } P^{\%}_{P} = \frac{\text{Equivalent } P^{\%}_{PU} + \text{Equivalent } P^{\%}_{PL}}{2}$$

$$\text{Equivalent } P^{\%}_{PK} = Min\{\text{Equivalent } P^{\%}_{PU},\ \text{Equivalent } P^{\%}_{PL}\}$$

$\Phi^{-1}(p)$ is the standard normal inverse CDF, evaluated in Excel with the =NORMSINV(p) function. These formulas presume that the tolerance limits are on the correct side of the center of the distribution at τ. For more general formulas, use the complete formula for $F_X()$, listed below, as the argument for $\Phi_X^{-1}()$. Table 17-1 lists selected values of equivalent $P_{PL}^{\%}$ and $P_{PU}^{\%}$ for exponential processes.

The bottom entry in Table 17-1 represents tolerance limits at 21 β, approximately 15 standard deviations away from the mean. At this point the defects are equivalent to a normal distribution with a tolerance limit at six standard deviations above the mean. Using conventional formulas for process capability, which assume a normal distribution, would seriously underestimate the defects.

Probability Density Function:

$$f_{Laplace(\beta,\tau)}(x) = \frac{\exp\left(\dfrac{-|x - \tau|}{\beta}\right)}{2\beta}$$

In an Excel worksheet, calculate $f_{Laplace(\beta,\tau)}(x)$ with the formula =GAMMADIST(ABS(x-τ),1,β,FALSE)/2.

Table 17-1   Selected Capability Values for Laplace Random Variables

LTL	Equivalent $P_{PL}^{\%}$	UTL	Equivalent $P_{PU}^{\%}$
$\tau - 3\beta$	0.654	$\tau + 3\mu$	0.654
$\tau - 5\beta$	0.903	$\tau + 5\mu$	0.903
$\tau - 7\beta$	1.105	$\tau + 7\mu$	1.105
$\tau - 9\beta$	1.280	$\tau + 9\mu$	1.280
$\tau - 11\beta$	1.435	$\tau + 11\mu$	1.435
$\tau - 13\beta$	1.576	$\tau + 13\mu$	1.576
$\tau - 15\beta$	1.707	$\tau + 15\mu$	1.707
$\tau - 17\beta$	1.828	$\tau + 17\mu$	1.828
$\tau - 19\beta$	1.943	$\tau + 19\mu$	1.943
$\tau - 21\beta$	2.051	$\tau + 21\mu$	2.051

Cumulative Distribution Function:

$$F_{Laplace(\beta,\tau)}(x) = \begin{cases} 1 - \dfrac{1}{2}\exp\left(\dfrac{-(x-\tau)}{\beta}\right) & x \geq \tau \\[2ex] \dfrac{1}{2}\exp\left(\dfrac{-(\tau-x)}{\beta}\right) & x < \tau \end{cases}$$

In an Excel worksheet, calculate $F_{Laplace(\beta,\tau)}(x)$ with the formula =IF(x<τ,(1-GAMMADIST(τ-x,1,β,TRUE))/2,1-(1-GAMMADIST(x-τ,1,β,TRUE))/2).

Inverse Cumulative Distribution Function:

$$F^{-1}_{Laplace(\beta,\tau)}(prob) = \begin{cases} -\beta\ln(2[1 - prob]) + \tau & prob \geq 0.5 \\[2ex] \beta\ln(2prob) + \tau & prob < 0.5 \end{cases}$$

In an Excel worksheet, calculate $F^{-1}_{Laplace(\beta,\tau)}(x)$ with =τ+β*IF($prob$<0.5, LN(2*$prob$),-LN(2*(1-$prob$))).

Random Number Generation: To generate exponential random numbers in an Excel worksheet, use this formula =τ+β*IF(RAND()<0.5,1,-1)* LN(RAND()).

With Crystal Ball and Excel software, use the formula =(2*CB.YesNo(0.5)-1)* CB.Exponential(1/β).

Survival Function:

$$R_{Laplace(\beta,\tau)}(x) = \begin{cases} \dfrac{1}{2}\exp\left(\dfrac{-(x-\tau)}{\beta}\right) & x \geq \tau \\[2ex] 1 - \dfrac{1}{2}\exp\left(\dfrac{-(\tau-x)}{\beta}\right) & x < \tau \end{cases}$$

Hazard Function:

$$h_{Laplace(\beta,\tau)}(x) = \begin{cases} \dfrac{1}{\beta} & x \geq \tau \\[3ex] \dfrac{\exp\left(\dfrac{-(\tau-x)}{\beta}\right)}{\beta\left[2 - \exp\left(\dfrac{-(\tau-x)}{\beta}\right)\right]} & x < \tau \end{cases}$$

Figure 17-3 illustrates the PDF, CDF, survival, and hazard functions for Laplace random variables with $\tau = 0$ and $\beta = 0.5$, 1.0 and 2.0.

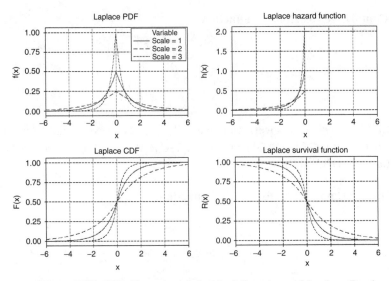

Figure 17-3   PDF, CDF, Hazard, and Survival Functions of Laplace Random Variables with $\tau = 0$ and Scale Parameter $\beta = 0.5$, 1, and 2.

Mean (Expected Value): $E[Laplace(\beta, \tau)] = \tau$

Median: τ

Mode: τ

Standard Deviation: $SD[Laplace(\beta, \tau)] = \beta\sqrt{2}$

Variance: $V[Laplace(\beta, \tau)] = 2\beta^2$

Coefficient of Skewness: $Skew[Laplace(\beta, \tau)] = 0$

The Laplace random variable is symmetric.

Coefficient of Kurtosis: $Kurt[Laplace(\beta, \tau)] = 6$

Coefficient of Excess Kurtosis: $ExKurt[Laplace(\beta, \tau)] = 3$

The Laplace random variable is leptokurtic.

Moment Generating Function: $M_{Laplace(\beta, \tau)}(t) = \dfrac{e^{\tau t}}{1 - \beta^2 t^2}$

18

Logistic Distribution Family

A logistic random variable is a continuous random variable with a symmetric, bell-shaped probability curve. The logistic probability curve is similar to a normal probability curve, except that the logistic central peak is taller and the tails are slightly heavier. Figure 18-1 compares logistic and normal distributions with the same mean and standard deviation.

The logistic distribution is most famous for its cumulative distribution function, which is of the form $F(z) = [1 + e^{-z}]^{-1}$. This function has many useful properties. One such property is that the slope of the curve is proportional to the product of its difference from its starting point 0 and its difference from its ending point 1. This property describes growth in many processes, such as autocatalytic chemical reactions, in which the reaction products catalyze the reaction.

As a model for random behavior, the logistic distribution is less popular than it once was. The formulas describing the logistic distribution are simpler than those describing a normal distribution. Decades ago, before computerized normal probability functions were widely available, some would use the logistic distribution as a simpler substitute for the normal distribution. Today, there is no need to do this.

However, there are some theoretical reasons for using logistic distribution models. One is that the difference between two independent extreme value random variables is a logistic random variable. Because of this property, the sample midrange, which is the average of the maximum and minimum values, follows a logistic distribution in many situations. Balakrishnan (1992) edited a book on the logistic distribution and its many applications.

In addition to any theoretical reasons, the distribution of observed data may suggest a logistic distribution model instead of a normal distribution model. Because of the sharper peak and heavier tails of their probability curves,

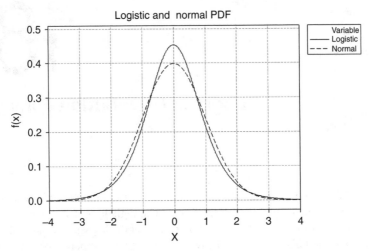

Figure 18-1 Logistic and Normal Probability Density Functions, with the Same Mean and Standard Deviation.

logistic random variables have an excess kurtosis of 1.2. If observed data has a symmetric distribution with consistently higher kurtosis than a normal distribution, the logistic distribution may be a good modeling choice. Laplace and Student's *t* distributions are other good modeling candidates for this situation.

When a process produces logistic data, applying normal-based tools will give satisfactory results in some cases, if the decisions rely on the middle of the distribution and not the extreme tails. The following paragraphs illustrate the impact of using normal-based control charts and capability metrics on logistic data. When logistic distributions occur in a Six Sigma project, process capability depends on tail probabilities. Because of this, it is particularly important to use equivalent capability metrics for logistic processes instead of normal-based capability metrics.

Later in this chapter, section 18.1 describes the loglogistic and three-parameter loglogistic families, which are popular models for reliability and waiting time data.

Parameters: The logistic distribution family has two parameters:

- μ is the mean of the random variable. μ can be any number.
- β is the scale parameter of the random variable. β can be any positive number.

Representation: $X \sim Logistic(\mu, \beta)$

Support: $(-\infty, \infty)$

Relationships to Other Distributions:

- The difference between two independent extreme value random variables is a logistic random variable. If $X_1, X_2 \stackrel{iid}{\sim} SEV(\eta, \beta)$ or if $X_1, X_2 \stackrel{iid}{\sim} LEV(\eta, \beta)$ then $X_1 - X_2 \sim Logistic(0, \beta)$. It also follows that the sum of a smallest extreme value and an independent largest extreme value, with the same parameter values, is logistic. $SEV(\eta_1, \beta) + LEV(\eta_2, \beta) \sim Logistic(\eta_1 + \eta_2, \beta)$
- The natural log of a loglogistic random variable is a logistic random variable. $\ln[Loglogistic(\mu, \beta)] \sim Logistic(\mu, \beta)$.

Process Control Tools: Because of the closeness of logistic and normal probabilities, most Shewhart normal-based control charts will work reasonably well for logistic data, with a somewhat higher rate of false alarms. This is particularly true for $\overline{X}$ charts, since the central limit theorem will cause the subgroup means to be closer to a normal distribution.

To plot individual observations from a logistic process on a logistic IX, MR chart, use the following formulas:

$$UCL_{IX(Logistic)} = X + 3.30\overline{MR}$$

$$CL_{IX(Logistic)} = X$$

$$LCL_{IX(Logistic)} = X - 3.30\overline{MR}$$

$$UCL_{MR(Logistic)} = 3.62\overline{MR}$$

$$CL_{MR(Logistic)} = \overline{MR}$$

$$LCL_{MR(Logistic)} = 0$$

Table 18-1 compares the average run length (ARL) of a standard IX, MR chart applied to normal data, a standard IX, MR chart applied to logistic data, and a logistic IX, MR chart applied to normal data. These values were determined by simulation.

Table 18-1 shows that plotting logistic data on a normal IX, MR chart lowers the ARL and increases the rate of false alarms. If the additional false alarms happen too often, the logistic IX, MR chart corrects the false alarm rate.

Table 18-1 ARL of IX, MR Charts Applied to Normal and Logistic Data

Distribution of Data	Type of Chart	ARL of IX Chart	ARL of MR Chart
Normal	Normal IX, MR	371	108
Logistic	Normal IX, MR	103	62
Logistic	Logistic IX, MR	376	112

Estimating Parameter Values: Simple estimators for the mean and scale parameters are:

$$\hat{\mu} = \overline{X}$$

$$\hat{\beta} = \frac{s\sqrt{3}}{\pi}$$

Maximum likelihood estimators have lower variance than these, but these estimators require iterative solutions of large equations. It is best to use a major statistical package, such as MINITAB or STATGRAPHICS to estimate logistic distribution parameters from observed data.

Capability Metrics: For a process producing logistic data, calculate equivalent capability metrics this way:

$$\text{Equivalent } P_{PU}^{\%} = \frac{-1}{3}\, \Phi^{-1}\!\left(\frac{\exp\!\left(\frac{-(UTL - \mu)}{\beta}\right)}{1 + \exp\!\left(\frac{-(UTL - \mu)}{\beta}\right)} \right)$$

$$\text{Equivalent } P_{PL}^{\%} = \frac{-1}{3}\, \Phi^{-1}\!\left(\frac{1}{1 + \exp\!\left(\frac{-(LTL - \mu)}{\beta}\right)} \right)$$

$$\text{Equivalent } P_{P}^{\%} = \frac{\text{Equivalent } P_{PU}^{\%} + \text{Equivalent } P_{PL}^{\%}}{2}$$

$$\text{Equivalent } P_{PK}^{\%} = Min\{\text{Equivalent } P_{PU}^{\%}, \text{Equivalent } P_{PL}^{\%}\}$$

$\Phi^{-1}(p)$ is the standard normal inverse CDF, evaluated in Excel with the =NORMSINV(p) function. Table 18-2 lists selected values of equivalent $P_{PL}^{\%}$ and $P_{PU}^{\%}$ for logistic random variables.

According to Table 18-2, the equivalent capability metrics achieve a level of 2.0 when the tolerance limits are nearly 12 standard deviations away from

Table 18-2 Equivalent Capability Metrics for Logistic Data. In This Table, the Standard Deviation $\sigma = \beta\pi/\sqrt{3} \cong 1.81\beta$

LTL	Equivalent $P_{PL}^{\%}$	UTL	Equivalent $P_{PU}^{\%}$
$\mu-1\sigma$	0.360	$\mu+1\sigma$	0.360
$\mu-2\sigma$	0.648	$\mu+2\sigma$	0.648
$\mu-3\sigma$	0.875	$\mu+3\sigma$	0.875
$\mu-4\sigma$	1.064	$\mu+4\sigma$	1.064
$\mu-5\sigma$	1.228	$\mu+5\sigma$	1.228
$\mu-6\sigma$	1.374	$\mu+6\sigma$	1.374
$\mu-7\sigma$	1.507	$\mu+7\sigma$	1.507
$\mu-8\sigma$	1.631	$\mu+8\sigma$	1.631
$\mu-9\sigma$	1.746	$\mu+9\sigma$	1.746
$\mu-10\sigma$	1.854	$\mu+10\sigma$	1.854
$\mu-11\sigma$	1.957	$\mu+11\sigma$	1.957
$\mu-12\sigma$	2.055	$\mu+12\sigma$	2.055

the mean. This means that a logistic process produces nearly as many units outside $\pm$ 12 standard deviation limits as a normal process produces outside $\pm$ 6 standard deviation limits. This is because the tails of the logistic distribution are much heavier than the tails of a normal distribution.

Probability Density Function:

$$f_{Logistic(\mu,\beta)}(x) = \frac{\exp\left(\frac{-(x-\mu)}{\beta}\right)}{\beta\left[1 + \exp\left(\frac{-(x-\mu)}{\beta}\right)\right]^2}$$

Cumulative Distribution Function:

$$F_{Logistic(\mu,\beta)}(x) = \frac{1}{1 + \exp\left(\frac{-(x-\mu)}{\beta}\right)}$$

Inverse Cumulative Distribution Function:

$$F_{Logistic(\mu,\beta)}^{-1}(p) = \mu - \beta\ln\left(\frac{1-p}{p}\right)$$

Random Number Generation: To generate logistic random numbers in an Excel worksheet, use the formula =μ-β*LN(1/RAND()-1).

To generate logistic random numbers with Excel and Crystal Ball software, use the formula =CB.Logistic(μ,β).

Survival Function:

$$R_{Logistic(\mu,\beta)}(x) = \frac{\exp\left(\frac{-(x-\mu)}{\beta}\right)}{1 + \exp\left(\frac{-(x-\mu)}{\beta}\right)}$$

Hazard Function:

$$h_{Logistic(\mu,\beta)}(x) = \frac{1}{\beta\left[1 + \exp\left(\frac{-(x-\mu)}{\beta}\right)\right]}$$

Figure 18-2 shows the PDF, CDF, survival and hazard functions of a logistic random variable with mean $\mu = 0$ and scale parameter $\beta = 1$.

Mean (Expected Value): $E[Logistic(\mu, \beta)] = \mu$

Median: μ

Mode: μ

Standard Deviation: $SD[Logistic(\mu, \beta)] = \frac{\beta\pi}{\sqrt{3}}$

Variance: $V[Logistic(\mu, \beta)] = \frac{\beta^2\pi^2}{3}$

Coefficient of Variation: $CV[Logistic(\mu, \beta)] = \frac{\beta\pi}{\mu\sqrt{3}}$

Coefficient of Skewness: $Skew[Logistic(\mu, \beta)] = 0$

Coefficient of Kurtosis: $Kurt[Logistic(\mu, \beta)] = 4.2$

Coefficient of Excess Kurtosis: $Kurt[Logistic(\mu, \beta)] = 1.2$

Moment Generating Function:

$$m_{Logistic(\mu,\beta)}(t) = e^{\mu t}B(1 - \beta t, 1 + \beta t)$$
$$= e^{\mu t}\Gamma(1 - \beta t)\Gamma(1 + \beta t)$$

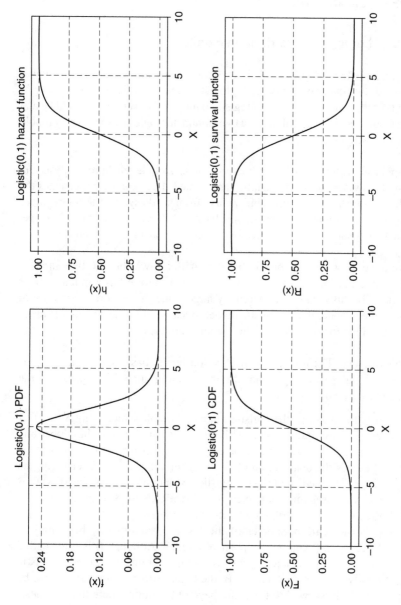

Figure 18-2 PDF, CDF, Survival and Hazard Function of a *Logistic*(0, 1) Random Variable.

In the above formula, $B(\alpha, \beta)$ is the beta function, and $\Gamma(\alpha)$ is the gamma function, which can be evaluated by the Excel formula =EXP(GAMMALN(α)).

18.1 Loglogistic Distribution Family

A loglogistic random variable is a continuous random variable with a lower bound. For the two-parameter loglogistic family, the lower bound is zero, but for the three-parameter loglogistic family, the lower bound could be any real number. The natural log of a two-parameter loglogistic random variable is a logistic random variable.

Applications for loglogistic random variables include reliability analysis and any field where positive quantities are studied. The parameter β changes the shape of the loglogistic probability curve from nearly symmetric at low values of β to exponential shapes at high values of β. Because of its flexible shape, the loglogistic family is an effective empirical model for many processes.

Relative to other distribution families with a lower bound, like gamma, Weibull, and lognormal, the loglogistic distribution has a much heavier right tail. This means that extremely large values have more probability of occurrence under the loglogistic model. Because of the heavy tail, the mean is undefined for some loglogistic random variables.

Parameters: The loglogistic distribution family has two parameters, plus an optional third parameter. In these paragraphs, X represents the loglogistic random variable.

- μ is the mean of the logistic random variable which is $\ln(X - \tau)$. μ can be any real number. Although μ is often called the "location" parameter, it is a scale parameter of the loglogistic family. Increasing μ stretches the distribution of X to higher values, without changing its shape. The median of the loglogistic distribution is e^{μ}. While MINITAB software uses this parameter, μ, STATGRAPHICS software specifies the loglogistic distribution by its median e^{μ}.
- β is the scale parameter of the logistic random variable which is $\ln(X - \tau)$. β can be any positive number. Although β is often called the "scale" parameter, it acts as a shape parameter of the loglogistic family. With large values of β, the loglogistic random variable can be extremely skewed to the right, beyond the point where the skewness, standard deviation, and mean are undefined. As β gets small, the skew decreases, until the shape is nearly symmetric and nearly logistic. While MINITAB software and most references call β a "scale"

parameter, STATGRAPHICS software calls it a "shape" parameter, using the symbol σ.

- τ is an optional threshold parameter, which serves as a lower bound. τ can be any real number, and if it is not specified, the default value is $\tau = 0$.

Representation: $X \sim Loglogistic(\mu, \beta)$ or $X \sim Loglogistic(\mu, \beta, \tau)$

Support: $[\tau, \infty)$

Relationships to Other Distributions:

- The natural log of a two-parameter loglogistic random variable is a logistic random variable. $\ln[Loglogistic(\mu, \beta)] \sim Logistic(\mu, \beta)$.

Process Control Tools: To plot two-parameter loglogistic data collected in rational subgroups on a control chart, plot the natural log of the data on a normal-based Shewhart control chart. To create a control chart for individual data, plot the natural log of the data on a logistic IX, MR chart described earlier in this chapter.

To plot three-parameter loglogistic data on a control chart, first subtract the threshold parameter τ, then follow the steps in the previous paragraph.

Estimating Parameter Values: It is best to use a major statistical package, such as MINITAB or STATGRAPHICS to estimate loglogistic distribution parameters from observed data.

Capability Metrics: For a process producing loglogistic data, calculate equivalent capability metrics using the following formulas. For the two-parameter loglogistic distribution, $\tau = 0$.

$$\text{Equivalent } P_{PU}^{\%} = \frac{-1}{3}\Phi^{-1}\left(\frac{\exp\left(\frac{-(\ln(UTL \, - \, \tau) \, - \, \mu)}{\beta}\right)}{1 + \exp\left(\frac{-(\ln(UTL \, - \, \tau) \, - \, \mu)}{\beta}\right)}\right)$$

$$\text{Equivalent } P_{PL}^{\%} = \frac{-1}{3}\Phi^{-1}\left(\frac{1}{1 + \exp\left(\frac{-(\ln(LTL \, - \, \tau) \, - \, \mu)}{\beta}\right)}\right)$$

$$\text{Equivalent } P_{P}^{\%} = \frac{\text{Equivalent } P_{PU}^{\%} + \text{Equivalent } P_{PL}^{\%}}{2}$$

$$\text{Equivalent } P_{PK}^{\%} = Min\{\text{Equivalent } P_{PU}^{\%}, \text{Equivalent } P_{PL}^{\%}\}$$

$\Phi^{-1}(p)$ is the standard normal inverse CDF, evaluated in an Excel worksheet with the =NORMSINV(p) function.

Probability Density Function:

Two-parameter loglogistic:

$$f_{Loglogistic(\mu,\beta)}(x) = \frac{\exp\left(\frac{-(\ln x - \mu)}{\beta}\right)}{\beta x\left[1 + \exp\left(\frac{-(\ln x - \mu)}{\beta}\right)\right]^2}, \text{ for } x > 0$$

Three-parameter loglogistic:

$$f_{Loglogistic(\mu,\beta,\tau)}(x) = \frac{\exp\left(\frac{-(\ln(x - \tau) - \mu)}{\beta}\right)}{\beta(x - \tau)\left[1 + \exp\left(\frac{-(\ln(x - \tau) - \mu)}{\beta}\right)\right]^2}, \text{ for } x > \tau$$

Cumulative Distribution Function:

Two-parameter loglogistic:

$$F_{Loglogistic(\mu,\beta)}(x) = \frac{1}{1 + \exp\left(\frac{-(\ln x - \mu)}{\beta}\right)}, \text{ for } x > 0$$

Three-parameter loglogistic:

$$F_{Loglogistic(\mu,\beta,\tau)}(x) = \frac{1}{1 + \exp\left(\frac{-(\ln(x - \tau) - \mu)}{\beta}\right)}, \text{ for } x > \tau$$

Inverse Cumulative Distribution Function:

Two-parameter loglogistic:

$$F^{-1}_{Loglogistic(\mu,\beta)}(p) = \exp\left[\mu - \beta\ln\left(\frac{1 - p}{p}\right)\right]$$

Three-parameter loglogistic:

$$F^{-1}_{Loglogistic(\mu,\beta,\tau)}(p) = \tau + \exp\left[\mu - \beta\ln\left(\frac{1 - p}{p}\right)\right]$$

Random Number Generation: To generate three-parameter loglogistic random numbers in an Excel worksheet, use the formula =τ+EXP(μ-β*LN (1/RAND()-1)).

To generate logistic random numbers with Excel and Crystal Ball software, use the formula =τ+EXP(CB.Logistic(μ,β)).

Survival Function:

Two-parameter loglogistic:

$$R_{Loglogistic(\mu,\beta)}(x) = \frac{\exp\left(\frac{-(\ln x - \mu)}{\beta}\right)}{1 + \exp\left(\frac{-(\ln x - \mu)}{\beta}\right)}, \text{ for } x > 0$$

Three-parameter loglogistic:

$$R_{Loglogistic(\mu,\beta,\tau)}(x) = \frac{\exp\left(\frac{-(\ln(x - \tau) - \mu)}{\beta}\right)}{1 + \exp\left(\frac{-(\ln(x - \tau) - \mu)}{\beta}\right)}, \text{ for } x > \tau$$

Hazard Function:

Two-parameter loglogistic:

$$h_{Loglogistic(\mu,\beta)}(x) = \frac{1}{\beta x\left[1 + \exp\left(\frac{-(\ln x - \mu)}{\beta}\right)\right]}$$

Three-parameter loglogistic:

$$h_{Loglogistic(\mu,\beta,\tau)}(x) = \frac{1}{\beta(x - \tau)\left[1 + \exp\left(\frac{-(\ln(x - \tau) - \mu)}{\beta}\right)\right]}$$

Figure 18-3 shows the PDF, CDF, survival and hazard functions of selected loglogistic random variables.

Note: In the following formulas, $\Gamma(\alpha)$ is the gamma function, which can be evaluated by the Excel formula =EXP(GAMMALN(α)).

Mean (Expected Value): If $\beta < 1$, then
$E[Loglogistic(\mu, \beta, \tau)] = \tau + e^{\mu}\Gamma(1 + \beta)\Gamma(1 - \beta)$. If $\beta \geq 1$, the mean does not exist.

Median: e^{μ}

Mode: If $\beta < 1$, then the mode is $\tau + \exp\left[\mu + \beta\ln\left(\frac{1 - \beta}{1 + \beta}\right)\right]$. If $\beta \geq 1$, the mode is τ.

Standard Deviation: If $\beta < 0.5$, then $SD[Loglogistic(\mu, \beta, \tau)] = e^{\mu}\sqrt{\Gamma(1 + 2\beta)\Gamma(1 - 2\beta) - [\Gamma(1 + \beta)\Gamma(1 - \beta)]^2}$. If $\beta \geq 0.5$, the standard deviation does not exist.

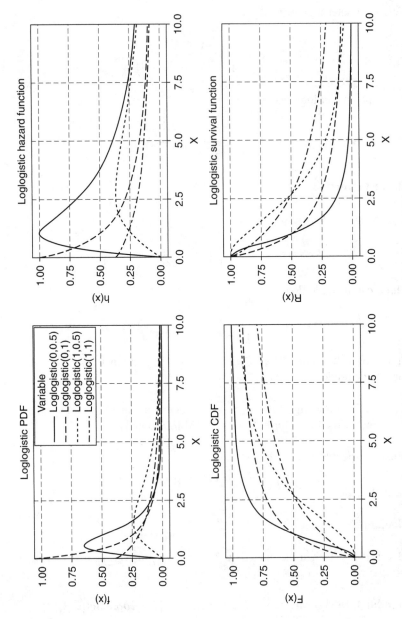

Figure 18-3 PDF, CDF, Survival and Hazard Function of Selected Loglogistic Random Variables.

19

Lognormal Distribution Family

A lognormal random variable is a continuous random variable with a lower bound. For the two-parameter lognormal family, the lower bound is zero, but for the three-parameter lognormal family, the lower bound could be any real number. The natural log of a two-parameter lognormal random variable is a normal random variable. Lognormal random variables are always skewed to the right, but as the parameter σ gets small, lognormal random variables become nearly symmetric and nearly normal in distribution.

The two parameters of the lognormal family represent the mean μ and the standard deviation σ of the normal random variable, which is the natural log of the lognormal random variable. In some software and books, the parameters μ and σ are often called "location" and "scale" parameters respectively, but they do not act like location and scale parameters for the lognormal family. Changing the μ and σ parameters change the scale and shape, respectively, of the lognormal distribution. The optional third parameter of a lognormal random variables X is a threshold parameter τ, so that $\ln(X - \tau)$ is normally distributed. τ is the lower bound of the three-parameter lognormal random variable, and $P[X \le \tau] = 0$.

Applications for lognormal random variables include reliability analysis and any field where positive quantities are studied. Since the lognormal distribution approaches a normal distribution as $\sigma \to 0$, the lognormal distribution is a useful model for processes with right skew caused by a fixed lower bound. Process distributions with a left skew may be modeled by the negative of the lognormal distribution.

Example 19.1

Buzz is a Black Belt at a company manufacturing RF amplifiers for satellite uplink and downlink applications. The gain of each amplifier is measured as a ratio of output voltage to input voltage: V_{Out}/V_{In}. More commonly, the gain is expressed in decibels, calculated this way: $dB_{Gain} = 20\log_{10}(V_{Out}/V_{In})$.

Buzz measures the gain of a prototype build of 350 amplifiers with a nominal gain of 20 dB. Experience with similar amplifiers suggests that dB_{Gain} has a normal distribution. Since dB_{Gain} is a scaled logarithm of V_{Out}/V_{In}, then the gain ratio V_{Out}/V_{In} should have a lognormal distribution.

Figure 19-1 shows two histograms of Buzz's measurements. As expected, dB_{Gain} has an apparently normal distribution, while a lognormal distribution fits the gain ratio V_{Out}/V_{In}.

Parameters: The lognormal distribution family has two parameters, plus an optional third parameter. In this discussion, X represents the lognormal random variable:

- μ is the mean of the normal random variable, which is $\ln(X - \tau)$. μ is often called a location parameter, but it acts as a scale parameter for the lognormal family. Increasing μ stretches the distribution of X out to higher numbers, without changing its shape. μ can be any number.
- σ is the standard deviation of the normal random variable, which is $\ln(X - \tau)$. σ is often called a scale parameter, but it acts as a shape parameter for the lognormal family. With large values of σ, the lognormal random variable can be extremely skewed to the right. As σ

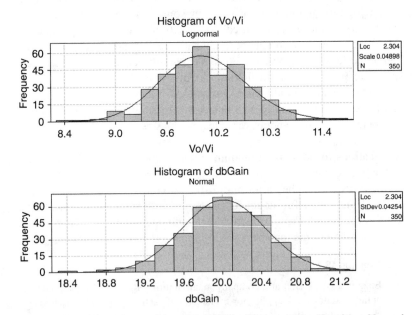

Figure 19-1   Histograms of Amplifier Gains, Expressed in dB with a Normal Distribution, or as a Gain Ratio with a Lognormal Distribution.

gets small, the skew decreases, until the shape is nearly symmetric and nearly normal. σ can be any positive number.

- τ is an optional threshold parameter, which serves as a lower bound for X. τ can be any real number, and if it is not specified, the default value is $\tau = 0$.

Representations of the lognormal family vary in books and software. MINITAB and Excel functions use the parameters described above, μ and σ, which are the mean and standard deviation of the antilog of the lognormal random variable. However, STATGRAPHICS functions and the Crystal Ball graphical interface use μ' and σ', which are the mean and standard deviation of the lognormal random variable. To calculate the mean and standard deviation of a lognormal random variable, use these formulas:

$$\mu' = E[LN(\mu, \sigma)] = \exp\left(\mu + \frac{\sigma^2}{2}\right)$$

$$\sigma' = SD[LN(\mu, \sigma)] = \sqrt{\exp(2\mu + \sigma^2)[e^{\sigma^2} - 1]}$$

To calculate μ and σ from μ' and σ', use these formulas:

$$\mu = \frac{1}{2} \ln\left[\frac{(\mu')^4}{(\mu')^2 + (\sigma')^2}\right]$$

$$\sigma = \ln\left[\frac{(\mu')^2 + (\sigma')^2}{2(\mu')^2}\right]$$

When using Crystal Ball Excel functions to generate random numbers, both options are available: =CB.Lognormal(μ,σ) or =CB.Lognormal2(μ',σ').

Representation: $X \sim LN(\mu, \sigma)$ or $X \sim LN(\mu, \sigma, \tau)$

Support: $[\tau, \infty)$

Relationships to Other Distributions:

- The natural log of a two-parameter lognormal random variable is a normal random variable. $\ln[LN(\mu, \sigma, \tau) - \tau] \sim N(\mu, \sigma^2)$.

Normalizing Transformation:

If $X \sim LN(\mu, \sigma)$, then $\ln X \sim N(\mu, \sigma^2)$.

If $X \sim LN(\mu, \sigma, \tau)$, then $\ln(X - \tau) \sim N(\mu, \sigma^2)$.

Process Control Tools: Because of the easy log transformation to a normal distribution, the best strategy is to plot the log of the data (minus τ for a three-parameter lognormal) on a normal Shewhart control chart, such as an $\bar{X}$, s chart for subgrouped data or an IX, MR chart for individual data.

Estimating Parameter Values: For a two-parameter lognormal distribution when $\tau = 0$, or for a three-parameter lognormal distribution, when τ is known, calculate $\ln(X - \tau)$, and use normal-based estimation methods. Refer to Chapter 21 for a summary of these techniques.

For a three-parameter lognormal distribution when τ is unknown, it is best to use a major statistical package, such as MINITAB or STATGRAPHICS to estimate lognormal distribution parameters from observed data.

Capability Metrics: For a process producing lognormal data, calculate equivalent capability metrics using the following formulas. For the two-parameter lognormal distribution, $\tau = 0$.

$$\text{Equivalent } P^{\%}_{PU} = \frac{\ln(UTL - \tau) - \mu}{3\sigma}$$

$$\text{Equivalent } P^{\%}_{PL} = \frac{\mu - \ln(LTL - \tau)}{3\sigma}$$

$$\text{Equivalent } P^{\%}_{P} = \frac{\text{Equivalent } P^{\%}_{PU} + \text{Equivalent } P^{\%}_{PL}}{2}$$

$$\text{Equivalent } P^{\%}_{PK} = Min\{\text{Equivalent } P^{\%}_{PU}, \text{Equivalent } P^{\%}_{PL}\}$$

Probability Density Function:

Two-parameter lognormal:

$$f_{LN(\mu,\sigma)}(x) = \begin{cases} \dfrac{1}{x\sigma\sqrt{2\pi}} \exp\left[\dfrac{-(\ln x - \mu)^2}{2\sigma^2}\right] & x \geq 0 \\ \\ 0 & \text{otherwise} \end{cases}$$

Three-parameter lognormal:

$$f_{LN(\mu,\sigma,\tau)}(x) = \begin{cases} \dfrac{1}{(x - \tau)\sigma\sqrt{2\pi}} \exp\left[\dfrac{-(\ln(x - \tau) - \mu)^2}{2\sigma^2}\right] & x \geq \tau \\ \\ 0 & \text{otherwise} \end{cases}$$

The lognormal PDF can be calculated in terms of $\phi(x)$, the standard normal PDF, using the formula $f_{LN(\mu,\sigma,\tau)} = \frac{1}{x-\tau}\phi\left(\frac{\ln(x-\tau)-\mu}{\sigma}\right)$

To calculate the lognormal PDF in an Excel worksheet, use the formula =NORMDIST(LN(x-τ),μ,σ,FALSE)/(x-τ).

Cumulative Distribution Function: The lognormal CDF is easy to calculate in terms of $\Phi(x)$, the standard normal CDF:

Two-parameter lognormal: $F_{LN(\mu,\sigma)}(x) = \Phi\left(\frac{\ln x - \mu}{\sigma}\right)$

Three-parameter lognormal: $F_{LN(\mu,\sigma,\tau)}(x) = \Phi\left(\frac{\ln(x-\tau)-\mu}{\sigma}\right)$

To calculate the lognormal CDF in an Excel worksheet, use the formula =LOGNORMDIST(x-τ,μ,σ).

Inverse Cumulative Distribution Function: The lognormal inverse CDF is easy to calculate in terms of $\Phi^{-1}(x)$, the standard normal inverse CDF:

Two-parameter lognormal: $F_{LN(\mu,\sigma)}^{-1}(p) = \exp[\mu + \sigma\Phi^{-1}(p)]$

Three-parameter lognormal: $F_{LN(\mu,\sigma,\tau)}^{-1}(p) = \tau + \exp[\mu + \sigma\Phi^{-1}(p)]$

To calculate the lognormal inverse CDF in an Excel worksheet, use the formula =τ+LOGINV(p,μ,σ).

Random Number Generation: To generate three-parameter lognormal random numbers in an Excel worksheet, use the formula =τ+LOGINV (RAND(p),μ,σ).

To generate three-parameter lognormal random numbers with Excel and Crystal Ball software, use the formula =CB.Lognormal(μ,σ). To use the alternate parameters $\mu' = E[X]$ and $\sigma' = SD[X]$, use the formula =CB.Lognormal2(μ',σ').

Survival Function:

Two-parameter lognormal: $R_{LN(\mu,\sigma)}(x) = 1 - \Phi\left(\frac{\ln x - \mu}{\sigma}\right)$

Three-parameter lognormal: $R_{LN(\mu,\sigma,\tau)}(x) = 1 - \Phi\left(\frac{\ln(x-\tau)-\mu}{\sigma}\right)$

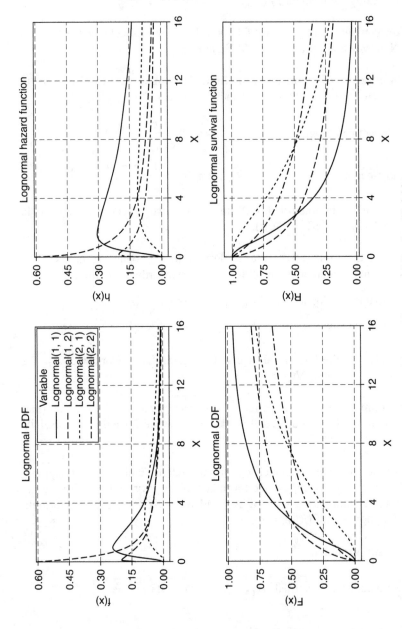

Figure 19-2 PDF, CDF, Survival and Hazard Function of Selected Lognormal Random Variables.

Hazard Function:

Two-parameter lognormal: $h_{LN(\mu,\sigma)}(x) = \dfrac{\phi\left(\frac{\ln x - \mu}{\sigma}\right)}{x\left[1 - \Phi\left(\frac{\ln x - \mu}{\sigma}\right)\right]}$

Three-parameter lognormal: $h_{LN(\mu,\sigma,\tau)}(x) = \dfrac{\phi\left(\frac{\ln(x - \tau) - \mu}{\sigma}\right)}{(x - \tau)\left[1 - \Phi\left(\frac{\ln(x - \tau) - \mu}{\sigma}\right)\right]}$

Figure 19-2 shows the PDF, CDF, survival and hazard functions of selected lognormal random variables.

Note: In the following formulas, $\tau = 0$ for the two-parameter lognormal.

Mean (Expected Value): $E[LN(\mu, \sigma, \tau)] = \tau + \exp\left(\mu + \frac{\sigma^2}{2}\right)$

Median: $\tau + e^{\mu}$

Mode: $\tau + \exp(\mu - \sigma^2)$

Standard Deviation: $SD[LN(\mu, \sigma, \tau)] = \sqrt{\exp(2\mu + \sigma^2)[e^{\sigma^2} - 1]}$

Variance: $V[LN(\mu, \sigma, \tau)] = \exp(2\mu + \sigma^2)[e^{\sigma^2} - 1]$

Coefficient of Variation: $CV[LN(\mu, \sigma, \tau)] = \dfrac{\sqrt{\exp(2\mu + \sigma^2)[e^{\sigma^2} - 1]}}{\tau + \exp(\mu + \frac{1}{2}\sigma^2)}$

Coefficient of Skewness: $Skew[LN(\mu, \sigma, \tau)] = (e^{\sigma^2} + 2)\sqrt{e^{\sigma^2} - 1}$

Lognormal random variables are always skewed to the right, but as σ gets small, the skewness approaches 0.

Coefficient of Kurtosis: $Kurt[LN(\mu, \sigma, \tau)] = e^{4\sigma^2} + 2e^{3\sigma^2} + 3e^{2\sigma^2} - 3$

Coefficient of Excess Kurtosis:

$$ExKurt[LN(\mu, \sigma, \tau)] = e^{4\sigma^2} + 2e^{3\sigma^2} + 3e^{2\sigma^2} - 6$$

Lognormal random variables are always leptokurtic, but as σ gets small, the kurtosis approaches normal kurtosis.

Negative Binomial Distribution Family

A negative binomial random variable is a discrete random variable representing the count of nondefective items before the kth defective item in a series of independent items, each with p probability of being defective. More generally, in any experiment consisting of a series of independent trials with two possible outcomes, if the probability of outcome "A" is p on every trial, the count of "B" outcomes before the kth "A" outcome is a negative binomial random variable.

An experiment consists of a series of Bernoulli trials if it meets these criteria:

- Every trial can have only two outcomes "A" and "B"
- The probability of outcome "A" is the same on every trial
- The trials are independent

In any series of Bernoulli trials, the count of outcome "B" before the kth outcome "A" is a negative binomial random variable.

Several alternative versions of the negative binomial random variable are in use, and two are described in this chapter. X_0 represents the number of "B" outcomes before the kth "A" outcome, and has a minimum value of 0. X_k represents the total number of trials up to and including the kth "A" outcome, and has a minimum value of k. A third version, not discussed here, counts the number of "B" and "A" outcomes up to but not including the kth "A" outcome, and has a minimum value of $k - 1$. When referring to publications or when using software, be careful to use the correct version. The negative binomial functions in STATGRAPHICS, JMP, and Excel use X_0. Crystal Ball software uses X_k. Starting in release 15, MINITAB will include negative binomial functions for both X_0 and X_k.

Figure 20-1   Blaise Pascal.

When k is an integer, the negative binomial distribution is also called the Pascal distribution, in honor of renowned mathematician, physicist, and philosopher Blaise Pascal, (1623–1662). Pascal described special cases of this distribution family in correspondence with Pierre de Fermat (1601–1665). Most often, the Pascal distribution refers to the X_k version of the negative binomial family, when k is an integer. In general, the negative binomial distribution family does not require k to be an integer.

The negative binomial distribution is a generalization of the geometric distribution, and includes it as a special case when $k = 1$.

The name "negative binomial" suggests a relationship to the binomial distribution, but the relationship is a bit abstract. In the same way that the binomial PMF consists of terms in the binomial expansion of $(p + q)^n$, where $q = 1 - p$, the negative binomial PMF consists of terms in the negative binomial expansion of $(Q - P)^{-k}$, where $P = \frac{1-p}{p}$ and $Q = 1 + P$.

Example 20.1

Ric is evaluating constraints in a printer production line. The final inspection station is automated, and when a printer fails any part of the test, the tester diverts the printer into a repair queue. However, the queue only has enough space for three printers. If another printer fails when the repair queue is full, the line shuts down. To predict how often this happens, Ric needs a model to predict the number of printers manufactured before the fourth failed printer causes the line to shut down. Ric selects a negative binomial random variable, with $k = 4$, corresponding to 4 failed printers.

On average, $p = 0.01$ is the probability that any one printer fails the final test. The repair technician processes one printer and empties the repair queue in the time it takes to produce 125 printers. If 4 printers fail in less than the time it takes to manufacture 125 printers, the line will shut down.

Using Ric's model, the number of printers manufactured up to and including the 4th failure is a negative binomial random variable, X_k version. Ric needs to know the probability that fewer than 125 printers will be manufactured up to and including the 4th failure. Ric estimates this probability using a simple Crystal Ball model. The model has one assumption, which has a negative binomial distribution with $p = 0.01$ and $k = 4$. In Ric's simulation, 3.75% of the trials have values of less than 125 in this assumption. Therefore, Ric estimates that the queue will fill up 3.75% of the time.

With STATGRAPHICS software, Ric can directly calculate the probability of queue overflow. Since STATGRAPHICS negative binomial functions use the X_0 version, Ric needs to know $P[NB_0(4, 0.01) < 122]$. This value is 0.0375.

For a different approach to modeling this same problem, see Example 14.1 in the chapter on the gamma distribution.

Example 20.2

Judge Julie will preside over the trial of a notorious murderer. She needs to decide how many potential jurors to summon for the jury pool. Julie wants to be 99% sure of having enough potential jurors to seat a jury of 12. Julie estimates the probability that any one potential juror will be seated on the jury is 0.1. Therefore, the total number of potential jurors required is a negative binomial, X_k version, with $k = 12$ and $p = 0.1$. The required size of the jury pool is the smallest x such that $F_{X_k}(x) > 0.99$.

Following the instructions in this chapter, Julie enters this formula for $F_{X_0}(x)$ into an Excel worksheet: =1-BINOMDIST(12-1,12+x,0.1,TRUE). Julie tries different values for x, and discovers that with $x = 198$, the formula returns a value of 0.99025. Since this is the X_0 version of the negative binomial, Julie must add 12 to include the 12 accepted jurors in the pool of potential jurors. Julie needs to summon at least 210 potential jurors.

Example 20.3

Andy likes ice cream with nuts, but on his new diet, he is limited to only 1.5 scoops of his favorite macadamia ripple ice cream. Andy is also cheap, and he only buys the store brand macadamia ripple ice cream, in which only 50% of the scoops contain a macadamia nut. If Andy scoops out a scoop without a nut, he puts that scoop in his wife's bowl and tries again. Assume that the number of nuts in every scoop is independent of the number of nuts in every other.

The total number of scoops Andy must scoop to get 1.5 scoops with a nut in each is approximately a negative binomial random variable with $p = 0.5$ and $k = 1.5$. Note that the total number of scoops is the X_k version of a negative binomial random variable. The mean total scoops for X_k is $\frac{k}{p} = \frac{1.5}{0.5} = 3$.

Parameters: The negative binomial distribution family has two parameters:

- k represents the number of outcomes "A" in a series of Bernoulli trials, for which X_0 represents the count of outcomes "B" before the kth occurrence of outcome "A." In Excel's NEGBINOMDIST function, k is the number_s parameter. In Crystal Ball functions, k is the Shape parameter.
- p represents the probability of observing outcome "A" in any one trial, when X_0 represents the count of outcomes "B" before the kth occurrence of outcome "A." $0 \leq p \leq 1$. In Excel's NEGBINOMDIST function, p is the probability_s parameter. In Crystal Ball, k is the Probability parameter.

Representation: $X_0 \sim NB_0(k, p)$ or $X_k \sim NB_k(k, p)$

Support:

The support of X_0 is $\{0, 1, 2 \ldots\}$

The support of X_k is $\{k, k + 1, k + 2 \ldots\}$

Relationships to Other Distributions:

- A Pascal random variable is the same as a negative binomial random variable, when the parameter k is an integer. Most often, Pascal random variables have a minimum value of k, corresponding to the X_k version of the negative binomial random variable.
- In a sequence of independent Bernoulli (Yes-No) random variables, the count of 0 values before the kth 1 value is a negative binomial random variable, X_0 version. If $X_i \overset{iid}{\sim} Bern(p)$, then the count of 0 values before the kth 1 value in the sequence is $NB_0(k, p)$.
- A geometric random variable is a special case of a negative binomial random variable with $k = 1$. $NB_0(1, p) \sim Geom_0(p)$ and $NB_k(1, p) \sim Geom_1(p)$.
- The sum of k independent geometric random variables is a negative binomial random variable. If $X_i \overset{iid}{\sim} Geom_0(p)$, then $\sum_{i=1}^{k} X_i \sim NB_0(k,p)$. If $X_i \overset{iid}{\sim} Geom_1(p)$, then $\sum_{i=1}^{k} X_i \sim NB_k(k, p)$.
- *Reproductive property.* The sum of j independent negative binomial random variables is a negative binomial random

variable. If $X_i \sim NB_0(k_i, p)$, and the X_i are independent, then $\sum_{i=1}^{j} X_i \sim NB_0(\sum_{i=1}^{j} k_i, p)$.

- Left tail probabilities of a negative binomial distribution are the same as right tail probabilities of a binomial distribution. If $X \sim NB_0(k, p)$ and $Y \sim Bin(k + x, p)$, then $P[X \le x] = P[Y \ge k]$. This identity is useful in programs (like Excel) which do not have a function for calculating the negative binomial CDF.

- The Poisson random variable with $\lambda = k(1 - p)$ is a limiting form for the negative binomial random variable. This means that as $k \to \infty$ and $p \to 1$, with $k(1 - p)$ constant, then $NB_0(k, p) \xrightarrow{D} Pois(k(1 - p))$.

Process Control Tools: Very few control charts have been proposed for negative binomial processes. Shore's general control charts for attributes (2000a) are effective in this application because they correct for the occasionally extreme skewness of attribute distributions. Based on Shore's method, here are formulas for a negative binomial control chart, based on observed values of X_0, with k known.

$$\hat{p} = \frac{k}{k + \overline{X}_0}$$

$$UCL_{NB} = \frac{k(1 - \hat{p})}{\hat{p}} + 3\sqrt{\frac{k(1 - \hat{p})}{\hat{p}^2}} + 1.324\left(\frac{2 - \hat{p}}{\hat{p}}\right) - 0.5$$

$$CL_{NB} = \frac{k(1 - \hat{p})}{\hat{p}}$$

$$LCL_{NB} = \frac{k(1 - \hat{p})}{\hat{p}} - 3\sqrt{\frac{k(1 - \hat{p})}{\hat{p}^2}} + 1.324\left(\frac{2 - \hat{p}}{\hat{p}}\right) + 0.5$$

Normalizing Transformation: Anscombe (1948) gives a relatively simple normalizing transformation for X_0, as follows:

$$Y = \sqrt{k - 0.5}\sinh^{-1}\sqrt{\frac{X + \frac{3}{8}}{k - \frac{3}{4}}}$$

Section 5.6 of Johnson *et al* (2005) lists several other more accurate and more complex normalizing transformations.

Estimating Parameter Values: When k is known, a maximum likelihood estimator of p, based on X_0, is $\hat{p} = \frac{k}{k + \overline{X}_0}$. When the observed count is X_k, the estimator is $\hat{p} = \frac{k}{\overline{X}_k}$.

Capability Metrics: When a negative binomial process with rate λ has tolerance limits LTL and UTL, calculate equivalent long-term capability metrics this way:

$$\text{Equivalent } P_{PL}^{\%} = \frac{-\Phi^{-1}(F_X(LTL - 1))}{3}$$

$$\text{Equivalent } P_{PU}^{\%} = \frac{-\Phi^{-1}(R_X(UTL))}{3}$$

$$\text{Equivalent } P_P^{\%} = (\text{Equivalent } P_{PL}^{\%} + \text{Equivalent } P_{PU}^{\%})/2$$

$$\text{Equivalent } P_{PK}^{\%} = Min\{\text{Equivalent } P_{PL}^{\%}, \text{Equivalent } P_{PU}^{\%}\}$$

If LTL is not an integer, replace $(LTL - 1)$ with LTL. In these formulas, $F_X(x)$ is the CDF of a negative binomial random variable with parameters k and p, and $R_X(x) = 1 - F_X(x)$ is the corresponding survival function. Assuming that the observed negative binomial values are the X_0 version, with 0 as a possible value, evaluate $R_{X_0}(x)$ in Excel with the formula =BINOMDIST(k-1,k+x,p,TRUE), and $F_X(x) = 1 - R_X(x)$. Also, $\Phi^{-1}(p)$ is the standard normal inverse CDF, evaluated in Excel with the =NORMSINV(p) function. Table 20-1 lists selected values of equivalent $P_{PL}^{\%}$ and equivalent $P_{PU}^{\%}$ for negative binomial processes.

Probability Mass Function (Pascal):

For the X_0 version:

$$f_{NB_0(k,p)}(x) = \begin{cases} \binom{k + x - 1}{k - 1} p^k(1 - p)^x & x \in \{0, 1, \ldots\} \\ 0 & \text{otherwise} \end{cases}$$

For the X_k version:

$$f_{NB_k(k,p)}(x) = \begin{cases} \binom{x - 1}{k - 1} p^k(1 - p)^{x-k} & x \in \{k, k + 1, \ldots\} \\ 0 & \text{otherwise} \end{cases}$$

To calculate $f_{NB_0(k,p)}(x)$ in Excel, use =NEGBINOMDIST(x,k,p).

Table 20-1 Selected Capability Values for Negative Binomial Random Variables

$k = 2, p = 0.5$		$k = 10, p = 0.5$				$k = 10, p = 0.01$			
UTL	Equiv. $P^\%_{PU}$	LTL	Equiv. $P^\%_{PL}$	UTL	Equiv. $P^\%_{PU}$	LTL	Equiv. $P^\%_{PL}$	UTL	Equiv. $P^\%_{PU}$
4	0.410	1	1.032	15	0.401	50	2.048	1400	0.420
8	0.766	2	0.840	20	0.675	100	1.699	1800	0.730
12	1.039	3	0.690	25	0.916	150	1.465	2200	0.999
16	1.267	4	0.561	30	1.133	200	1.285	2600	1.239
20	1.466	5	0.447	35	1.330	250	1.135	3000	1.457
24	1.645	6	0.344	40	1.513	300	1.006	3400	1.659
28	1.809	7	0.249	45	1.685	350	0.891	3800	1.846
32	1.960	8	0.161	50	1.846	400	0.788	4200	2.022
36	2.102	9	0.078	55	1.998	450	0.694	4600	2.188

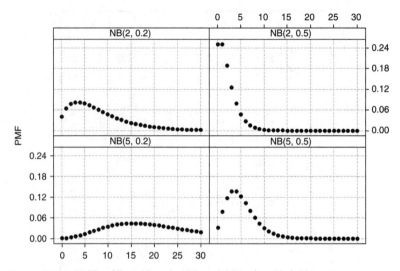

Figure 20-2 PMFs of Four Negative Binomial Random Variables.

Figure 20-2 illustrates the PMF for four negative binomial random variables, $NB_0(2, 0.2)$, $NB_0(2, 0.5)$, $NB_0(5, 0.2)$, and $NB_0(5, 0.5)$. The negative binomial PMF can assume variety of shapes, all with a heavy right tail. The geometric PMFs illustrated in Figure 15-2 are also negative binomial PMFs.

Probability Mass Function (negative binomial):

For the X_0 version:

$$f_{NB_0(k,p)}(x) = \begin{cases} \dfrac{\Gamma(k+x)}{\Gamma(k)x!}\,p^k(1-p)^x & x \in \{0, 1, \ldots\} \\ 0 & \text{otherwise} \end{cases}$$

For the X_k version:

$$f_{NB_k(k,p)}(x) = \begin{cases} \dfrac{(x-1)!}{\Gamma(k)\Gamma(x-k+1)}\,p^k(1-p)^{x-k} & x \in \{k, k+1, \ldots\} \\ 0 & \text{otherwise} \end{cases}$$

In this formula, $\Gamma(\ )$ is the gamma function. Note that Excel will not correctly calculate this PMF using the =NEGBINOMDIST function when k is not an integer. Expecting only integer values, Excel simply rounds

k down. However, Excel does calculate $\ln\Gamma(\ )$ with the =GAMMALN function. To calculate $f_{NB_0(k,p)}(x)$ in Excel, use this formula: =EXP(GAMMALN($k+x$)-GAMMALN(k))/FACT(x)*($p^\wedge k$)*(($1-p$)$^\wedge x$)

Cumulative Distribution Function (Pascal):

For the X_0 version: $F_{NB_0(k,p)}(x) = \sum_{i=0}^{x} \binom{k+i-1}{k-1} p^k(1-p)^i$

For the X_k version: $F_{NB_k(k,p)}(x) = \sum_{i=k}^{x} \binom{i-1}{k-1} p^k(1-p)^{i-k}$

Except in the geometric special case, when $k = 1$, there is no easy formula to simplify the CDF. JMP or STATGRAPHICS software can calculate the negative binomial CDF for the X_0 version, but Excel only provides the PMF as a built-in function, not the CDF. Starting in release 15, MINITAB has negative binomial functions for both X_0 and X_k versions. Earlier releases of MINITAB do not have negative binomial functions.

The identity relating binomial and negative binomial probabilities is useful here: If $X_0 \sim NB_0(k, p)$ and $Y \sim Bin(k + x, p)$, then $F_{X_0}(x) = P[X_0 \leq x] = P[Y \geq k] = 1 - F_Y(k - 1)$. Therefore, this Excel formula will return a value of $F_{X_0}(x)$: =1-BINOMDIST(k-1,$k+x$,p, TRUE).

Inverse Cumulative Distribution Function: Except in the geometric special case, when $k = 1$, no closed-form expression for the inverse CDF is available. Negative binomial quantiles may be computed by solving the CDF iteratively, as illustrated in Example 20.2.

Random Number Generation: One way to generate negative binomial random numbers is to generate k geometric random numbers, by either method listed in Chapter 15, and sum them.

With Excel and Crystal Ball software, random values for $NB_k(k, p)$ are returned by =CB.NegBinomial(p,k).

Mean (Expected Value):

For the X_0 version: $E[NB_0(k, p)] = \dfrac{k(1-p)}{p}$

For the X_k version: $E[NB_k(k, p)] = \dfrac{k}{p}$

Mode: When $m = \frac{(k - 1)(1 - p)}{p}$ is an integer, then X_0 has two equal modes at m and $m - 1$. When m is not an integer, then X_0 has a single mode at m, rounded down.

Standard Deviation: $SD[NB(k, p)] = \frac{\sqrt{k(1 - p)}}{p}$ for either version.

Variance: $V[NB(k, p)] = \frac{k(1 - p)}{p^2}$ for either version.

Coefficient of Variation:

For the X_0 version: $CV[NB_0(k, p)] = \dfrac{1}{\sqrt{k(1 - p)}}$

For the X_k version: $CV[NB_k(k, p)] = \sqrt{k(1 - p)}$

Coefficient of Skewness: $Skew[NB(k, p)] = \dfrac{2 - p}{\sqrt{k(1 - p)}}$ for either version.

A negative binomial random variable is always skewed to the right.

Coefficient of Kurtosis: $Kurt[NB(k, p)] = \dfrac{p^2}{k(1 - p)} + \dfrac{6}{k} + 3$ for either version.

Coefficient of Excess Kurtosis: $ExKurt[NB(k, p)] = \dfrac{p^2}{k(1 - p)} + \dfrac{6}{k}$ for either version.

A negative binomial random variable is always leptokurtic.

Moment Generating Function:

For the X_0 version: $M_{NB_0(k,p)}(t) = \left[\dfrac{p}{1 - (1 - p)e^t} \right]^k$

For the X_k version: $M_{NB_k(k,p)}(t) = \left[\dfrac{pe^t}{1 - (1 - p)e^t} \right]^k$

21

Normal (Gaussian) Distribution Family

A normal or Gaussian random variable is a continuous random variable with a symmetric, bell-shaped probability function. According to the central limit theorem, sums or averages of many independent random variables tend to have normal distributions, regardless of the distribution of the individual random variables. Although there are some limitations on this result, it is true for most practical situations. Because of this fact, characteristics of many natural and manufactured processes tend to have normal distributions. The normal model is easily the most widely applied distribution model ever devised.

The standard normal random variable Z, has mean $\mu = 0$ and standard deviation $\sigma = 1$. The normal distribution family has two parameters, μ and σ. Any normal random variable X may be expressed as a function of Z by the formula $X = \mu + \sigma Z$. Conversely, $Z = \frac{X - \mu}{\sigma}$ standardizes any normal random variable into what is called a "z-score."

Figure 21-1 illustrates the bell-shaped probability function for the normal distribution. The mean, median, and mode are all located at μ. The standard deviation σ is the distance between the mean and the point of inflection on either side of the curve. Figure 21-1 labels selected probabilities in percentages or PPM. Of particular interest in Six Sigma applications is the probability of observing values more than 4.5 standard deviations away from the mean, which is 3.4 PPM on either side. If a tolerance or specification limit is located 4.5 standard deviations away from the mean, then parts outside that limit are defects, and the defect rate is 3.4 DPM.

Historically, the normal distribution was first published as an approximation to the binomial (deMoivre, 1733) and hypergeometric (Laplace, 1774) distributions. In 1809, Carl Friedrich Gauss (1777–1855) published *Theoria Motus* (*Theory of Motion*) on the orbits of celestial bodies. In this work and

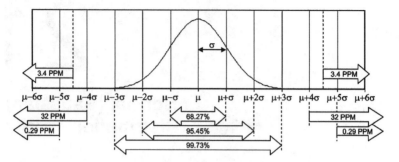

Figure 21-1 Normal Probability Function with Selected Probabilities.

later, while supervising a geodesic survey of Bavaria, Gauss observed that the most likely value of an uncertain quantity is the mean of many measurements, and that the mean tends to have a bell-shaped distribution. Further, Gauss observed that the variability of the sample mean decreases as the sample size increases. In 1816, Gauss proved that the sum of a large number of independent errors has a normal distribution. In Gauss's honor, the normal distribution is often called the Gaussian distribution.

Today, the wealth of statistical theory based on the normal distribution provides many practical tools for those to make decisions from data. Even

Figure 21-2 Carl Friedrich Gauss.

when normal-based tools are applied to mildly nonnormal data, most of these tools have proven to be robust and effective.

One very important result is that maximum likelihood estimators are asymptotically normal, regardless of the distribution family or the parameter of interest. Combining this result with the flexibility of maximum likelihood estimation means that nearly every characteristic of every distribution can be estimated with an approximate confidence interval expressing the precision of the estimate. While these calculations may not be easy, many statistical programs provide them. These confidence intervals not only express the uncertainty in estimates, but they also provide tests to decide whether two or more populations have different distribution characteristics.

This chapter describes several versions of normal random variables. The standard normal random variable and normal family are described first. Section 21.1 describes half-normal distributions. Section 21.2 describes a more general family of truncated normal distributions.

Example 21.1

Lori tested a batch of 100 springs that should have a spring rate of 6.0 ± 1.0 N/cm. The statistics from Lori's sample are $\overline{X} = 5.880$ and $s = 0.3484$. Figure 21-3 shows a histogram of Lori's measurements with a symmetric, bell-shaped distribution. An Anderson-Darling test of normality, not shown here,

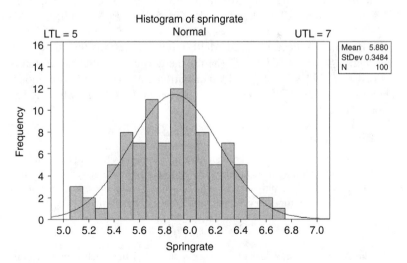

Figure 21-3 Histogram of 100 Spring Rates, with a Normal Probability Density Function that Fits the Data.

gives a P-value of 0.875, which means that there is no reason to reject the normal distribution model. Therefore, Lori decides to use a normal distribution model to decide whether the process producing the springs is acceptable.

Based on the normal distribution model, what is the probability that a spring rate will be unacceptable? Lori uses the Excel function NORMDIST, which calculates the CDF, or the probability that $X \leq x$ for any normal random variable X and any value x. To calculate the probability of being below the lower tolerance limit, $P[X \leq 5.0]$, Lori enters =NORMDIST(5.0,5.88, 0.3484,TRUE), which returns the value 0.005771. To calculate the probability of being above the upper tolerance limit, $P[X > 5.0]$, Lori enters =1-NORMDIST(7.0,5.88,0.3484,TRUE), which returns the value 0.000653. Summing these two probabilities, Lori predicts that these springs have a 0.006424 probability of a spring rate outside the tolerance limits.

Even though none of the 100 springs tested by Lori were unacceptable, she predicts that 1 spring out of 156 ($1/0.006424 = 156$) in the population of springs will be unacceptable.

In the above example, some may wonder why Lori calculated $P[X \leq 5.0]$ for one limit and $P[X > 5.0]$ for the other? Should the calculation for the lower limit be $P[X < 5.0]$ to be fair, since $P[X > 5.0]$ excludes the probability of being exactly equal to 5.0?

For discrete distributions, like binomial and Poisson, the difference between $<$ and $\leq$ is extremely important, but a continuous random variable has zero probability of taking any single value. If X is continuous, then $P[X = x] = 0$ for all x. Therefore, $P[X \leq 5.0] = P[X < 5.0]$.

In an Excel worksheet, the NORMDIST function returns the cumulative distribution function (CDF) $F_X(x)$ for the normal distribution. $F_X(x)$ is defined to be $P[X \leq x]$ for all distributions, discrete and continuous. Therefore, 1-NORMDIST returns $1 - F_X(x) = P[X > x]$, where X has a normal distribution.

Example 21.2

Jay is a mechanical engineer designing packages for electronic products. A typical product may have hundreds of opportunities for interferences, which would obstruct assembly or function. To avoid interferences, Jay computes a transfer function for each potential interference of this form:

$$Y_i = \sum_{j=1}^{k} b_j X_j$$

The X_j are random values of component characteristics, each with a mean μ_j and standard deviation σ_j. Assuming that the X_j are independent, Jay calculates the mean and standard deviation of Y_i:

$$\mu_{Y_i} = \sum_{j=1}^{k} b_j \mu_j$$

$$\sigma_{Y_i} = \sqrt{\sum_{j=1}^{k} (b_j \sigma_j)^2}$$

Jay does not know the distribution shape for each of the X_j, but this is less important than the mean and standard deviation. If each Y_i is a sum of many independent X_j with similar standard deviations, then Jay can assume that each Y_i is normally distributed. Under this assumption, the probability that the interference will cause a problem is

$$P[Y_i \leq 0] = \Phi\left(\frac{0 - \mu_{Y_i}}{\sigma_{Y_i}}\right)$$

where $\Phi( )$ is the standard normal CDF, evaluated in Excel with the NORMSDIST function.

For one potential interference, Jay calculates that the average clearance is $\mu_Y = 0.0045''$ and the standard deviation is $\sigma_Y = 0.0010''$. The clearance has one tolerance limit, a lower limit of 0. The standardized version (also called the z-score) of 0 is

$$\frac{0 - \mu_Y}{\sigma_Y} = \frac{0 - 0.0045}{0.0010} = -4.5$$

To calculate the probability of an interference instead of a clearance, Jay calculates $\Phi(-4.5)$ with the Excel formula =NORMSDIST(-4.5). The answer is 3.4×10^{-6}, which is in line with Six Sigma standards of quality.

Parameters: The normal distribution family has two parameters:

- μ represents the mean of the random variable, and can be any real number.
- σ represents the standard deviation of the random variable, and can be any positive real number.

Representation:

Z is the standard normal random variable with $\mu = 0$ and $\sigma = 1$.

The standard notation for a normal random variable in statistical literature is $X \sim N(\mu, \sigma^2)$, which includes *mean* μ and the *variance* σ^2, not the *standard deviation* σ. Among non-statisticians, the standard deviation is more meaningful than the variance because it has the same units of measure as the random variable. A clear but less concise notation is $X \sim N(\mu = 1, \sigma = 2)$, which describes a normal random variable with mean $\mu = 1$ and standard deviation $\sigma = 2$.

Support: $(-\infty, +\infty)$

Relationships to Other Distributions:

- *Central limit theorem.* If X_1, X_2, . . . , X_n are mutually independent, identically distributed random variables with finite mean μ and standard deviation σ, then the distribution of the standardized sum $\left[\left(\sum_{i=1}^{n} X_i\right) - n\mu\right]/\sigma\sqrt{n}$ tends to have a standard normal distribution as n goes to infinity. Similarly, the distribution of the standardized sample mean $\left[\left(\frac{1}{n}\sum_{i=1}^{n} X_i\right) - \mu\right]/\frac{\sigma}{\sqrt{n}}$ tends to have a standard normal distribution as n goes to infinity.
- The standard normal random variable $Z \sim N(0, 1)$ and the normal random variable $X \sim N(\mu, \sigma^2)$ are related by $Z = \frac{X - \mu}{\sigma}$ and $X = \mu + \sigma Z$.
- Any linear function of mutually independent normal random variables is also a normal random variable. If X_1, X_2, . . . , X_n are mutually independent normal random variables with mean μ_i and standard deviation σ_i, and b_0, b_1, . . . ,b_n are real constants, then $b_0 + \sum_{i=1}^{n} b_i X_i$ is a normal random variable with mean $\mu = b_0 + \sum_{i=1}^{n} b_i\mu_i$ and standard deviation $\sigma = \sqrt{\sum_{i=1}^{n}(b_i\sigma_i)^2}$.
- The natural log of a lognormal random variable is a normal random variable. If $X \sim LN(\mu, \sigma)$, then $\ln X \sim N(\mu, \sigma^2)$. Note that the standard parameters for the lognormal distribution family are the mean and standard deviation of the natural log of X, and not the mean and standard deviation of X.
- The sum of the squares of n mutually independent standard normal random variables is a chi-squared random variable with n degrees of freedom. If $Z_i \overset{iid}{\sim} N(0, 1)$, then $\sum_{i=1}^{n} Z_i \sim \chi^2(n)$.
- The absolute value of a normal random variable with mean 0 is a half-normal random variable. If $X \sim N(0, \sigma^2)$, then $|X| \sim HN(\sigma)$. See section 21.1 for more information on this distribution.
- The normal random variable is the limiting distribution for the standardized forms $\frac{X - E(X)}{SD(X)}$ of many other random variables, including these:
  - Beta $\beta(\alpha, \beta)$ as $\alpha, \beta \to \infty$ with $\alpha = \beta$
  - Binomial $Bin(n, p)$ as $n \to \infty$
  - Chi-squared $\chi^2(v)$ as $v \to \infty$
  - Gamma $\gamma(\alpha, \beta)$ as $\alpha \to \infty$
  - Hypergeometric $Hypergeom(N,D,n)$ as $n \to \infty$ and $N \to \infty$ such that D/N is fixed and $N > n$
  - Lognormal $LN(\mu, \sigma)$ as $\sigma \to 0$
  - Poisson $Pois(\lambda)$ as $\lambda \to \infty$
  - Student's $t(v)$ as $t \to \infty$

Process Control Tools: All standard Shewhart control charts for variable data are designed for normally distributed data. These charts include the

IX, *MR* chart, $\overline{X}$, *R* chart, $\overline{X}$, *s* chart, and many others. For more information on these tools, see Montgomery (2005), Sleeper (2006), AIAG (2005) or any book of standard SPC tools.

Estimating Parameter Values: To estimate the mean μ, the sample mean $\hat{\mu} = \overline{X} = \frac{1}{n}\sum_{i=1}^{n}X_i$ is an unbiased, maximum likelihood estimator, calculated in Excel with the =AVERAGE function. A $100(1-\alpha)\%$ confidence interval for μ is $\overline{X} \pm T_7\left(n, \frac{\alpha}{2}\right)s$, where *s* is the sample standard deviation $s = \sqrt{\frac{1}{n-1}\sum_{i=1}^{n}(X_i - \overline{X})^2}$, calculated in Excel with the =STDEV function. $T_7\left(n, \frac{\alpha}{2}\right)$ is a shorthand notation introduced by Bothe (2002a), defined as $T_7\left(n, \frac{\alpha}{2}\right) = t_{\alpha/2,n-1}/\sqrt{n}$. Bothe (2002a) and Sleeper (2006) contain tabulated values for $T_7\left(n, \frac{\alpha}{2}\right)$. To calculate $T_7\left(n, \frac{\alpha}{2}\right)$ in Excel, use this formula: =TINV(α,n-1)/SQRT(n). $t_{\alpha/2,n-1}$ is the $1-\alpha/2$ quantile of the Student's *t* distribution with $n-1$ degrees of freedom, calculated in Excel by the formula =TINV(α,n-1). Be careful when using the TINV function, because the probability parameter is different from other inverse CDF functions in Excel. To calculate a 95% confidence interval, $\alpha = 0.05$. Since the α risk is split between two ends of the confidence interval, the correct *t* quantile is $t_{0.025,n-1}$ which Excel calculates with =TINV(0.05,n-1).

To estimate the long-term standard deviation σ_{LT} from a sample representing long-term variation, an unbiased estimator is $\hat{\sigma}_{LT} = \frac{s}{c_4(n)}$, where *s* is the sample standard deviation as defined above, and $c_4(n)$ is an unbiasing constant defined by

$$c_4(n) = \frac{\Gamma\left(\frac{n}{2}\right)}{\Gamma\left(\frac{n-1}{2}\right)}\sqrt{\frac{2}{n-1}}$$

In this formula, $\Gamma()$ is the gamma function. In an Excel worksheet, $c_4(n)$ may be calculated by the formula =EXP(GAMMALN(n/2)-GAMMALN((n-1)/2))*SQRT(2/(n-1)). As *n* gets large, $c_4(n)$ becomes very close to 1, and it is often ignored. Bothe (1997), Sleeper (2006), Montgomery (2005), AIAG (2005), and most other books on SPC contain tabulated values for $c_4(n)$.

To calculate a $100(1-\alpha)\%$ confidence interval for long-term standard deviation σ_{LT}, the lower limit is $L_{\sigma_{LT}} = s/T_2\left(n, 1 - \frac{\alpha}{2}\right)$ and the upper limit is $U_{\sigma_{LT}} = s/T_2\left(n, \frac{\alpha}{2}\right)$, where

$$T_2(n, \tfrac{\alpha}{2}) = \sqrt{\frac{\chi^2_{\alpha/2,n-1}}{n-1}}$$

evaluated in Excel with the formula =SQRT(CHIINV($1-\alpha/2,n\text{-}1$)/($n\text{-}1$)). T_2 is shorthand notation introduced by Bothe (2002a).

Other estimators are also used for σ_{LT}. Most common is simply $\hat{\sigma}_{LT} = s$, which is biased low. As an estimator for the variance σ_{LT}^2, $\hat{\sigma}_{LT}^2 = s^2$ is unbiased, but taking the square root of s^2 introduces bias. A different estimator is

$$s_n = s\sqrt{\frac{n-1}{n}} = \sqrt{\tfrac{1}{n}\sum_{i=1}^{n}(X_i - \overline{X})^2}$$

which is a maximum likelihood estimator for σ_{LT}. In Excel, the **STDEVP** function calculates s_n. The **P** in **STDEVP** stands for population. If every value in the population were available, then s_n is the population standard deviation. This situation is rare. More often, only a sample is available. As a sample statistic, $s_n < s$, so s_n is biased even more low than s. Therefore, for estimating standard deviation from a sample, s_n is not recommended.

To estimate short-term standard deviation σ_{ST} requires a sample containing rational subgroups collected over a long time. The variation within each subgroup is used to estimate σ_{ST}. Four unbiased estimators for σ_{ST} are listed here in order from most precise to least precise. In these formulas, k represents the count of subgroups, and n represents the count of observations in each subgroup.

- Pooled standard deviation: $\hat{\sigma}_{ST} = \dfrac{1}{c_4(k(n-1)+1)}\sqrt{\dfrac{\sum_{i=1}^{k}s_i^2}{k}}$

- Average standard deviation: $\hat{\sigma}_{ST} = \dfrac{\bar{s}}{c_4(n)}$

- Average range: $\hat{\sigma}_{ST} = \dfrac{\overline{R}}{d_2(n)}$ Tabulated values of d_2 are listed in many SPC books, including Bothe (1997), Montgomery (2005), Sleeper (2006) and AIAG (2005).

- Average moving range, where $MR_i = |X_i - X_{i-1}|$: $\hat{\sigma}_{ST} = \dfrac{\overline{MR}}{1.128}$

Only the first three of these estimators have confidence intervals, which are described in Sleeper (2006). Since the pooled standard deviation is the most precise estimate, its confidence interval is the most narrow.

Capability Metrics: Section 3.2 covers many well-established methods for measuring process capability for normally distributed processes.

Probability Density Function:

For the standard normal random variable: $f_Z(x) = \phi(x) = \dfrac{e^{-x^2/2}}{\sqrt{2\pi}}$

The notation $\phi(x)$ commonly represents the standard normal PDF.

For a normal random variable with parameters μ and σ:

$$f_{N(\mu,\sigma^2)}(x) = \frac{1}{\sigma}\,\phi\!\left(\frac{x-\mu}{\sigma}\right) = \frac{1}{\sigma\sqrt{2\pi}}\exp\!\left(\frac{-(x-\mu)^2}{2\sigma^2}\right)$$

To calculate $f_{N(\mu,\sigma^2)}(x)$ in an Excel worksheet, use the formula =NORMDIST(x,μ,σ,FALSE).

Cumulative Distribution Function: For the standard normal random variable:

$$F_Z(x) = \Phi(x) = \int_{-\infty}^{x}\frac{e^{-u^2/2}}{\sqrt{2\pi}}du$$

The notation $\Phi(x)$ commonly represents the standard normal CDF. To calculate $\Phi(x)$ in an Excel worksheet, use the formula =NORMSDIST(x).

For a normal random variable with parameters μ and σ:
$$F_{N(\mu,\sigma^2)}(x) = \Phi\!\left(\frac{x-\mu}{\sigma}\right)$$

To calculate $F_{N(\mu,\sigma^2)}(x)$ in an Excel worksheet, use the formula =NORMDIST(x,μ,σ,TRUE).

Inverse Cumulative Distribution Function: For a standard normal random variable, the symbol $\Phi^{-1}(prob)$ represents the inverse CDF. That is, $\Phi^{-1}(prob)$ is x which solves the equation $F_{N(0,1)}(x) = prob$. For a $N(\mu, \sigma^2)$ random variable, $F_{N(\mu,\sigma^2)}^{-1}(prob) = \mu + \sigma\Phi^{-1}(prob)$. There is no simple formula to calculate $\Phi^{-1}(prob)$, and it must be calculated by an iterative numerical procedure. Excel, MINITAB, JMP, STATGRAPHICS, and most other statistical programs provide built-in algorithms to calculate the normal inverse CDF.

To calculate $F_{N(\mu,\sigma^2)}^{-1}(prob)$ in an Excel worksheet, use the formula =NORMINV($prob,\mu,\sigma$). To calculate $\Phi^{-1}(prob)$ in an Excel worksheet, use the formula =NORMSINV($prob$).

Random number generation: To generate standard normal random numbers in an Excel worksheet, use the formula =NORMSINV(RAND()). To generate normal random numbers with mean μ and standard deviation σ in an Excel worksheet, use the formula =NORMINV(RAND(),μ,σ). To generate normal random numbers with Excel and Crystal Ball software, use =CB.Normal(μ,σ).

A different algorithm can be used in a programming environment lacking a function for $\Phi^{-1}(prob)$. Generate two independent uniform random numbers between 0 and 1, $U_1 \sim Unif(0, 1)$ and $U_2 \sim Unif(0, 1)$. Then the following two functions compute two independent standard normal random numbers:

$$Z_1 = \sqrt{-2\ln U_1}\, \sin(2\pi U_2)$$

$$Z_2 = \sqrt{-2\ln U_1}\, \cos(2\pi U_2)$$

Survival Function: For the standard normal random variable:

$$R_Z(x) = 1 - \Phi(x)$$

For a normal random variable with parameters μ and σ:

$$R_{N(\mu,\sigma^2)}(x) = 1 - \Phi\left(\frac{x - \mu}{\sigma}\right)$$

Hazard Function: The hazard function for normal random variables is always increasing.

For the standard normal random variable:

$$h_Z(x) = \frac{\phi(x)}{1 - \Phi(x)}$$

For a normal random variable with parameters μ and σ:

$$h_{N(\mu,\sigma^2)}(x) = \frac{\phi\left(\frac{x - \mu}{\sigma}\right)}{\sigma\left[1 - \Phi\left(\frac{x - \mu}{\sigma}\right)\right]}$$

Figure 21-4 illustrates the CDF, PDF, Survival, and hazard functions for a standard normal random variable.

Mean (Expected Value): $E[N(\mu, \sigma^2)] = \mu$

Median: μ

Mode: μ

Standard Deviation: $SD[N(\mu, \sigma^2)] = \sigma$

Variance: $V[N(\mu, \sigma^2)] = \sigma^2$

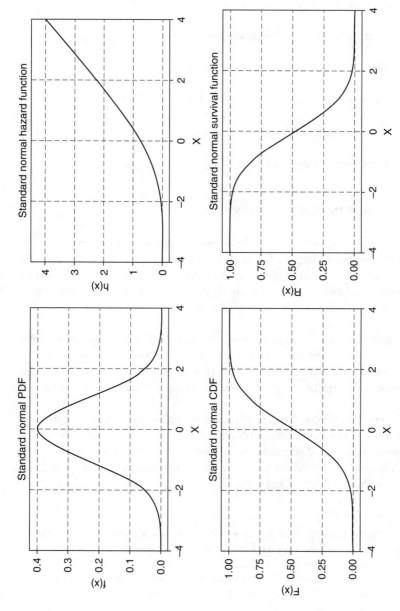

Figure 21-4 PDF, CDF, Hazard, and Survival Functions for the Standard Normal Random Variable.

Coefficient of Variation: $CV[N(\mu, \sigma^2)] = \frac{\sigma}{\mu}$

Coefficient of Skewness: $Skew[N(\mu, \sigma^2)] = 0$

A normal random variable is always symmetric.

Coefficient of Kurtosis: $Kurt[N(\mu, \sigma^2)] = 3$

Coefficient of Excess Kurtosis: $ExKurt[N(\mu, \sigma^2)] = 0$

A normal random variable is always mesokurtic.

Moment Generating Function: $M_{N(\mu,\sigma^2)}(t) = \exp\left(t\mu + \frac{t^2\sigma^2}{2}\right)$

21.1   Half-Normal Distribution Family

The absolute value of a normal random variable with mean zero is a half-normal random variable. The half-normal distribution family has a single parameter σ, representing the standard deviation of the underlying normal random variable. If needed, the family could also have a location parameter, but this generalization is not described here.

If two points on a line have independent normally distributed positions on the line, with the same mean position, the distance between the two points is a half-normal random variable. For this reason, the half-normal is a reasonable statistical model for many geometric dimensions of manufactured items.

Because it is limited to positive values, the half-normal random variable is also a useful model in reliability engineering for times to failure. The hazard function of a half-normal random variable is always increasing, and compared to Weibull models with increasing hazard function, the half-normal hazard function starts at a higher value.

Example 21.3

Thad is a mechanical engineer responsible for camshaft designs in engines. The relative position of each cam is critical to efficient and smooth engine operation. Figure 21-5 illustrates two cams on a shaft. For illustration purposes, each cam is rotated on the shaft so that its major axis is vertical. X describes the difference in extension of each cam away from the axis of the shaft.

If the maximum extension of each cam is normally distributed with the same mean value, then X, the difference between extensions, has a half-normal distribution.

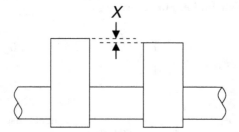

Figure 21-5 Two Cams on a Camshaft, Rotated so that the Maximum Cam Extension is Vertical.

Parameters: The half-normal distribution family has one parameter:

- σ represents the standard deviation of a normal random variable with mean 0, of which X is the absolute value. σ can be any positive real number.

Representation: $X \sim HN(\sigma)$

Support: $[0, +\infty)$

Relationships to Other Distributions:

- If $X \sim N(0, \sigma)$, then $|X| \sim HN(\sigma)$.
- A normal random variable with mean zero and truncated on the left at zero is a half-normal random variable. $TN(0, \sigma, 0, \infty) \sim HN(\sigma)$.
- A standard half-normal distribution, with $\sigma = 1$, is a chi distribution with one degree of freedom. $HN(1) \sim \chi(1)$.
- A folded normal random variable is a generalized version of a half-normal random variable, formed by the absolute value of a normal random variable with any mean value.

Probability Density Function:

$$f_{HN(\sigma)}(x) = \begin{cases} \dfrac{2}{\sigma}\phi\left(\dfrac{x}{\sigma}\right) = \dfrac{2}{\sigma\sqrt{2\pi}}\exp\left(\dfrac{-x^2}{2\sigma^2}\right) & x \geq 0 \\ \\ 0 & x < 0 \end{cases}$$

To calculate $f_{HN(\sigma)}(x)$ in an Excel worksheet, use the formula =IF(x>=0,2*NORMDIST(x,0,σ,FALSE),0).

Cumulative Distribution Function:

$$F_{HN(\sigma)}(x) = \begin{cases} 2\left(\Phi\left(\frac{x}{\sigma}\right) - 0.5\right) & x \geq 0 \\ 0 & x < 0 \end{cases}$$

To calculate $F_{HN(\sigma)}(x)$ in an Excel worksheet, use the formula =IF(x>=0,2*(NORMDIST(x,0,σ,TRUE)-0.5),0).

Inverse Cumulative Distribution Function:

$$F^{-1}_{HN(\sigma)}(prob) = \sigma\Phi^{-1}\left(\frac{prob + 1}{2}\right)$$

To calculate $F^{-1}_{HN(\sigma)}(prob)$ in an Excel worksheet, use the formula =σ*NORMSINV(($prob$+1)/2).

Random number generation: To generate half-normal random numbers in an Excel worksheet, use the formula =σ*NORMSINV((RAND()+1)/2). To generate half-normal random numbers with Excel and Crystal Ball software, use =CB.Normal(μ,σ,0).

Survival Function:

$$R_{HN(\sigma)}(x) = \begin{cases} 2 - 2\Phi\left(\frac{x}{\sigma}\right) & x \geq 0 \\ 1 & x < 0 \end{cases}$$

Hazard Function: The hazard function for half-normal random variables is always increasing.

$$h_{HN(\sigma)}(x) = \frac{\frac{1}{\sigma}\phi\left(\frac{x}{\sigma}\right)}{1 - \Phi\left(\frac{x}{\sigma}\right)}$$

Figure 21-6 illustrates the CDF, PDF, Survival, and hazard functions for a standard normal random variable.

Mean (Expected Value): $E[HN(\sigma)] = \sigma\sqrt{\frac{2}{\pi}}$

Median: $\sigma\Phi^{-1}\left(\frac{3}{4}\right) \cong 0.675\sigma$

Mode: 0

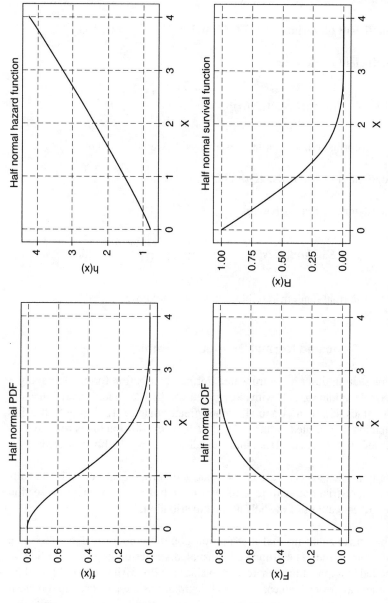

Figure 21-6 PDF, CDF, Hazard, and Survival Functions for the Standard Half-Normal Random Variable.

Standard Deviation: $SD[HN(\sigma)] = \sigma\sqrt{1 - \frac{2}{\pi}}$

Variance: $V[HN(\sigma)] = \sigma^2\left(1 - \frac{2}{\pi}\right)$

Coefficient of Variation: $CV[HN(\sigma)] = \sqrt{\frac{\pi - 2}{2}} \cong 0.7555$

Coefficient of Skewness:

$$Skew[HN(\sigma)] = \frac{\sqrt{\frac{2}{\pi}}\left(\frac{4}{\pi} - 1\right)}{\left(1 - \frac{2}{\pi}\right)^{3/2}} \cong 0.995$$

A half-normal random variable is always skewed to the right.

Coefficient of Kurtosis: $Kurt[HN(\sigma)] \cong 0.869$

Coefficient of Excess Kurtosis:

$$ExKurt[HN(\sigma)] = \frac{-3\left(\frac{2}{\pi}\right)^2 - 2\left(\frac{2}{\pi}\right) + 3}{\left(1 - \frac{2}{\pi}\right)^2} \cong 3.869$$

A half-normal random variable is always leptokurtic.

21.2 Truncated Normal Distribution Family

This section describes a truncated normal distribution formed by taking a normal random variable with mean μ and standard deviation σ, and truncating all values less than A and all values greater than B. If $A = -\infty$, the distribution is truncated on the right. If $B = \infty$, the distribution is truncated on the left. If both A and B are finite, the distribution is doubly truncated.

An important special case of the truncated normal distribution is the half-normal distribution with $\mu = 0$, $A = 0$, and $B = \infty$. The previous section describes features of the half-normal distribution.

The untruncated normal distribution could have any real number as a possible value. The truncated normal distribution is often applied to modeling applications where some values are impossible or would result in numerical errors. Truncated normal distributions are also used to represent inspection or screening processes which are intended to remove units that are outside acceptable limits.

Often, truncation is applied inappropriately. If values above or below a certain threshold cause numerical problems for a model, perhaps this is because those values never occur in the physical system being modeled. Distribution families other than the normal distribution include lower bounds, upper bounds or both, and one of these may be a more appropriate model for naturally bounded processes than a truncated normal distribution. Section 1.6.1 discusses the misuse of truncation in modeling applications, and provides an example.

Parameters: The truncated normal distribution family has four parameters:

- μ represents the mean of a normal random variable, before truncation. μ could be any real number. μ is not the mean of the truncated random variable.
- σ represents the standard deviation of a normal random variable, before truncation. σ could be any positive real number. σ is not the standard deviation of the truncated random variable.
- A represents the left truncation limit. A could be any real number or $-\infty$ to represent no left truncation.
- B represents the right truncation limit. B could be any real number or $+\infty$ to represent no right truncation.

Representation: $X \sim TN(\mu, \sigma, A, B)$

Support: (A, B)

Relationships to Other Distributions: $HN(\sigma) \sim TN(0, \sigma, 0, \infty)$

Probability Density Function:

$$
f_{TN(\mu,\sigma,A,B)}(x) = \begin{cases} \dfrac{\phi\left(\frac{x - \mu}{\sigma}\right)}{\sigma\left[\Phi\left(\frac{B - \mu}{\sigma}\right) - \Phi\left(\frac{A - \mu}{\sigma}\right)\right]} & A \leq x \leq B \\ \\ 0 & \text{otherwise} \end{cases}
$$

To calculate $f_{TN(\mu,\sigma,A,B)}(x)$ in an Excel worksheet, use the formula =IF(AND(x>=A,x<=B),NORMDIST(x,μ,σ,FALSE)/(NORMDIST(B,μ, σ,TRUE)-NORMDIST(A,μ,σ,TRUE)),0).

Figure 21-7 illustrates the PDF of four standard normal random variables truncated at different points.

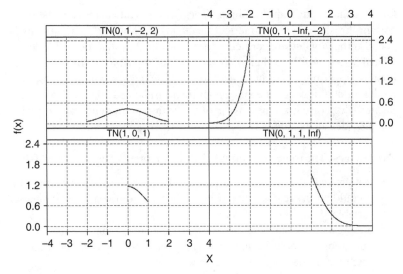

Figure 21-7   PDFs of Four Truncated Versions of the Standard Normal Random Variable.

Cumulative Distribution Function:

$$
F_{TN(\mu,\sigma,A,B)}(x) = \begin{cases} 0 & x < A \\ \dfrac{\Phi\left(\dfrac{x-\mu}{\sigma}\right) - \Phi\left(\dfrac{A-\mu}{\sigma}\right)}{\Phi\left(\dfrac{B-\mu}{\sigma}\right) - \Phi\left(\dfrac{A-\mu}{\sigma}\right)} & A \le x \le B \\ 1 & x > B \end{cases}
$$

To calculate $F_{TN(\mu,\sigma,A,B)}(x)$ in an Excel worksheet, use =MIN(1,MAX(0,(NORMDIST(x,μ,σ,TRUE)-NORMDIST(A,μ,σ, TRUE))/(NORMDIST(B,μ,σ,TRUE)-NORMDIST(A,μ,σ,TRUE)))).

Inverse Cumulative Distribution Function:

$$
F_{TN(\mu,\sigma,A,B)}^{-1}(prob) = \mu + \sigma\Phi^{-1}\left(\Phi\left(\frac{A-\mu}{\sigma}\right)\right.
$$

$$
\left. + prob \times \left[\Phi\left(\frac{B-\mu}{\sigma}\right) - \Phi\left(\frac{A-\mu}{\sigma}\right)\right]\right)
$$

To calculate $F_{TN(\mu,\sigma,A,B)}^{-1}(prob)$ in an Excel worksheet, use the formula =μ+σ*NORMSINV(NORMDIST(A,μ,σ,TRUE)+$prob$*(NORMDIST(B,μ,σ,TRUE)-NORMDIST(A,μ,σ,TRUE))).

Random number generation: To generate truncated normal random numbers in an Excel worksheet, use the formula =μ+σ*NORMSINV(NORMDIST(A,μ,σ,TRUE)+RAND()*(NORMDIST(B,μ,σ,TRUE)-NORMDIST(A,μ,σ,TRUE))). To generate truncated normal random numbers in Excel with Crystal Ball software, use the formula =CB.Normal(μ,σ,A,B).

Mean (Expected Value):

$$E[TN(\mu, \sigma, A, B)] = \mu + \sigma\left[\frac{\phi\left(\frac{A-\mu}{\sigma}\right) - \phi\left(\frac{B-\mu}{\sigma}\right)}{\Phi\left(\frac{B-\mu}{\sigma}\right) - \Phi\left(\frac{A-\mu}{\sigma}\right)}\right]$$

For right truncated normal random variables, when $A = -\infty$,

$$E[TN(\mu, \sigma, -\infty, B)] = \mu + \sigma\left[\frac{-\phi\left(\frac{B-\mu}{\sigma}\right)}{\Phi\left(\frac{B-\mu}{\sigma}\right)}\right]$$

For left truncated normal random variables, when $B = \infty$,

$$E[TN(\mu, \sigma, A, \infty)] = \mu + \sigma\left[\frac{\phi\left(\frac{A-\mu}{\sigma}\right)}{1 - \Phi\left(\frac{A-\mu}{\sigma}\right)}\right]$$

Standard Deviation:

$SD[TN(\mu, \sigma, A, B)] =$

$$\sigma\sqrt{1 + \frac{\frac{A-\mu}{\sigma}\phi\left(\frac{A-\mu}{\sigma}\right) - \frac{B-\mu}{\sigma}\phi\left(\frac{B-\mu}{\sigma}\right)}{\Phi\left(\frac{B-\mu}{\sigma}\right) - \Phi\left(\frac{A-\mu}{\sigma}\right)} - \left[\frac{\phi\left(\frac{A-\mu}{\sigma}\right) - \phi\left(\frac{B-\mu}{\sigma}\right)}{\Phi\left(\frac{B-\mu}{\sigma}\right) - \Phi\left(\frac{A-\mu}{\sigma}\right)}\right]^2}$$

For right truncated normal random variables, when $A = -\infty$,

$$SD[\,TN(\mu, \sigma, -\infty, B)] = \sigma\sqrt{1 - \frac{\frac{B-\mu}{\sigma}\phi\left(\frac{B-\mu}{\sigma}\right)}{\Phi\left(\frac{B-\mu}{\sigma}\right)} - \left[\frac{\phi\left(\frac{B-\mu}{\sigma}\right)}{\Phi\left(\frac{B-\mu}{\sigma}\right)}\right]^2}$$

For left truncated normal random variables, when $B = \infty$,

$$SD[\,TN(\mu, \sigma, A, \infty)] = \sigma\sqrt{1 + \frac{\frac{A-\mu}{\sigma}\phi\left(\frac{A-\mu}{\sigma}\right)}{1 - \Phi\left(\frac{A-\mu}{\sigma}\right)} - \left[\frac{\phi\left(\frac{A-\mu}{\sigma}\right)}{1 - \Phi\left(\frac{A-\mu}{\sigma}\right)}\right]^2}$$

Sugiura and Gomi (1985) provide formulas to calculate the coefficients of skewness and kurtosis for doubly truncated normal random variables.

22

Pareto Distribution Family

A Pareto random variable is a continuous random variable with an extremely heavy right tail and a lower bound at some number $\beta > 0$. Pareto random variables are common models for distributions of income and other financial measurements.

The name of the Pareto distribution honors the economist who developed it, Vilfredo Federico Damaso Pareto (1848–1923). As published by Pareto (1897), Pareto's law states that if X is a random variable representing income over a population, then the distribution of X obeys this law: $P[X > x] \propto x^{-\alpha}$ Random variables obeying this law are now called Pareto random variables. Economists continue to debate the applicability or, as claimed by Pareto, the universality of this law. Pareto distributions remain important models for many economic quantities. Economists who have applied Pareto's law to income distributions report that the parameter α falls within a relatively narrow range, between 1.5 (Bresciani-Turroni, 1939) and 2.1 (Cramer, 1971).

For Six Sigma professionals, the name Pareto is well known for the Pareto principle and Pareto chart, both devised by Joseph M. Juran and named in Pareto's honor. Pareto observed that the majority (80%) of the wealth belonged to the minority (20%) of people. With proper scaling, Pareto distributions display this same relationship. As applied to quality phenomena by Dr. Juran, the Pareto principle states that 80% of quality problems are the effects of 20% of the causes.

For modelers and Six Sigma practitioners, Pareto distributions are both easy and difficult to use. They are easy to use because the formulas describing the distribution are mathematically quite simple. Relatively simple formulas for estimating the parameters of the distribution are also available. At the same time, it is difficult to apply many standard statistical methods to Pareto distributions, because familiar parameters like the mean and standard

Figure 22-1   Vilfredo Pareto.

deviation are undefined for small values of the shape parameter α. Instead of measuring the location and variation of Pareto distributions by the mean and standard deviation, it is better to use the median $[F^{-1}(0.5)]$ and interquartile range $[F^{-1}(0.75) - F^{-1}(0.25)]$, or other measures based on quantiles.

For an entire book on Pareto distributions and their applications, see Arnold (1983).

Parameters: The Pareto distribution family has one shape parameter, plus one optional scale parameter.

- α is a shape parameter, describing the rate of decrease in the probability curve for larger values. α can be any positive number.
- β is an optional scale parameter. β can be any positive number. If β is not specified, the default value is $\beta = 1$. β is a scale parameter because $\beta \times Pareto(\alpha) \sim Pareto(\alpha, \beta)$ β is also a lower bound for the Pareto random variable. Because of this fact, β is also called a location parameter in Crystal Ball and STATGRAPHICS software.

Representation: $X \sim Pareto(\alpha)$ or $X \sim Pareto(\alpha, \beta)$

Support: $[\beta, \infty)$

Relationships to Other Distributions:

- A uniform random variable between 0 and 1, raised to a negative power, is a Pareto random variable. If $U \sim Unif(0, 1)$, then $U^{-\alpha} \sim Pareto(\alpha)$.
- The natural antilog of an exponential random variable is a Pareto random variable. If $E \sim Exp(1)$, then $\beta \exp\left(\frac{E}{\alpha}\right) \sim Pareto(\alpha, \beta)$
- Conversely, the natural log of a Pareto random variable is an exponential random variable. If $X \sim Pareto(\alpha, \beta)$, then $\alpha \ln\left(\frac{X}{\beta}\right) \sim Exp(1)$
- There are many other versions of the Pareto distributions, adapted for different applications, including these:
  - A third parameter τ added to a one-parameter or a two-parameter Pareto random variable provides flexibility to locate the lower bound at any value $\tau + \beta$.
  - The Pareto distribution described here is called the Pareto distribution of the first kind.
  - The Pareto distribution of the second kind, sometimes called a Lomax distribution, has a CDF of this form:

$$F(x) = 1 - \frac{C^{\alpha}}{(x + C)^{\alpha}}, \text{ for } x > 0$$

  - The Pareto distribution of the third kind has a CDF of this form:

$$F(x) = 1 - \frac{Ce^{-bx}}{(x + C)^{\alpha}}, \text{ for } x > 0$$

  - For additional variations and generalizations, see Chapter 20 of Johnson *et al* (1994).

Normalizing Transformation: A $Pareto(\alpha, \beta)$ random variable can be transformed into an approximately normal distribution by this transformation: $[\alpha \ln(X/\beta)]^{0.2654}$. As with other normalizing transformations, the success of this method depends on how accurately the parameters α and β are known.

Process Control Tools: For processes producing Pareto distributions, it is extremely important not to apply normal-based techniques without correcting for the nonnormality. When $\alpha \leq 2$, as in many economic models, the standard deviation of Pareto random variables is infinite, or more formally, undefined. This detail means that the central limit theorem does not apply, and sample means will not tend to a normal distribution. When individual data has undefined standard deviation, $\overline{X}$ charts provide no protection from nonnormality.

There are at least two viable control chart alternatives for processes following a Pareto distribution. One alternative is to create a Pareto individual X chart with false alarm risk ε, using these formulas:

$$UCL = \frac{\beta}{(\varepsilon/2)^{1/\alpha}}$$

$$CL = 2^{1/\alpha}\beta$$

$$LCL = \frac{\beta}{(1 - \varepsilon/2)^{1/\alpha}}$$

For parity with Shewhart control charts, set $\varepsilon = 0.0027$.

Another method is to transform the Pareto data into an approximately normal distribution, using the transformation $[\alpha \ln(X/\beta)]^{0.2654}$. Then, plot the transformed data on any Shewhart control chart for normal data.

Estimating Parameter Values: If β is known, a maximum likelihood estimator for α is

$$\hat{\alpha} = \frac{1}{\ln\left(\frac{1}{\beta}\left[\prod_{i=1}^{n} X_i\right]^{\frac{1}{n}}\right)}$$

A $100(1-\alpha)\%$ confidence interval for α is

$$\left(\frac{\hat{\alpha}\chi^2_{2n,1-\alpha/2}}{2n}, \frac{\hat{\alpha}\chi^2_{2n,\alpha/2}}{2n}\right)$$

where $\chi^2_{v,\alpha}$ is the $1-\alpha$ quantile of the χ^2 distribution with v degrees of freedom, calculated by the Excel function =CHIINV(α,v).

If α and β are both unknown, unbiased estimates can be calculated by these formulas:

$$X_{(1)} = \underset{i}{Min}X_i$$

$$\alpha' = \frac{1}{\ln\left(\frac{1}{X_{(1)}}\left[\prod_{i=1}^{n} X_i\right]^{\frac{1}{n}}\right)}$$

$$\hat{\alpha} = \left(1 - \frac{2}{n}\right)\alpha'$$

$$\hat{\beta} = \left(1 - \frac{1}{(n-1)\alpha'}\right)X_{(1)}$$

Capability Metrics: Since the mode of a Pareto process is the same as its lower bound, generally there will only be an upper tolerance limit UTL. Calculate equivalent long-term capability metrics this way:

$$\text{Equivalent } P_P^\% = \text{Equivalent } P_{PK}^\% = \frac{-\Phi^{-1}\left(\left[\frac{\beta}{UTL}\right]^\alpha\right)}{3}$$

$\Phi^{-1}(p)$ is the standard normal inverse CDF, evaluated in Excel with the =NORMSINV(p) function. Table 22-1 lists selected values of equivalent $P_{PK}^\%$ for selected Pareto processes.

Probability Density Function:

$$f_{Pareto(\alpha,\beta)}(x) = \begin{cases} \dfrac{\alpha\beta^\alpha}{x^{\alpha+1}} & x \geq \beta \\ 0 & x < \beta \end{cases}$$

Table 22-1   Selected Capability Values for Pareto Random Variables

	Equivalent $P_{PK}^\%$			
UTL	Pareto (0.5, β)	Pareto (1, β)	Pareto (1.5, β)	Pareto (2, β)
100 β	0.427	0.775	1.030	1.240
1000 β	0.619	1.030	1.333	1.584
10,000 β	0.775	1.240	1.584	1.871
100,000 β	0.910	1.422	1.803	2.120
10^6 β	1.030	1.584	1.999	
10^7 β	1.139	1.733	2.179	
10^8 β	1.240	1.871		
10^9 β	1.333	1.999		
10^{12} β	1.584	2.344		
10^{15} β	1.803			
10^{18} β	1.999			

Note: In the following formulas, when β is unspecified, the default value is $\beta = 1$.

Cumulative Distribution Function:

$$F_{Pareto(\alpha,\beta)}(x) = \begin{cases} 1 - \left(\dfrac{\beta}{x}\right)^{\alpha} & x \geq \beta \\[2mm] 0 & x < \beta \end{cases}$$

Inverse Cumulative Distribution Function:

$$F^{-1}_{Pareto(\alpha,\beta)}(p) = \frac{\beta}{(1 - p)^{1/\alpha}}$$

Random number generation: To generate $Pareto(\alpha, \beta)$ random numbers in an Excel worksheet, use the formula =β/(RAND()^(1/α)).

To generate $Pareto(\alpha, \beta)$ random numbers with Excel and Crystal Ball software, use the formula =CB.Pareto(β,α).

Survival Function:

$$R_{Pareto(\alpha,\beta)}(x) = \begin{cases} \left(\dfrac{\beta}{x}\right)^{\alpha} & x \geq \beta \\[2mm] 1 & x < \beta \end{cases}$$

Hazard Function:

$$h_{Pareto(\alpha,\beta)}(x) = \frac{\alpha}{x}$$

Figure 22-2 illustrates the PDF, CDF, hazard and survival functions for selected Pareto random variables.

Mean (Expected Value): When $\alpha > 1$, $E[X] = \frac{\alpha\beta}{\alpha - 1}$

Median: $2^{1/\alpha}\beta$

Mode: β

Standard Deviation: When $\alpha > 2$, $SD[X] = \frac{\beta}{\alpha - 1}\sqrt{\frac{\alpha}{\alpha - 2}}$

Variance: When $\alpha > 2$, $V[X] = \frac{\beta^2\alpha}{(\alpha - 1)^2(\alpha - 2)}$

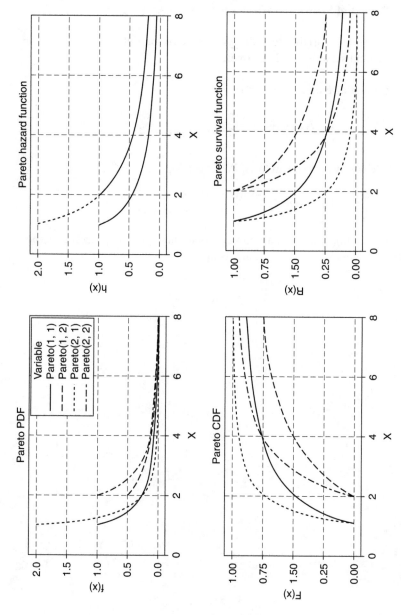

Figure 22-2 PDF, CDF, Hazard and Survival Functions for Selected Pareto Random Variables.

Coefficient of Variation: When $\alpha > 2$, $CV[X] = \sqrt{\frac{1}{\alpha(\alpha - 2)}}$

Coefficient of Skewness: When $\alpha > 3$, $Skew[X] = \frac{2(\alpha + 1)}{\alpha - 3}\sqrt{\frac{\alpha - 2}{\alpha}}$

Pareto random variables are always skewed to the right, and approach an exponential skewness of 2, as α gets large.

Coefficient of Kurtosis: When $\alpha > 4$,

$$Kurt[X] = \frac{3(\alpha - 2)(3\alpha^2 + \alpha + 2)}{\alpha(\alpha - 3)(\alpha - 4)}$$

Coefficient of Excess Kurtosis: When $\alpha > 4$,

$$ExKurt[X] = \frac{3(\alpha - 2)(3\alpha^2 + \alpha + 2)}{\alpha(\alpha - 3)(\alpha - 4)} - 3$$

Pareto random variables are always leptokurtic, and approach an exponential excess kurtosis of 6 as α gets large.

23

Poisson Distribution Family

A Poisson random variable is a discrete random variable representing the count of independent events occurring per unit of time, space, or product. Poisson random variables may represent defects per sheet of film, customers arriving per hour, or contaminant particles per liter. This type of application is so common that the Poisson distribution has been called the "per" distribution.

The Poisson family is one of the simplest distribution models, with a single parameter λ representing the expected count of events. The variance of a Poisson random variable is also λ, so the standard deviation is $\sqrt{\lambda}$.

Not all counts of events per unit are Poisson random variables. The Poisson model assumes that events occur independently of each other. Customers arriving per hour might meet this requirement, if customers make individual and independent decisions to arrive.

A Poisson random variable may have any nonnegative integer value, possibly a very large number. In most real situations, there is a practical upper limit to possible values. Most stores cannot accommodate one million customers at once, but according to the Poisson model, such large numbers are possible. If this creates a problem for simulations or predictions, a Poisson random variable may be truncated to prevent impractically large values. Section 23.1 describes the right-truncated Poisson distribution.

Section 23.2 describes a positive Poisson distribution, with zero values truncated. For more general cases of left truncation or double truncation, see Chapter 4 of Johnson *et al* (2005).

The name of the Poisson distribution honors Siméon-Denis Poisson (1781–1840), who made enormous contributions to mathematics, physics, electromechanical theory, and other fields.

Figure 23-1   Siméon-Denis Poisson.

Example 23.1

Jane runs a regional service center for Sup-R-Dup copiers. On average, her department receives 30 service requests per day, sometimes more, sometimes less. Jane is building a simulation model to help optimize labor costs. In her model, Jane chooses to model the number of service requests per day by a Poisson distribution with the rate parameter $\lambda = 30$.

Example 23.2

Fern is a Black Belt in a fine furniture factory. Rework of the finish on hardwood desks creates significant expense. The finishing shop is always a constraint in the process flow, and rework makes the problem worse. In a recent lot of 100 desks, inspectors found 20 defects requiring rework, and some desks had more than one defect. To estimate the costs of this problem, Fern uses a Poisson distribution with $\lambda = 0.2$ to represent the number of finish defects per desk. Based on this model, the probability that a desk will have zero defects is $e^{-\lambda} = e^{-0.2} = 0.82$. Therefore, the yield is 82%, and 18% of desks will have one or more finish defects.

Example 23.2 illustrates an important application of the Poisson distribution in the calculation of rolled throughput yield (RTY). For a process with many steps, RTY represents the probability that one unit will flow through the process with zero stops for defects, rework, or other activity that does not add value to a product. If one unit has many opportunities for defects, and defects are independent, then the count of defects per unit is a Poisson random variable. Using this model, if the average number of defects per unit (DPU) is known, then RTY is the probability of zero defects, that is, $RTY = e^{-\text{DPU}}$

Parameters: The Poisson distribution family has one parameter:

- λ represents the mean or expected value of the random variable. $\lambda \geq 0$. λ is called the Poisson rate parameter.

Representation: $X \sim Pois(\lambda)$

Support: $\{0, 1, \dots \}$

Relationships to Other Distributions:

- *Reproductive property.* The sum of independent Poisson random variables is also a Poisson random variable. If $X_i \sim Pois(\lambda_i)$ and the $\{X_i\}$ are mutually independent, then $\sum_i X_i \sim Pois(\sum_i \lambda_i)$.
- The converse of the above statement is also true. If a sum of mutually independent random variables is Poisson, then each of the individual random variables is also Poisson. The Poisson family may be the only one with this unusual property.
- If a Poisson random variable represents a count of events per unit of time (or along a vector in space), then the time (or distance) between events is an exponential random variable with the same rate parameter. In reliability applications, if the number of failures per unit time is $Pois(\lambda)$, then the time between failures is $Exp(\lambda)$. Therefore, the mean time (or distance) between Poisson events is $1/\lambda$.
- Poisson left-tail probabilities are related to right-tail probabilities of a chi-squared distribution. Specifically, if $X \sim Pois(\lambda)$ and $Y \sim \chi^2(v)$, then $P\left[X < \frac{v}{2}\right] = P[Y > 2\lambda]$. This fact is used to construct exact confidence intervals and tests for the Poisson rate parameter.
- The Poisson random variable may be approximated by a normal random variable with $\mu = \lambda$ and $\sigma = \sqrt{\lambda}$. As a general rule of thumb, use this approximation only when $\lambda \geq 5$.
- The Poisson random variable is a limiting form for the binomial random variable. This means that when p is small and np is large, a binomial random variable may be approximated by a Poisson random variable with $\lambda = np$. That is, when $X \sim Bin(n, p)$, then $X \xrightarrow{D} Pois(n\,p)$. As a general rule of thumb, use this approximation only when $p < 0.1$ and $np > 20$.
- The Poisson random variable is a limiting form for the hypergeometric random variable. Because of this, a hypergeometric random variable may be approximated by a Poisson random variable with $\lambda = nD/N$. As a general rule, use this approximation only when $N/n \geq 10$, $D/N < 0.1$ and $nD/N > 20$.

Normalizing Transformation: When applying regression models or hypothesis tests to Poisson data, a transformation is useful to stabilize the variation between groups, and to normalize the residuals. The simplest and most common transformation is $2\sqrt{X}$, which is approximately normally distributed with mean $2\sqrt{\lambda}$ and standard deviation 1.

To calculate Poisson tail probabilities, especially when λ is large, a crude but simple normal approximation is available. If $X \sim Pois(\lambda)$, then $P[X \leq x] \cong \Phi\left(\frac{x - \lambda}{\sqrt{\lambda}}\right)$ where $\Phi(z)$ is the standard normal CDF. For greater accuracy and flexibility to handle smaller values of λ, Peizer and Pratt (1968) recommend the approximation that,

$$P[X \leq x] \cong \Phi(z), \text{ where } z = \left(x - \lambda + \frac{2}{3} + \frac{0.02}{x + 1}\right)\sqrt{\frac{1 + T\left(\frac{x + 1/2}{\lambda}\right)}{\lambda}}$$

and
$$T(y) = \frac{1 - y^2 + 2y \ln y}{(1 - y)^2}$$

Process Control Tools: Two standard control charts, the c-chart and the u-chart, are designed for Poisson processes. When the subgroup size n is constant, the c-chart is simpler and easier to apply than the u-chart. When the subgroup size varies, the u-chart must be used, and the control limits will change depending on the size of each subgroup.

Shore's general control charts for attributes (2000a) compensate for skewness in the distribution. Applying Shore's method, the corrected control limits for a c-chart are:

$$UCL_c = \bar{c} + 3\sqrt{\bar{c}} + 0.824$$

$$LCL_c = \bar{c} - 3\sqrt{\bar{c}} + 1.824$$

Similarly, the corrected control limits for a u-chart are:

$$UCL_u = \bar{u} + 3\sqrt{\frac{\bar{u}}{n_i}} + \frac{0.824}{n_i}$$

$$LCL_u = \bar{u} - 3\sqrt{\frac{\bar{u}}{n_i}} + \frac{1.824}{n_i}$$

For processes where defects occur very rarely, consider counting the number of conforming units between each defect. This count will have an approximate geometric distribution, X_0 version. To create a control chart for geometric counts, perform a double square root transformation $Y = X^{1/4}$ and plot the transformed data on an individual X chart, with control limits determined by the moving ranges. Only a single control chart is required. See the process control section of Chapter 15 for more information.

Estimating Parameter Values: The sample mean $\overline{X}$ is an unbiased, maximum likelihood estimator for λ. $\hat{\lambda} = \overline{X} = \frac{1}{n}\sum_{i=1}^{n}X_i$. An exact $(1 - \alpha)100\%$ confidence interval for λ may be calculated from an observed Poisson count $X = n\,\overline{X}$ by using the relationship with the chi-squared distribution. The lower limit of the confidence interval is $L_\lambda = 0.5\chi^2_{1-\alpha/2,2X}$, and the upper limit is $U_\lambda = 0.5\chi^2_{\alpha/2,2(X+1)}$, where $\chi^2_{p,\nu}$ is the $(1 - p)$ quantile of the chi-squared distribution with ν degrees of freedom. In Excel, $\chi^2_{p,\nu}$ is calculated by the =CHIINV(p,ν) function.

It is common to use a normal approximation to the Poisson for confidence intervals and hypothesis tests, but this is unnecessary when an exact method is available.

Capability Metrics: When a Poisson process with rate λ has tolerance limits LTL and UTL, calculate equivalent long-term capability metrics this way:

$$\text{Equivalent } P^\%_{PL} = \frac{-\Phi^{-1}(F_X(LTL - 1))}{3}$$

$$\text{Equivalent } P^\%_{PU} = \frac{-\Phi^{-1}(R_X(UTL))}{3}$$

$$\text{Equivalent } P^\%_P = (\text{Equivalent } P^\%_{PL} + \text{Equivalent } P^\%_{PU})/2$$

$$\text{Equivalent } P^\%_{PK} = Min\{\text{Equivalent } P^\%_{PL}, \text{Equivalent } P^\%_{PU}\}$$

If LTL is not an integer, replace $(LTL - 1)$ with LTL. In these formulas, $F_X(x)$ is the CDF of a Poisson random variable with rate parameter λ, evaluated in Excel with the =POISSON$(x,\lambda,$TRUE$)$ function, and $R_X(x) = 1 - F_X(x)$ is the corresponding survival function. Also, $\Phi^{-1}(p)$ is the standard normal inverse CDF, evaluated in Excel with the =NORMSINV(p) function. Table 23-1 lists selected values of equivalent $P^\%_{PL}$ and equivalent $P^\%_{PU}$ for Poisson processes.

In the UTL columns of Table 23-1, the values increase by one standard deviation in each row. A normal process achieves $P_{PU} = 2.00$ when UTL is six standard deviations above the mean. For a Poisson process to achieve equivalent $P^\%_{PU} = 2.00$, UTL must be ten standard deviations above the

Table 23-1 Selected Capability Values for Poisson Random Variables

Rate $\lambda = 1$		Rate $\lambda = 9$				Rate $\lambda = 25$			
UTL	Equiv. $P_{PU}^{\%}$	LTL	Equiv. $P_{PL}^{\%}$	UTL	Equiv. $P_{PU}^{\%}$	LTL	Equiv. $P_{PL}^{\%}$	UTL	Equiv. $P_{PU}^{\%}$
2	0.468	1	1.222	12	0.385	1	2.219	30	0.365
3	0.692	2	1.009	15	0.671	2	2.054	35	0.668
4	0.894	3	0.833	18	0.939	3	1.914	40	0.958
5	1.081	4	0.676	21	1.192	4	1.788	45	1.234
6	1.255	5	0.533	24	1.432	6	1.562	50	1.501
7	1.420	6	0.399	27	1.663	8	1.359	55	1.758
8	1.577	7	0.273	30	1.884	10	1.171	60	2.008
9	1.726	8	0.152	33	2.098	15	0.748	65	2.250
10	1.870								
11	2.009								

mean when $\lambda = 1$, eight standard deviations above the mean when $\lambda = 9$, or seven standard deviations above the mean when $\lambda = 25$. A Poisson random variable has a heavier right tail than a normal distribution, so applying normal capability metrics to Poisson data would underestimate the rate of defects.

Probability Mass Function (PMF): $f_{Pois(\lambda)}(x) = \dfrac{e^{-\lambda}\lambda^x}{x!}$, for $x = 0, 1, 2, \ldots$

To calculate $f_{Pois(\lambda)}(x)$ in Excel, use =POISSON(x,λ,FALSE).

Figure 23-2 illustrates the PMF for Poisson random variables with mean values of 0.5, 1, 2, and 5. To create tables or for use in programs, it is useful to know that

$$f_{Pois(\lambda)}(0) = e^{-\lambda} \quad \text{and} \quad f_{Pois(\lambda)}(x) = \frac{\lambda}{x} f_X(x - 1)$$

Cumulative Distribution Function (CDF): $F_{Pois(\lambda)}(x) = \sum_{i=0}^{x}\dfrac{e^{-\lambda}\lambda^i}{i!}$

To calculate $F_{Pois(\lambda)}(x)$ in Excel, use =POISSON(x,λ,TRUE).

The POISSON algorithm used by Excel changed significantly in Excel 2003. In Excel 97 through XP, POISSON returned correct cumulative probabilities in the extreme left tail of the distribution, but sometimes returned

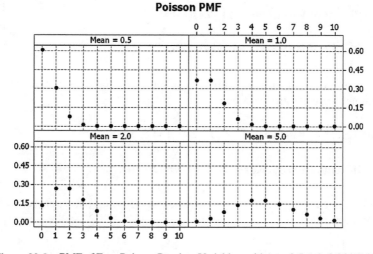

Figure 23-2   PMF of Four Poisson Random Variables, with $\lambda = 0.5, 1.0, 2.0,$ and 5.0.

error values in the middle section. The algorithm used in Excel 2003 returns correct answers in the middle of the distribution, but in the extreme left tail, when $F_{Pois(\lambda)}(x) < 10^{-14}$, the POISSON function returns zero. See Knüsel (2005) for more information.

Inverse Cumulative Distribution Function: $F_{Pois(\lambda)}^{-1}(prob) = x$ if $prob > \sum_{i=0}^{x-1} \frac{e^{-\lambda}\lambda^i}{i!}$ and $prob \le \sum_{i=0}^{x} \frac{e^{-\lambda}\lambda^i}{i!}$.

There is no simple way to calculate the inverse CDF in Excel, although MINITAB or STATGRAPHICS can perform this calculation.

Random Number Generation: To generate Poisson random numbers with Excel and Crystal Ball, use =CB.Poisson(λ).

Mean (Expected Value): $E[Pois(\lambda)] = \lambda$

Mode: The mode of a Poisson random variable is $\lfloor \lambda \rfloor$, where $\lfloor\ \rfloor$ denotes rounding down. If λ is an integer, the random variable has two modes at $\lambda - 1$ and λ.

Standard Deviation: $SD[Pois(\lambda)] = \sqrt{\lambda}$

Variance: $V[Pois(\lambda)] = \lambda$

Coefficient of Variation: $CV[Pois(\lambda)] = \frac{1}{\sqrt{\lambda}}$

Coefficient of Skewness: $Skew[Pois(\lambda)] = \frac{1}{\sqrt{\lambda}}$

A Poisson random variable is always skewed to the right, but approaches a symmetric distribution as λ gets large.

Coefficient of Kurtosis: $Kurt[Pois(\lambda)] = \frac{1}{\lambda} + 3$

Coefficient of Excess Kurtosis: $ExKurt[Pois(\lambda)] = \frac{1}{\lambda}$

A Poisson random variable is always leptokurtic, but approaches a mesokurtic distribution as λ gets large.

Moment Generating Function: $M_{Pois(\lambda)}(t) = \exp(\lambda[e^t - 1])$

23.1 Right-Truncated Poisson Distribution Family

In many applications, the Poisson random variable must be truncated to represent natural limits in the implementation of a process. This section describes a Poisson random variable truncated at a positive integer B so that $P[X > B] = 0$.

Estimating Parameter Values: Assuming that B is known, Moore (1954) published this unbiased estimator for λ: $\hat{\lambda} = \frac{1}{m}\sum_{i=1}^{N} X_i$, where m is the count of observations less than $B - 1$.

Example 23.3

Shawn is a black belt working to improve customer satisfaction metrics for a call center. The average call ends in 5 minutes, freeing the line for another call, but the call center has only 20 lines. Shawn collects data for the number of calls arriving in 122 five-minute periods and makes a histogram as shown in Figure 23-3. How can Shawn model this distribution in a call center simulation?

If the call center had an unlimited number of lines, a Poisson model would be reasonable, but in this case, some callers get busy signals and hang up. Shawn would like to estimate how many callers get busy signals.

The average number of calls in five minutes is 14.7. However, this is a truncated Poisson distribution, so 14.7 is not a good estimate for λ. Using Moore's formula, $\hat{\lambda} = \frac{1}{m}\sum_{i=1}^{N} X_i$, where m is the count of observations less than $B - 1$. In Shawn's process, $B = 20$. The sum of all X_i is 1793, and $m = 109$,

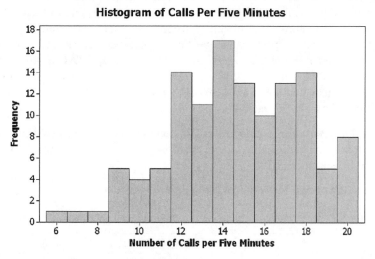

Figure 23-3 Histogram of Observed Counts of Calls Received During a Five-Minute Period.

excluding 13 values of 19 or 20. Therefore, $\hat{\lambda} = \frac{1793}{109} = 16.45$. This is an estimate of the rate parameter, which would be the mean calls per five minutes, if the distribution were not truncated.

In Excel, Shawn calculates $P[X > 20]$ which is the probability that at least one caller is bumped in a five minute period. Shawn enters the formula =1-POISSON(20,16.45,TRUE) which evaluates to 0.158. In nearly 16% of the 5-minute periods, at least one caller is bumped.

To model the number of calls per five minute period in Crystal Ball, Shawn would define a Poisson assumption with rate parameter = 16.45 and upper truncation limit = 20.

Note: In the formulas below, $F_{Pois(\lambda)}(x)$ refers to the CDF of an untruncated Poisson random variable with rate parameter λ, specifically $F_{Pois(\lambda)}(x) = \sum_{i=0}^{x} e^{-\lambda} \lambda^i / i!$. This can be calculated in Excel with the =POISSON(x,λ,TRUE) function.

Probability Mass Function (PMF): $f_X(x) = \dfrac{e^{-\lambda} \lambda^x / x!}{F_{Pois(\lambda)}(B)}$, for $x = 0, 1, 2, \ldots , B$

To calculate $f_X(x)$ in Excel, use =POISSON(x,λ,FALSE)/POISSON (B,λ,TRUE).

Cumulative Distribution Function (CDF): $F_X(x) = \dfrac{F_{Pois(\lambda)}(x)}{F_{Pois(\lambda)}(B)}$

To calculate $F_X(x)$ in Excel, use =POISSON(x,λ,TRUE)/POISSON (B,λ,TRUE).

Inverse Cumulative Distribution Function: There is no simple way to calculate the inverse CDF of any Poisson random variable in Excel. To calculate a quantile at probability *prob* for a right-truncated Poisson random variable, first calculate the cumulative probability for an untruncated Poisson $F_{Pois(\lambda)}(B)$ using MINITAB or STATGRAPHICS functions. Then multiply this by *prob* and use the inverse probability function to calculate $F_{Pois(\lambda)}^{-1}(prob \times F_{Pois(\lambda)}(B))$. This is the *prob*-quantile of a right-truncated Poisson random variable.

Random Number Generation: To generate right-truncated Poisson random numbers with Excel and Crystal Ball software, use =CB.Poisson(λ,0,B).

Mean (Expected Value): $E[X] = \lambda \dfrac{F_{Pois(\lambda)}(B-1)}{F_{Pois(\lambda)}(B)}$

Standard Deviation:

$$SD[X] = \sqrt{\lambda \frac{F_{Pois(\lambda)}(B-1)}{F_{Pois(\lambda)}(B)} + \lambda^2 \left[\frac{F_{Pois(\lambda)}(B-2)}{F_{Pois(\lambda)}(B)} - \left(\frac{F_{Pois(\lambda)}(B-1)}{F_{Pois(\lambda)}(B)} \right)^2 \right]}$$

Variance:

$$V[X] = \lambda \frac{F_{Pois(\lambda)}(B-1)}{F_{Pois(\lambda)}(B)} + \lambda^2 \left[\frac{F_{Pois(\lambda)}(B-2)}{F_{Pois(\lambda)}(B)} - \left(\frac{F_{Pois(\lambda)}(B-1)}{F_{Pois(\lambda)}(B)} \right)^2 \right]$$

23.2   Positive (Zero-Truncated) Poisson Distribution Family

Sometimes, zero values are missing from a set of Poisson data. This may arise when the zero values are undetectable, or when the count of such cases are not recorded. To reflect this real situation in a simulation model, the value zero is truncated from the Poisson random variable, creating a positive Poisson random variable.

Estimating Parameter Values: When zero cases are not observed, a maximum likelihood estimator for λ is the solution to the equation $\overline{X} = \dfrac{\lambda}{1 - e^{-\lambda}}$. The Excel solver or any numerical solution tool can solve this equation. Moore (1954) published this simple estimator for λ, which is simply the average of all the observations greater than 1: $\hat{\lambda} = \frac{1}{n} \sum_{i=1}^{N} (X_i | X_i > 1)$

Example 23.4

Vic works for Varistor Village, a leading electronics retailer. Vic's job is to set prices for extended warranties on products like plasma monitors. A major cause of plasma monitor returns is bad or missing cells. Even on a monitor with over one million cells, one bad cell leaves an annoying black spot, and customers tend to complain about that. To review warranty pricing, Vic has collected data on the number of bad cells in 100 returned monitors. Of these 100 monitors, 70 had one bad cell, 25 had two bad cells, and 5 had three bad cells. Because the distribution network through Varistor Villages worldwide is enormous, Vic does not have accurate numbers for the number of monitors sold during the same time as the 100 returned monitors. He also does not know if these were the only ones returned. Therefore, the number of monitors with zero bad cells is unknown.

Vic needs to estimate the rate of returns from this data. Vic's first challenge is to select a distribution model. A plasma monitor has a finite number of cells, but relatively few cells fail. A binomial distribution would be appropriate, but since the parameter n would be in the millions, the binomial model may not be practical. Instead, Vic decides to use a Poisson model as an approximation to the binomial. Either way, Vic must truncate the zero value from the distribution.

Vic needs to estimate the rate parameter λ. Applying Moore's formula,

$$\hat{\lambda} = \frac{(25 \times 2) + (5 \times 3)}{100} = 0.65$$

Vic could also calculate the sample mean, which is 1.35, and use the Excel solver to solve the equation $\overline{X} = \frac{\lambda}{1-e^{-\lambda}}$. This gives $\hat{\lambda} = 0.634$, almost the same answer.

Using $\hat{\lambda} = 0.65$, Vic calculates that the probability of a monitor having zero defects is $\frac{e^{-0.65}0.65^0}{0!} = e^{-0.65} = 0.522$, and that 47.8% of the monitors will have one or more bad cells.

Note: In the formulas below, $f_{Pois(\lambda)}(x)$ and $F_{Pois(\lambda)}(x)$ refer to the PMF and CDF, respectively, of the untruncated Poisson random variable with rate parameter λ. To calculate $f_{Pois(\lambda)}(x)$ in Excel, use =POISSON(x,λ,FALSE). To calculate $F_{Pois(\lambda)}(x)$ in Excel, use =POISSON(x,λ,TRUE)

Probability Mass Function (PMF): $f_X(x) = \frac{f_{Pois(\lambda)}(x)}{1-e^{-\lambda}}$, for $x = 1, 2, \ldots$
To calculate $f_X(x)$ in Excel, use =POISSON(x,λ,FALSE)/(1-EXP(-λ)).

Cumulative Distribution Function (CDF): $F_X(x) = \frac{F_{Pois(\lambda)}(x) - e^{-\lambda}}{1-e^{-\lambda}}$

To calculate $F_X(x)$ in Excel, use =(POISSON(x,λ,TRUE)-EXP(-λ))/(1-EXP(-λ)).

Inverse Cumulative Distribution Function: As with other versions of Poisson random variables, the inverse CDF of the positive Poisson random variable is difficult to evaluate in Excel, it can be done with MINITAB or STATGRAPHICS software. To evaluate $F_X^{-1}(prob)$, first calculate $u = prob \times (1 - e^{-\lambda}) + e^{-\lambda}$, and then calculate $F_{Pois(\lambda)}^{-1}(u)$.

Random Number Generation: To generate zero-truncated or positive Poisson random numbers with Excel and Crystal Ball software, use =CB.Poisson(λ,1).

Mean (Expected Value): $E[X] = \dfrac{\lambda}{1-e^{-\lambda}}$

Standard Deviation: $SD[X] = \sqrt{\dfrac{\lambda}{1-e^{-\lambda}} - \dfrac{\lambda^2 e^{-\lambda}}{(1-e^{-\lambda})^2}}$

Variance: $V[X] = \dfrac{\lambda}{1-e^{-\lambda}} - \dfrac{\lambda^2 e^{-\lambda}}{(1-e^{-\lambda})^2}$

Rayleigh Distribution Family

A Rayleigh random variable is a continuous random variable representing the square root of the sum of two squared independent normal random variables, each with mean 0 and the same standard deviation.

Since the distance between the origin and any point (X, Y) on a plane is $\sqrt{X^2 + Y^2}$, the Rayleigh distribution can model the distance between a point and its average position on a plane. Since the Rayleigh distribution has a lower bound of zero and is skewed to the right, it is also a popular model for waiting times or failure times in reliability applications.

The most common application of the Rayleigh distribution in Six Sigma applications is to model the time when defects are found in a software development project. By combining a Rayleigh model with an estimate of the code complexity, project managers can predict the time required to find a given percentage of defects and the number of defects remaining at release time. For more information on this application, see Hallowell (2006) or Kan (2003).

The name of the Rayleigh distribution honors John William Strutt, the third Baron Rayleigh (1842–1919). In addition to deriving the Rayleigh distribution to solve problems in acoustic theory, Lord Rayleigh earned the 1904 Nobel Prize for Physics for the discovery of the element argon.

Example 24.1

Ray is designing a motor housing. The upper housing has several holes that must line up with studs on the lower housing for the product to be assembled. If any hole is more than 1.0 mm away from its specified position, the motor might not be assembled. The true position of each hole is determined by its X and Y coordinates. The boring machine that makes these holes controls X and Y so that each has a mean equal to the target value and standard deviation $\sigma = 0.3$ mm. Based on a scatter plot and histograms of measured hole locations, Ray

Figure 24-1   Lord Rayleigh.

concludes that X and Y are independent and normally distributed. Ray needs to know the probability that any hole will be off target by more than 1.0 mm.

For each hole, suppose the target position is (0, 0) and the true position is (X, Y). From the above information, $X, Y \overset{iid}{\sim} N(0, \sigma = 0.3 \text{ mm})$. Therefore, the distance between each hole's true position and the target position is $\sqrt{X^2 + Y^2}$. Using the information in this chapter, Ray concludes that $\sqrt{X^2 + Y^2}$ is Rayleigh distributed with scale parameter $\beta = \sqrt{2} \times 0.3 \text{ mm} \cong 0.4243 \text{ mm}$. Therefore, the probability that $\sqrt{X^2 + Y^2}$ will exceed 1.0 mm is

$$P[Rayleigh(0.4243) > 1.0] = R_{Rayleigh(0.4243)}(1.0)$$

$$= \exp\left[-\left(\frac{1}{0.4243}\right)^2\right] = 0.00387$$

Parameters: The Rayleigh family of distributions has one scale parameter, plus one optional threshold parameter:

- β is a scale parameter, which must be a positive real number
- τ is an optional threshold parameter, which may be any real number. If τ is missing, assume that $\tau = 0$.

Alternative parameters are in use for the Rayleigh distribution family. In Chapter 18 of Johnson *et al* (1994), the Rayleigh scale parameter is $\sigma = \beta/\sqrt{2}$. Using σ as the scale parameter is consistent with the Rayleigh being a special case of the chi family. The use of β here is consistent with the Rayleigh being a special case of the Weibull family, and also consistent with Rayleigh functions in STATGRAPHICS software. To interpret Rayleigh applications correctly, be sure to check the definitions of the parameters.

Representation: $X \sim Rayleigh(\beta)$ or $X \sim Rayleigh(\beta, \tau)$

Support: $[\tau, \infty)$

Relationships to Other Distributions:

- The Rayleigh distribution family is a special case of the Weibull distribution family with shape parameter $\alpha = 2$. $Rayleigh(\beta) \sim Weibull(2, \beta)$ and $Rayleigh(\beta, \tau) \sim Weibull(2, \beta, \tau)$
- A chi, or χ, random variable with 2 degrees of freedom is also a Rayleigh random variable with $\tau = 0$. As used in this book, the scale parameters for the two families are related by $\beta = \sigma\sqrt{2}$. Therefore,

$$\chi(2, \sigma) \sim Rayleigh(\sigma\sqrt{2}) \text{ and } Rayleigh(\beta) \sim \chi(2, \beta/\sqrt{2})$$

- The square of a Rayleigh random variable with $\tau = 0$ is an exponential random variable with mean $\mu = \beta^2$. If $X \sim Rayleigh(\beta)$, then $X^2 \sim Exp(\mu = \beta^2)$.
- The square root of the sum of two squared normal random variables with the mean 0 and the same standard deviation is a Rayleigh random variable. If $X_1, X_2 \overset{iid}{\sim} N(0, \sigma^2)$, then $\sqrt{X_1^2 + X_2^2} \sim Rayleigh(\sigma\sqrt{2})$.

Normalizing Transformations: When the threshold parameter $\tau = 0$, the transformation $Y = X^{0.5308}$ transforms a Rayleigh distribution into an approximately normal distribution.

When the threshold parameter $\tau \neq 0$, the transformation $Y = (X - \tau)^{0.5308}$ transforms a Rayleigh distribution into an approximately normal distribution. In this case, any error in the estimation of τ will damage the effectiveness of this transformation.

Process Control Tools: There are two control chart options for a process producing Rayleigh data. One option is to plot the Rayleigh data on a Rayleigh individual X chart. Here are the formulas:

$$UCL = \tau + \beta\sqrt{-\ln(\alpha/2)}$$
$$CL = \tau + \beta\sqrt{-\ln 0.5}$$
$$LCL = \tau + \beta\sqrt{-\ln(1 - \alpha/2)}$$

In these formulas, α represents the false alarm rate for the control chart, which is usually 0.0027 to match the performance of standard Shewhart charts.

The second option is to apply a normalizing transformation $Y = (X - \tau)^{0.5308}$ to the Rayleigh data and plot the transformed data on any standard Shewhart control chart designed for normal distributions.

Estimating Parameter Values: The formulas in section applies only to a complete, uncensored dataset, when $\tau = 0$, or when τ is known and subtracted from the observed data. When both τ and β are unknown, or when the data is censored, it is best to use algorithms in a major statistical package such as STATGRAPHICS or MINITAB to estimate them.

A maximum likelihood estimator for the scale parameter β is $\hat{\beta} = \sqrt{\frac{1}{n}\sum_{i=1}^{n} X_i^2}$. This is the root-mean-square (RMS) combination of the measured values. A $100(1-\alpha)\%$ confidence interval for β is $\left(\hat{\beta}\sqrt{n/\chi^2_{2n,1-\alpha/2}}, \hat{\beta}\sqrt{n/\chi^2_{2n,\alpha/2}}\right)$, where $\chi^2_{p,\nu}$ is the $(1-p)$ quantile of the chi-squared distribution with ν degrees of freedom. In Excel, $\chi^2_{p,\nu}$ is calculated by the =CHIINV(p,ν) function.

Capability Metrics: For a process producing a Rayleigh distribution, calculate equivalent capability metrics using the Rayleigh CDF. Here are the formulas:

$$\text{Equivalent } P^{\%}_{PU} = \frac{-\Phi^{-1}\left(\exp\left[-\left(\frac{UTL}{\beta}\right)^2\right]\right)}{3}$$

$$\text{Equivalent } P^{\%}_{PL} = \frac{-\Phi^{-1}\left(1 - \exp\left[-\left(\frac{LTL}{\beta}\right)^2\right]\right)}{3}$$

$$\text{Equivalent } P^{\%}_{P} = \frac{\text{Equivalent } P^{\%}_{PU} + \text{Equivalent } P^{\%}_{PL}}{2}$$

$$\text{Equivalent } P^{\%}_{PK} = Min\{\text{Equivalent } P^{\%}_{PU}, \text{Equivalent } P^{\%}_{PL}\}$$

Table 24-1   Selected Capability Values for Rayleigh Random Variables

LTL	Equivalent $P^{\%}_{PL}$	*UTL*	Equivalent $P^{\%}_{PU}$
0.5β	0.256	1.33β	0.317
0.4β	0.349	1.5β	0.417
0.3β	0.455	1.66β	0.508
0.25β	0.517	2β	0.697
0.2β	0.587	2.5β	0.963
0.1β	0.776	3β	1.222
0.01β	1.240	3.5β	1.476
0.001β	1.584	4β	1.726
0.0001β	1.871	4.5β	1.973
0.00001β	2.120	5β	2.219

$\Phi^{-1}(p)$ is the standard normal inverse CDF, evaluated in Excel with the =NORMSINV(p) function. Table 24-1 lists selected values of equivalent $P^{\%}_{PL}$ and $P^{\%}_{PU}$ for Rayleigh random variables.

Probability Density Function (PDF):

$$f_{Rayleigh(\beta,\tau)}(x) = \begin{cases} \dfrac{2(x-\tau)}{\beta^2} \exp\left[-\left(\dfrac{x-\tau}{\beta}\right)^2\right] & x \geq \tau \\ 0 & x < \tau \end{cases}$$

To calculate the Rayleigh PDF in an Excel worksheet, use the formula =WEIBULL(x-τ,2,β,FALSE).

Cumulative Distribution Function (CDF):

$$F_{Rayleigh(\beta,\tau)}(x) = \begin{cases} 1 - \exp\left[-\left(\dfrac{x-\tau}{\beta}\right)^2\right] & x \geq \tau \\ 0 & x < \tau \end{cases}$$

To calculate the Rayleigh CDF in an Excel worksheet, use the formula =WEIBULL(x-τ,2,β,TRUE).

Inverse Cumulative Distribution Function:

$$F^{-1}_{Rayleigh(\beta,\tau)}(prob) = \tau + \beta \sqrt{-\ln(1 - prob)}$$

To calculate Rayleigh quantiles using the inverse CDF in an Excel worksheet, use the formula =τ+β*SQRT(-LN(1-*prob*))

Random number generation: To generate $Rayleigh(\beta, \tau)$ random numbers in an Excel worksheet, use the formula =τ+β*SQRT(-LN(RAND())).

To generate $Rayleigh(\beta, \tau)$ random numbers with Crystal Ball and Excel software, use the formula =CB.Weibull(τ,β,2).

Survival Function:

$$R_{Rayleigh(\beta,\tau)}(x) = \begin{cases} \exp\left[-\left(\dfrac{x - \tau}{\beta}\right)^2\right] & x \geq \tau \\ \\ 1 & x < \tau \end{cases}$$

Hazard Function:

$$h_{Rayleigh(\beta,\tau)}(x) = \frac{2(x - \tau)}{\beta^2}$$

The Rayleigh hazard function is always increasing.

Figure 24-2 illustrates the PDF, CDF, hazard, and survival functions for Rayleigh random variables with τ = 0 and β = 0.5, 1.0, and 2.0.

Mean (Expected Value):

$$E[Rayleigh(\beta, \tau)] = \tau + \beta \frac{\sqrt{\pi}}{2} \cong \tau + 0.8862\beta$$

Median: $\tau + \beta \sqrt{\ln 2}$

Mode: $\tau + \dfrac{\beta}{\sqrt{2}}$

Standard Deviation: $SD[Rayleigh(\beta, \tau)] = \beta \dfrac{\sqrt{4 - \pi}}{2} \cong 0.4633\beta$

Variance: $V[Rayleigh(\beta, \tau)] = \beta^2\left(\dfrac{4 - \pi}{4}\right) \cong 0.2146\beta^2$

Coefficient of Variation: $CV[Rayleigh(\beta, \tau)] = \dfrac{\beta\sqrt{4 - \pi}}{2\tau + \beta\sqrt{\pi}}$

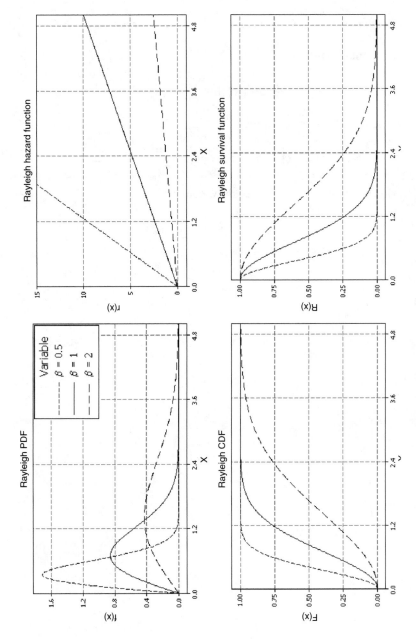

Figure 24-2 PDF, CDF, Hazard, and Survival Functions for Rayleigh Random Variables with Scale Parameters of 0.5, 1.0 and 2.0.

Coefficient of Skewness: $Skew[Rayleigh(\beta,\tau)] = \frac{2(\pi - 3)\sqrt{\pi}}{(4 - \pi)^{3/2}} \cong 0.6311$

Rayleigh random variables are always skewed to the right.

Coefficient of Kurtosis:

$$Kurt[Rayleigh(\beta, \tau)] = \frac{32 - 3\pi^2}{(4 - \pi)^2} \cong 3.2451$$

Coefficient of Excess Kurtosis:

$$ExKurt[Rayleigh(\beta, \tau)] = \frac{32 - 3\pi^2}{(4 - \pi)^2} - 3 \cong 0.2451$$

Rayleigh random variables are slightly leptokurtic.

25

Student's t Distribution Family

A Student's t random variable is a continuous random variable representing the distribution of $\frac{\sqrt{n}(\overline{X} - \mu)}{s}$, when $\overline{X}$ is the sample mean and s is the sample standard deviation of a random sample from a stable normal distribution, $N(\mu, \sigma^2)$. Because of this fact, the t distribution is used in many statistical tests designed to detect changes in the population mean, or differences between two population means. Sleeper (2006) describes many t tests with Six Sigma applications.

The distribution of a Student's t random variable is symmetric, with heavier tails than a normal distribution. The heaviness of the tails, measured by the excess kurtosis, ranges from infinite (more formally, undefined), approaching normal excess kurtosis of 0, as the degrees of freedom parameter gets large. For modelers, t random variables are useful models for processes producing a symmetric distribution with heavy tails. These distributions arise naturally when the short-term distribution is normal, with a mean that varies over time. Other distribution families for this situation, Laplace and logistic, have fixed values for excess kurtosis, 3 and 1.2, respectively, but the t family spans the full range of leptokurtic distributions.

Student was the curious pen name of William Sealy Gosset (1876–1937), who derived an early form of the t distribution (Student, 1908). As a chemist and statistician, Gosset worked for the brewery of Arthur Guinness & Son in Dublin. After a publication of Guiness trade secrets by another employee, Guiness forbade the publication of any papers by its employees. To avoid trouble at work, Gosset published under the pseudonym Student.

His choice of pen name and the lower case t are not the only evidence of Gosset's unusual modesty. In his time, few people appreciated the importance of Gosset's work as much as Sir Ronald A. Fisher, who developed the design of experiments and other tools with revolutionary

Figure 25-1   William Sealy Gosset, Who Wrote Under the Pseudonym "Student."

impact. In fact, it was Fisher (1925) who modified Gosset's work into the form of Student's t distribution we know today. Gossett wrote to Fisher, saying, "I am sending you a copy of Student's Tables, as you are the only man that's ever likely to use them!" (Pearson, 1990).

Section 25.1 describes the noncentral t distribution family, which is used to calculate power and sample sizes for statistical tests involving t statistics.

Parameters: The Student's t distribution family has only one parameter:

- ν is the degrees of freedom. When the t random variable represents the ratio $\frac{\sqrt{n}(\bar{X} - \mu)}{s}$, $\nu = n - 1$. Ordinarily, ν is restricted to positive integer values. However, the PDF for the t distribution is a valid PDF for any positive real value ν. Crystal Ball software allows noninteger ν values and adjusts the distribution accordingly. Excel TINV and TDIST functions allow noninteger ν values, but round ν down to the next lower integer. Because of this behavior, Excel TINV and TDIST functions will not return correct values for noninteger values of ν.

Crystal Ball software implements the Student's t distribution family with two additional parameters, a midpoint parameter m and a scale parameter s. m can be any number, and s can be any positive number. The use of non-integer ν values with the addition of midpoint and scale parameters gives modelers more flexibility to adapt the t distribution to the widest variety of leptokurtic distributions. To adjust the formulas in this chapter to use the additional parameters, substitute $\frac{x - m}{s}$ for x.

Representation: $X \sim t(\nu)$

Support: $(-\infty, \infty)$

Relationships to Other Distributions:

- After translation and scaling, the sample mean divided by the sample standard deviation of a random sample of size n from a normal distribution has a t distribution with $n - 1$ degrees of freedom. When $X_i \overset{iid}{\sim} N(\mu, \sigma^2)$, the sample mean $\overline{X} = \frac{1}{n}\sum_{i=1}^{n} X_i$, and the sample standard deviation $s = \sqrt{\frac{1}{n-1}\sum_{i=1}^{n}(X_i - \overline{X})^2}$, then $\frac{\sqrt{n}(\overline{X} - \mu)}{s} \sim t(n - 1)$.
- As the degrees of freedom parameter increases, the t distribution approaches the standard normal distribution. $t(\nu) \xrightarrow[\nu \to \infty]{D} N(0, 1)$.
- After scaling, the ratio of a normal random variable to a chi random variable is a t random variable. If $Z \sim N(0, 1)$, $X \sim \chi(\nu)$, and Z and X are independent, then $\frac{Z\sqrt{\nu}}{X} \sim t(\nu)$. Similarly, if $Y \sim \chi^2(\nu)$ and Z and Y are independent, then $\frac{Z\sqrt{\nu}}{\sqrt{Y}} \sim t(\nu)$.
- An F random variable with one numerator degree of freedom is also a t random variable. $F(1, \nu) \sim t(\nu)$.

Normalizing Transformation: Because of the importance of the t distribution and relative complexity of its mathematics, normalizing transformations have been researched extensively. Bailey (1980) derived one of the best functions for normalizing $X \sim t(\nu)$:

$$Y = Sgn(X)\left[\frac{8\nu + 1}{8\nu + 9}\right]\sqrt{\left(\nu + \frac{19}{12}\right)\ln\left(1 + \frac{X^2}{\nu + \frac{1}{12}}\right)}$$

where $Sgn(X)$ is either $+1$ or -1 depending on the sign of X.

Process Control Tools: A process producing data with a Student's t distribution may be unstable. If the process has a short-term normal distribution with a wandering mean, the long-term distribution might look like a t distribution. Therefore, before instituting a control chart for a process with a t distribution, Six Sigma practitioners should investigate whether the process can be further stabilized. If it is not practical or economical to remove special causes of variation and normalize the distribution, then a special control chart be implemented for the t distribution.

After estimating the degrees of freedom parameter v, transform the observed data into a normal distribution using the transformation described above. Then, plot the transformed data on any Shewhart control chart designed for normal data.

Estimating Parameter Values: When v is the only unknown parameter, it may be estimated from the sample excess kurtosis $\hat{K}$ by this formula: $\hat{v} = \frac{6}{\hat{K}} + 4$. Since the sample kurtosis has an extremely wide sampling distribution, this method is not precise unless sample sizes are extremely large.

It is better to use maximum likelihood estimation. Of the statistical programs mentioned in this book, few provide this functionality. STATGRAPHICS distribution fitting functions will fit Student's t distributions when v is the only parameter. Crystal Ball functions will fit Student's t distributions to data using three unknown parameters, v, m and s.

Capability Metrics: For a process producing a Student's t distribution, calculate equivalent capability metrics using right-tail t probabilities, also known as the survival function $R_{t(v)}(x)$. Here are the formulas:

$$\text{Equivalent } P_{PU}^{\%} = \frac{-\Phi^{-1}(R_{t(v)}(UTL))}{3}$$

$$\text{Equivalent } P_{PL}^{\%} = \frac{\Phi^{-1}(R_{t(v)}(-LTL))}{3}$$

$$\text{Equivalent } P_{P}^{\%} = \frac{\text{Equivalent } P_{PU}^{\%} + \text{Equivalent } P_{PL}^{\%}}{2}$$

$$\text{Equivalent } P_{PK}^{\%} = Min\{\text{Equivalent } P_{PU}^{\%}, \text{Equivalent } P_{PL}^{\%}\}$$

In an Excel worksheet, calculate $R_{t(v)}(x)$ with the formula =TDIST($x,v,1$). $\Phi^{-1}(p)$ is the standard normal inverse CDF, evaluated in Excel with the =NORMSINV(p) function. Table 25-1 lists selected values of equivalent $P_{PL}^{\%}$ and $P_{PU}^{\%}$ for processes with a t distribution. The rightmost column in Table 25-1 shows capability metrics for a normal process for comparison.

Note: In the formulas that follow, $\Gamma(\alpha)$ is the gamma function, which can be calculated in an Excel spreadsheet by the formula =EXP(GAMMALN(α)).

Probability Density Function:

$$f_{t(v)}(x) = \frac{\Gamma\left(\frac{v+1}{2}\right)}{\sqrt{v\pi}\,\Gamma\left(\frac{v}{2}\right)}\left(1 + \frac{x^2}{v}\right)^{-\frac{(v+1)}{2}}$$

Table 25-1   Equivalent Capabiltiy Metrics for Student's t Processes

$-LTL$ or UTL	Equivalent $P_{PL}^\%$ or $P_{PU}^\%$					
	$t(1)$	$t(2)$	$t(5)$	$t(10)$	$t(30)$	$N(0, 1)$
1	0.225	0.267	0.303	0.317	0.328	0.333
2	0.349	0.443	0.545	0.597	0.641	0.667
3	0.423	0.556	0.723	0.825	0.928	1.000
4	0.473	0.634	0.855	1.007	1.184	1.333
5	0.510	0.693	0.957	1.154	1.410	1.667
6	0.540	0.739	1.038	1.274	1.609	2
8	0.585	0.809	1.162	1.461	1.937	2.667
10	0.619	0.860	1.253	1.600	2.195	3.333
20	0.716	1.008	1.512	1.995	2.960	6.667
50	0.830	1.180	1.806	2.440	3.823	16.67
100	0.909	1.297	2.002	2.733	4.379	33.33
1000	1.139	1.631	2.552	3.539	5.870	333.3
10,000	1.333	1.910	3.006	4.196	7.055	3333
100,000	1.505	2.156	3.401	4.764	8.068	33,333
10^6	1.660	2.377	3.756	5.272	8.967	333,333
10^9	2.060	2.945	4.662	6.564		

Excel functions do not include a function for calculating the t PDF, so it must be calculated using the formula above, like this: =EXP(GAMMALN$((\nu+1)/2)$-GAMMALN$(\nu/2)$)/SQRT$(\nu$*PI())*$((1+x^*x/\nu)$^(-$(\nu+1)/2$)).

Cumulative Distribution Function and Survival Function: The t CDF and survival function (left-tail and right-tail probabilities, respectively) must be calculated by numerical integration. However, Excel and all statistical programs have built-in functions to perform this calculation.

The Excel function TDIST only calculates the right-tail probability. Since the t distribution is symmetric, both tails are the same.

When $x < 0$, to calculate the left-tail probability $F_{t(\nu)}(x)$ in an Excel worksheet, use the formula =-TDIST(-x,ν,1).

When $x \geq 0$, to calculate the right-tail probability $R_{t(\nu)}(x)$ in an Excel worksheet, use the formula =TDIST(x,ν,1).

For two-tailed statistical tests, it is often necessary to calculate the sum of left and right tail probabilities. In symbols, this is $F_{t(\nu)}(-x) + R_{t(\nu)}(x)$. To calculate this quantity in an Excel worksheet, use the formula =TDIST(x,ν,2).

Figure 25-2 shows the PDF, CDF, survival, and hazard functions for selected Student's t random variables and a standard normal random variable.

Inverse Cumulative Distribution Function and Inverse Survival Function: The inverse CDF and survival function for the t distribution must be calculated iteratively. The Excel TINV function has different parameters than other Excel inverse CDF functions, and care is required to calculate the correct value.

To calculate critical values for a statistical test requires one of two calculations. One-tailed tests (for example, $H_0:\mu = 0$ versus $H_A:\mu > 0$) require a t quantile in the right tail, which is the inverse survival function $R_{t(\nu)}^{-1}(\alpha)$. This quantile is more commonly referred to as $t_{\nu,\alpha}$. To calculate this quantile in an Excel worksheet, use the formula =TINV($2^*\alpha,\nu$).

For two-tailed tests (for example, $H_0:\mu = 0$ versus $H_A:\mu \neq 0$), the Type I error probability α must be split evenly between the left and right tails. The required t quantile is the inverse survival function $R_{t(\nu)}^{-1}(\alpha/2)$ or $t_{\nu,\alpha/2}$ To calculate this quantile in an Excel worksheet, use the formula =TINV(α,ν).

Random number generation: Random numbers with a $t(\nu)$ distribution may be calculated by generating $\nu + 1$ $N(0, 1)$ random numbers. Calculate the sample mean $\overline{X}$ and the sample standard deviation s. Finally, calculate $\frac{\overline{X}\sqrt{\nu + 1}}{s}$, which has a $t(\nu)$ distribution.

When using the Excel TINV function to generate random numbers, care is required because TINV only represents the right tail of the t distribution.

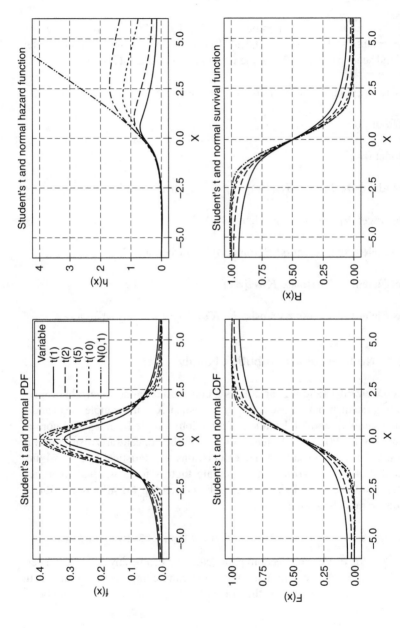

Figure 25-2 PDF, CDF, Survival and Hazard Functions of Selected Student's *t* Random Variables and a Standard Normal Random Variable.

One way to generate $t(v)$ random numbers with a single Excel formula is: =IF(RAND()>0.5,1,-1)*TINV(RAND(),v).

To generate $t(v)$ random numbers with Excel and Crystal Ball software, use the formula =CB.StudentT(0,1,v). To use the optional midpoint parameter m and scale parameter s, use the formula =CB.StudentT(m,s,v)

Mean (Expected Value): $E[t(v)] = 0$ when $v > 1$

Median: 0

Mode: 0

Standard Deviation: $SD[t(v)] = \sqrt{\frac{v}{v-2}}$ when $v > 2$

Variance: $V[t(v)] = \frac{v}{v-2}$ when $v > 2$

Coefficient of Skewness: $Skew[t(v)] = 0$ when $v > 3$

Coefficient of Kurtosis: $Kurt[t(v)] = \frac{6}{v-4} + 3$ when $v > 4$

Coefficient of Excess Kurtosis: $ExKurt[t(v)] = \frac{6}{v-4}$ when $v > 4$

25.1 Noncentral t Distribution Family

A noncentral t random variable is a continuous random variable representing the distribution of $\frac{\sqrt{n}(\overline{X} - \mu + \delta)}{s}$, when $\overline{X}$ is the sample mean and s is the sample standard deviation of a random sample from a stable normal distribution, $N(\mu, \sigma^2)$. Because of this fact, the noncentral t distribution is used for power and sample size calculations for statistical tests involving t statistics. It is also used to calculate factors for statistical tolerance intervals. Because of its complexity, the noncentral t distribution is rarely used for modeling applications.

Parameters: The noncentral t distribution family has two parameters:

- v is the degrees of freedom parameter, and can be any positive integer.
- δ is the noncentrality parameter, and can be any positive number. In some older references, the noncentrality parameter might also be δ^2 or $\frac{1}{2}\delta^2$.

Representation: $X \sim NCt(v, \delta)$

Support: $(-\infty, \infty)$

Relationships to Other Distributions:

- After translation and scaling, the sample mean plus δ and divided by the sample standard deviation of a random sample of size n from a normal distribution has a noncentral t distribution with $n - 1$ degrees of freedom and noncentrality parameter δ. When $X_i \overset{iid}{\sim} N(\mu, \sigma^2)$, the sample mean $\overline{X} = \frac{1}{n}\sum_{i=1}^{n} X_i$, and the sample standard deviation $s = \sqrt{\frac{1}{n-1}\sum_{i=1}^{n}(X_i - \overline{X})^2}$, then $\frac{\sqrt{n}(\overline{X} - \mu + \delta)}{s} \sim NCt(n - 1, \delta)$
- After scaling, the ratio of a normal random variable with mean δ to a chi random variable is a noncentral t random variable. If $Z \sim N(\delta, 1)$, $X \sim \chi(\nu)$, and Z and X are independent, then $\frac{Z\sqrt{\nu}}{X} \sim NCt(\nu, \delta)$. Similarly, if $Y \sim \chi^2(\nu)$ and Z and Y are independent, then $\frac{Z\sqrt{\nu}}{\sqrt{Y}} \sim NCt(\nu, \delta)$

Probability Density Function:

$$f_{NCt(\nu)}(x) = \sum_{j=0}^{\infty} \frac{e^{\frac{-\delta^2}{2}}\Gamma\!\left(\frac{\nu + j + 1}{2}\right)\nu^{\frac{\nu+1}{2}}\left(x\delta\sqrt{2}\right)^j}{j!\sqrt{\nu\pi}\,\Gamma\!\left(\frac{\nu}{2}\right)(\nu + x^2)^{\frac{\nu+j+1}{2}}}$$

Calculating this function is not a job for Excel functions. Instead, use a major statistical program such as JMP or STATGRAPHICS.

Cumulative Distribution Function and Survival Function: To calculate left-tail probabilities (CDF) or right-tail probabilities (survival function) for noncentral t random variables, use a major statistical package such as JMP, MINITAB, or STATGRAPHICS.

Inverse Cumulative Distribution Function and Inverse Survival Function: To calculate p-quantiles (inverse CDF) for noncentral t random variables, use a major statistical package such as JMP, MINITAB, or STATGRAPHICS.

Random number generation: Random numbers with a $NCt(\nu, \delta)$ distribution may be calculated by generating $\nu + 1$ $N(\delta, 1)$ random numbers and then calculating $\frac{\overline{X}\sqrt{\nu + 1}}{s}$.

Mean (Expected Value):

$$E[NCt(\nu, \delta)] = \delta\sqrt{\frac{\nu}{2}}\frac{\Gamma\!\left(\frac{\nu + 1}{2}\right)}{\Gamma\!\left(\frac{\nu}{2}\right)} \text{ when } \nu > 1$$

Standard Deviation:

$$SD[NCt(\nu, \delta)] = \sqrt{\frac{\nu}{\nu - 2}(1 + \delta^2) - \frac{\nu\delta^2}{2}\left(\frac{\Gamma\left(\frac{\nu + 1}{2}\right)}{\Gamma\left(\frac{\nu}{2}\right)}\right)^2} \quad \text{when } \nu > 2$$

Variance:

$$V[NCt(\nu, \delta)] = \frac{\nu}{\nu - 2}(1 + \delta^2) - \frac{\nu\delta^2}{2}\left(\frac{\Gamma\left(\frac{\nu + 1}{2}\right)}{\Gamma\left(\frac{\nu}{2}\right)}\right)^2 \quad \text{when } \nu > 2$$

26

Triangular Distribution Family

A triangular random variable is a continuous random variable with a lower bound A, an upper bound B, and a mode, or most likely value M. The probability density function for a triangular random variable has a simple triangular shape.

Triangular distributions are common models for random variables, when observed data is limited or nonexistent. When expert opinion suggests that the random variable has a lower bound, an upper bound, and a most likely value, a triangular distribution is a reasonable representation of that expert opinion.

Real processes very rarely follow a triangular distribution exactly. The sharp peak, straight lines and definitive ends of the triangular distribution are unusual features for real process distributions. When fitting distribution models to real data, other distribution families are more likely to be useful models for prediction. If the process has a natural upper and lower bound, consider the beta distribution family as an alternative model. Many other bounded distribution families are available beyond the ones in this book. Kotz and van Dorp (2004) describe the triangular and many other bounded distribution families in great depth.

Example 26.1

Elle is simulating a business plan for a proposed new law firm. One of the assumptions in the simulation is the judgment amount for cases ending in a favorable judgment. The partners have established guidelines for which cases to accept. Based on these guidelines, Elle expects a minimum judgment amount of $100,000, a maximum judgment of $3,000,000, and a most likely judgment of $1,000,000. With no data to fit a distribution model, Elle decides to use a triangular distribution to reflect this expert opinion.

Figure 26-1 shows a Crystal Ball assumption form, illustrating a triangular distribution representing judgment amount. Of course, not all cases end in a favorable judgment. In the business plan model, Elle uses a Yes-No assumption

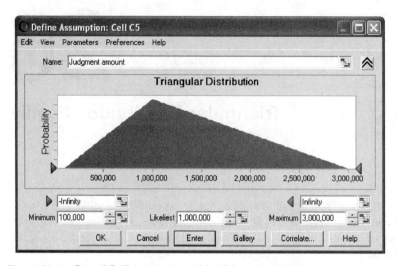

Figure 26-1 Crystal Ball Assumption with a Triangular Distribution.

to represent favorable judgments with the value 1 and unfavorable judgments with the value 0. The judgment for each case brought to trial is the product of the Yes-No assumption and the triangular judgment amount assumption.

Parameters: The triangular distribution family has three parameters:

- A is the lower bound for the triangular random variable.
- M is the mode, or most likely value, for the triangular random variable. M can be any number between A and B.
- B is the upper bound for the triangular random variable. B can be any number greater than A.

The term "standard triangular distribution" refers to the special case where $A = 0$ and $B = 1$.

Representation: $X \sim Tri(A, M, B)$

Support: $[A, B]$

Relationships to Other Distributions:

- The sum or average of two independent and identically distributed uniform random variables is a triangular random variable. If $X, Y \overset{iid}{\sim} Unif(A, B)$, then $X + Y \sim Tri(2A, A + B, 2B)$ and $\frac{X + Y}{2} \sim Tri\left(A, \frac{A + B}{2}, B\right)$

- The beta distribution family includes triangular random variables with the mode equal to the upper or lower bound. $\beta(1, 2, A, B) \sim Tri(A, A, B)$ and $\beta(2, 1, A, B) \sim Tri(A, B, B)$

Process Control Tools: When a process produces data with a triangular distribution, a control chart for individual data can be constructed from the triangular PDF, with the following formulas. The false alarm rate α in these formulas is typically 0.0027, for the same false alarm rate as Shewhart control charts.

$$UCL = B - \sqrt{\left(\tfrac{\alpha}{2}\right)(B - M)(B - A)}$$

$$CL = \begin{cases} B - \sqrt{(B - M)\dfrac{B - A}{2}} & M < \dfrac{A + B}{2} \\[2ex] M & M = \dfrac{A + B}{2} \\[2ex] A + \sqrt{(M - A)\dfrac{B - A}{2}} & M > \dfrac{A + B}{2} \end{cases}$$

$$LCL = A + \sqrt{\left(\tfrac{\alpha}{2}\right)(M - A)(B - A)}$$

When data is collected in rational subgroups, plot triangular data on a Shewhart control chart such as an $\overline{X}$, s control chart. As subgroup size increases, these charts will work almost as well for triangular data as they do for normal data. This is especially true when the distribution is nearly symmetric, but when M is near to A or B, Shewhart charts may not be as effective.

Estimating Parameter Values: Although the triangular distribution is conceptually simple, estimating parameter values from observed data is not simple. Chapter 1 of Kotz and van Dorp (2004) contains formulas, pseudocode, and examples illustrating this estimation process.

Crystal Ball distribution fitting functions fit a triangular distribution to observed data, using maximum likelihood methods.

Capability Metrics: For a process with a triangular distribution, the formulas for calculating equivalent capability metrics are listed below. These formulas assume that the tolerance limits LTL and UTL are on the correct side of the mode M.

$$\text{Equivalent } P^{\%}_{PL} = \begin{cases} \infty & LTL \leq A \\ \dfrac{-1}{3}\Phi^{-1}\left(\dfrac{(LTL - A)^2}{(M - A)(B - A)}\right) & A < LTL < M \end{cases}$$

$$\text{Equivalent } P^{\%}_{PU} = \begin{cases} \dfrac{-1}{3}\Phi^{-1}\left(\dfrac{(B - UTL)^2}{(B - M)(B - A)}\right) & M < UTL < B \\ \infty & B \leq UTL \end{cases}$$

$$\text{Equivalent } P^{\%}_{P} = \frac{\text{Equivalent } P^{\%}_{PU} + \text{Equivalent } P^{\%}_{PL}}{2}$$

$$\text{Equivalent } P^{\%}_{PK} = Min\{\text{Equivalent } P^{\%}_{PU}, \text{Equivalent } P^{\%}_{PL}\}$$

In these formulas, $\Phi^{-1}(p)$ is the standard normal inverse CDF, evaluated in an Excel worksheet with the formula =NORMSINV(p)

Probability Density Function:

$$f_{Tri(A,M,B)}(x) = \begin{cases} \dfrac{2(x - A)}{(M - A)(B - A)} & A \leq x < M \\ \dfrac{2}{B - A} & x = M \\ \dfrac{2(B - x)}{(B - M)(B - A)} & M < x \leq B \\ 0 & \text{otherwise} \end{cases}$$

Cumulative Distribution Function:

$$F_{Tri(A,M,B)}(x) = \begin{cases} 0 & x < A \\ \dfrac{(x - A)^2}{(M - A)(B - A)} & A \leq x < M \\ \dfrac{M - A}{B - A} & x = M \\ 1 - \dfrac{(B - x)^2}{(B - M)(B - A)} & M < x \leq B \\ 1 & x > B \end{cases}$$

Inverse Cumulative Distribution Function:

$$F^{-1}_{Tri(A,M,B)}(p) = \begin{cases} A + \sqrt{p(M - A)(B - A)} & 0 \le p \le \dfrac{M - A}{B - A} \\[3mm] B - \sqrt{(1 - p)(B - M)(B - A)} & \dfrac{M - A}{B - A} < p \le 1 \end{cases}$$

Survival Function:

$$R_{Tri(A,M,B)}(x) = \begin{cases} 1 & x < A \\[2mm] 1 - \dfrac{(x - A)^2}{(M - A)(B - A)} & A \le x < M \\[3mm] \dfrac{B - M}{B - A} & x = M \\[3mm] \dfrac{(B - x)^2}{(B - M)(B - A)} & M < x \le B \\[2mm] 0 & x > B \end{cases}$$

Hazard Function:

$$h_{Tri(A,M,B)}(x) = \begin{cases} \dfrac{2(x - A)}{MB - AB - AM - x^2 + 2Ax} & A < x < M \\[3mm] \dfrac{2}{B - x} & M < x < B \end{cases}$$

Figure 26-2 shows the PDF, CDF, hazard, and survival functions of several triangular random variables, all bounded between 0 and 1.

Mean (Expected Value): $E[Tri(A, M, B)] = \dfrac{A + M + B}{3}$

Median:

$$\text{The median is} \begin{cases} B - \sqrt{(B - M)\dfrac{B - A}{2}} & M < \dfrac{A + B}{2} \\[3mm] M & M = \dfrac{A + B}{2} \\[3mm] A + \sqrt{(M - A)\dfrac{B - A}{2}} & M > \dfrac{A + B}{2} \end{cases}$$

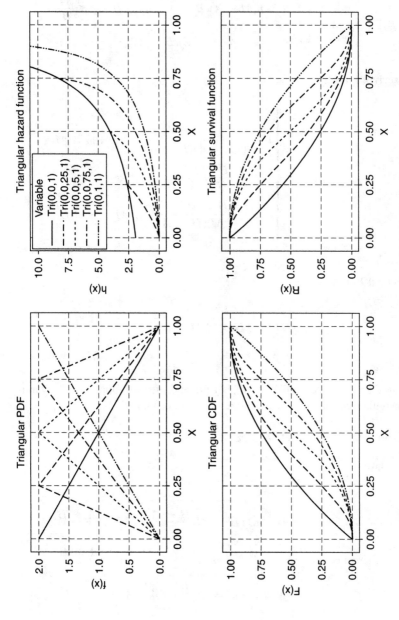

Figure 26-2 PDF, CDF, Hazard, and Survival Functions of Selected Triangular Random Variables between 0 and 1.

Mode: M

Standard Deviation:

$$SD[Tri(A, M, B)] = \sqrt{\frac{A^2 + M^2 + B^2 - AM - AB - BM}{18}}$$

Variance:

$$V[Tri(A, M, B)] = \frac{A^2 + M^2 + B^2 - AM - AB - BM}{18}$$

Coefficient of Skewness:

$$Skew[Tri(A, M, B)] = \frac{\sqrt{2}(1 - 2\theta)(2 - \theta)(1 - \theta)}{5[1 - \theta(1 - \theta)]^{3/2}}$$

where

$$\theta = \frac{M - A}{B - A}$$

Triangular random variables are skewed to the left if $M > \frac{A + B}{2}$ and skewed to the right if $M < \frac{A + B}{2}$.

Coefficient of Kurtosis: $Kurt[Tri(A, M, B)] = 2.4$

Coefficient of Excess Kurtosis: $ExKurt[Tri(A, M, B)] = -0.6$

Triangular random variables are always platykurtic.

Moment Generating Function:

$$m_{Tri(A,M,B)}(t) = \frac{2e^{tA}\left[e^{t(M-A)} - \frac{M - A}{B - A}e^{t(B-A)} - \left(\frac{B - M}{B - A}\right)\right]}{(M - A)(B - M)t^2}$$

27

Uniform Distribution Family

A uniform random variable is a continuous random variable that is equally likely to have any value between A and B. Uniform random variables are also called rectangular random variables, because the probability density function has a rectangular shape.

The most important use of uniform random variables is in the generation of random numbers. The inverse cumulative distribution function (inverse CDF) of any random variable transforms uniform random numbers between 0 and 1 into random numbers with the selected inverse CDF. If a computer program can generate random values of $U \sim Unif(0, 1)$, and if it can compute $F_X^{-1}()$ for any distribution X, then $F_X^{-1}(U)$ is distributed according to the CDF $F_X(x)$. This property is called the "probability integral transform."

Many modelers use a uniform model to represent a quantity that is believed to be between A and B, but without any other knowledge about its distribution. When no data is available to fit a different distribution model, a uniform model expresses an expert opinion that the quantity is bounded between upper and lower limits, without saying anything else about its distribution.

Modelers and Six Sigma practitioners should realize that real measured quantities very rarely have a uniform distribution. The sharp edges of a uniform distribution are extremely unusual in natural processes. Many process distributions do have a sharp edge at a single physical boundary. For example, an exponential distribution model for waiting time has a sharp edge at the lower physical boundary of zero. It is rare to have both an upper and a lower physical boundary combined with equal probability of falling anywhere between those boundaries.

The above observation has consequences for modelers. Here are some rules of thumb about applications for uniform distribution:

- When selecting a distribution model based on its fit to observed data, do not include a uniform distribution among candidate distribution models. If the process is physically bounded on upper and lower ends, consider instead the beta distribution, which includes uniform as a special case.
- Do not calculate process capability metrics or prepare control charts based on a uniform distribution, without strong theoretical justification that the uniform distribution is a reasonable model for the process.

Example 27.1

Lee is developing a business plan for an apartment complex. The business plan requires many assumptions, all of which must be conservative and justifiable to pass the scrutiny of potential investors. One critical assumption is vacancy rate. Lee surveyed other apartment complexes of similar quality in the same region, and found vacancy rates between 2% and 12%. Lee believes that the 12% represents poor management, and that his company will certainly do better than that.

To simulate the financial performance of his project, Lee assumes that the vacancy rate is uniformly distributed between 2% and 12%. Lee believes that a triangular distribution might be more appropriate, but he has no data to support that choice, nor does he know what the most likely value will be. To create a business plan that is thorough, conservative, and data-driven, he chooses a uniform distribution as a model for vacancy rate.

Example 27.2

Roger designs analog circuitry for an instrumentation manufacturer. The accuracy of a voltage measurement input circuit depends on several components, including a voltage reference, amplifiers, and numerous resistors. Each of these has an upper and a lower tolerance limit.

To simulate the accuracy of this circuit, Roger could assume that each component is normally distributed with some arbitrary level of capability, for instance, $C_{PK} = 1$. However, Roger has no measurement data for any of these components to support such a choice. Further, Roger has been an engineer long enough to know that many components have skewed or bimodal distributions, some with C_{PK} values far less than 1.

Therefore, Roger makes a more conservative assumption that each component is uniformly distributed between its tolerance limits. Roger believes that each component value is most likely within its tolerance limits, but his choice of a uniform distribution assumes nothing beyond that.

Since the standard deviation of a uniform random variable is $\frac{B - A}{2\sqrt{3}}$, the uniform assumption is equivalent to assuming that $C_{PK} = \frac{1}{\sqrt{3}} \cong 0.577$. This is worse capability than most, but not all components.

After Roger simulates the accuracy of his design with uniformly distributed components, he can perform a sensitivity analysis. A sensitivity analysis almost always shows that one or two inputs dominate the variation of the output. Once Roger knows which components are most important, he can gather more information to fit a better model for those components, if needed.

Parameters: The uniform family of distributions has two parameters:

- A represents the lower bound, and can be any real number.
- B represents the upper bound, and can be any real number greater than A.

Representation: $X \sim Unif(A, B)$

Support: $[A, B]$

Relationships to Other Distributions:

- The $Unif(0, 1)$ random variable can be converted into $Unif(A, B)$ by the formula $A + (B - A) \times Unif(0, 1)$.
- *Probability integral transform.* If $U \sim Unif(0, 1)$, and X is a random variable with CDF $F_X(x)$ and inverse CDF $F_X^{-1}(p)$, then $F_X^{-1}(U) \sim X$.
- An important special case of the above property is that the negative natural log of a $Unif(0, 1)$ random variable is an exponential random variable. If $U \sim Unif(0, 1)$, then $-\ln U \sim Exp(1)$ and $-\mu \ln U \sim Exp(\mu)$, where μ represents the mean parameter. It also follows that $-2\ln U \sim \chi^2(2)$.
- The sum or average of two independent uniform random variables is a triangular random variable. If $X_1, X_2 \overset{iid}{\sim} Unif(A, B)$, then

$$X_1 + X_2 \sim Tri(2A, A + B, 2B) \text{ and } \frac{X_1 + X_2}{2} \sim Tri\left(A, \frac{A + B}{2}, B\right)$$

Normalizing Transformation: The standard normal inverse CDF function $\Phi^{-1}(p)$ converts a $Unif(0, 1)$ random variable into a standard normal random variable. Therefore, if $X \sim Unif(A, B)$, then $\Phi^{-1}\left(\frac{U - A}{B - A}\right) \sim N(0, 1)$. In Excel software, the =NORMSINV function calculates $\Phi^{-1}(p)$.

Process Control Tools: If a process produces $Unif(A, B)$ data, then a simple individual X control chart with false alarm rate α would have control limits and centerline:

$$UCL = B - \frac{\alpha(B - A)}{2}$$

$$CL = \frac{A + B}{2}$$

$$LCL = A + \frac{\alpha(B - A)}{2}$$

Another alternative is to take averages of several values and plot them on a normal-based Shewhart control chart for subgrouped data, like the $\overline{X}$, s control chart. With subgroup sizes of six or more, the false alarm rate of the $\overline{X}$ chart with uniform data is similar to what it is with normal data.

Estimating Parameter Values: Maximum likelihood estimators for the lower and upper bound are the sample minimum and maximum: $\hat{A} = Min\{X_i\}$ and $\hat{B} = Max\{X_i\}$.

Unbiased versions of these estimators are $\hat{A} + (n + 1)(\hat{B} - \hat{A})$ and $\hat{B} - (n + 1)(\hat{B} - \hat{A})$.

Sometimes, instead of the upper and lower bound, one needs to estimate the midpoint $\eta = \frac{A + B}{2}$ and the half-width $\beta = \frac{B - A}{2}$. Unbiased estimators for these alternative parameters are $\hat{\eta} = \frac{(Min\{X_i\} + Max\{X_i\})}{2}$ and $\hat{\beta} = \frac{(n + 1)(Max\{X_i\} - Min\{X_i\})}{2(n - 1)}$

Capability Metrics: For every other distribution family in this book, the recommended method of calculating capability metrics is to first calculate the probability of defects and then calculate the equivalent capability metrics for a normal distribution with the same probability of defects. Applying this method to a uniform distribution creates a confusing situation. Whenever the tolerance limits are at or outside the bounds of the uniform distribution, the probability of defects is 0 and equivalent $P_{PK}^{\%} = \infty$.

For the uniform distribution only, it is less confusing and more realistic to use the same formulas to calculate capability metrics as for a normal distribution. Here are the calculations for a $Unif(A, B)$ random variable:

$$P_{PL} = \frac{\frac{A + B}{2} - LTL}{3\frac{B - A}{2\sqrt{3}}} = \frac{(A + B) - 2LTL}{\sqrt{3}(B - A)}$$

$$P_{PU} = \frac{UTL - \frac{A + B}{2}}{3\frac{B - A}{2\sqrt{3}}} = \frac{2UTL - (A + B)}{\sqrt{3}(B - A)}$$

$$P_{PK} = Min\{P_{PL}, P_{PU}\}$$

$$P_P = \frac{UTL - LTL}{\sqrt{3}(B - A)}$$

When $LTL = A$ and $UTL = B$, these formulas simplify to $P_P = P_{PK} = \frac{1}{\sqrt{3}} \cong 0.577$

Probability Density Function:

$$f_{Unif(A,B)}(x) = \begin{cases} \dfrac{1}{B - A} & A \leq x \leq B \\ 0 & \text{otherwise} \end{cases}$$

Cumulative Distribution Function:

$$F_{Unif(A,B)}(x) = \begin{cases} 0 & x < A \\ \dfrac{x - A}{B - A} & A \leq x \leq B \\ 1 & x > B \end{cases}$$

Inverse Cumulative Distribution Function:

$$F^{-1}_{Unif(A,B)}(p) = A + (B - A)p$$

Random number generation: Although every software product with mathematical functions claims to generate uniform random numbers, this task is more difficult than it might seem. A fundamental limitation of digital computers is that they are sequential machines, incapable of random behavior. Random number generators are properly called pseudo-random number generators, because the computer calculates the next value from the previous value in a long sequence. The starting value in the sequence may be fixed, specified by the user, or derived from the system clock. Eventually, the sequence will repeat itself.

Chapter 2 discusses known issues with the RAND function in Microsoft Office Excel software which generates pseudo-random values of $Unif(0, 1)$. Versions earlier than Excel 2003 should not be used, and Excel 2003 should have all available updates to correct a known defect in the RAND function. The random number generator in the Analysis ToolPak provided with Excel software should not be used at any release level.

Random number generators in Crystal Ball, MINITAB, JMP, and STATGRAPHICS software products have been extensively tested and verified. Before using the random number generator in other software, inquire about how it was tested. For more information on this subject, see Chapter 2.

To generate $Unif(A, B)$ random numbers in Microsoft Office Excel 2003 or later, use $= A + (B - A)*\text{RAND}()$. To generate $Unif(A, B)$ random numbers with Crystal Ball and Excel software, use $=\text{CB.Uniform}(A, B)$.

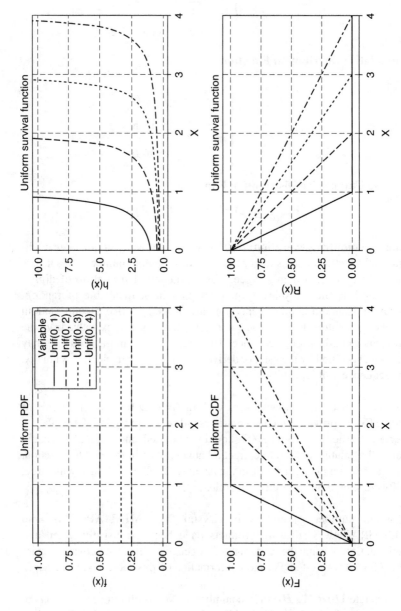

Figure 27-1 PDF, CDF, Survival, and Hazard Functions for Selected Uniform Random Variables.

Survival Function:

$$R_{Unif(A,B)}(x) = \begin{cases} 1 & x < A \\ \dfrac{B - x}{B - A} & A \leq x \leq B \\ 0 & x > B \end{cases}$$

Hazard Function:

$$h_{Unif(A,B)}(x) = \frac{1}{B - x}$$

Figure 27-1 illustrates the PDF, CDF, survival, and hazard functions for selected uniform random variables.

Mean (Expected Value): $E[\,Unif(A, B)] = \frac{A + B}{2}$

Median: $\frac{A + B}{2}$

Standard Deviation: $SD[\,Unif(A, B)] = \frac{B - A}{2\sqrt{3}}$

Variance: $V[\,Unif(A, B)] = \frac{(B - A)^2}{12}$

Coefficient of Variation: $CV[\,Unif(A, B)] = \frac{B - A}{\sqrt{3}(A + B)}$

Coefficient of Skewness: $Skew[\,Unif(A, B)] = 0$

Coefficient of Kurtosis: $Kurt[\,Unif(A, B)] = 1.8$

Coefficient of Excess Kurtosis: $ExKurt[\,Unif(A, B)] = -1.2$

All uniform random variables are unskewed and platykurtic.

28

Weibull Distribution Family

A Weibull random variable is a continuous random variable with a lower bound of zero. The family of Weibull random variables includes many important special cases, including the exponential and Rayleigh distributions, described in Chapters 11 and 24, respectively. The Weibull probability curve has a wide variety of shapes, determined by the shape parameter α. Most Weibull distributions are skewed to the right. However, a Weibull random variable with shape parameter $\alpha = 3.602$ is nearly symmetric, and is comparable to a normal distribution in many ways.

The Weibull distribution is a popular model for reliability applications, as a model for time to failure. There are several reasons for its importance in reliability applications, including these:

- By changing the shape parameter α, a Weibull random variable may have decreasing hazard function ($\alpha < 1$), constant hazard function ($\alpha = 1$), or increasing hazard function ($\alpha > 1$). Therefore, the Weibull family may be used to model infant mortality, useful life, or wearout portions of a product life. Figure 28-1 illustrates how the changes in hazard function over a product lifetime form a "bathtub curve."
- An exponential random variable raised to a positive power is a Weibull random variable.
- The Weibull distribution and the extreme value distribution are two of three families that may describe the minimum value in a sample with large sample sizes, as proved by Gnedenko (1943) and others. Because the time to failure of certain systems is the failure of a "weakest link," the Weibull or smallest extreme value distributions are plausible models for failure times. Section 12.1 discusses extreme value theory at a high level for practitioners of Six Sigma and reliability methods.
- Because of the wide variety of Weibull probability curves, the Weibull distribution fits a wide variety of observed datasets.

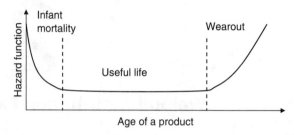

Figure 28-1 Bathtub Curve, Showing Three Portions of Product Life According to Trends in the Hazard Function.

The name of the Weibull family honors Ernst Hjalmar Waloddi Weibull (1887–1979), who applied the distribution as a model for breaking strength of materials and other applications (1939, 1939a, 1951). On occasion, the distribution has also been called the Fréchet distribution, after Maurice René Fréchet (1878–1973), who was the first to identify this distribution family as a model for extreme values (1927). However, the term Fréchet distribution more often refers to a very different distribution family applying to

Figure 28-2 Ernst Hjalmar Waloddi Weibull, Photograph by Sam C. Saunders, Used by Permission. Photograph Taken c. 1970 during Dr. Weibull's Visit to Boeing Scientific Research Laboratories in Seattle.

extreme values for certain heavy-tailed random variables. See Section 12.1 for more details.

The Weibull distribution family has two parameters, a shape parameter α and a scale parameter β. Section 28.1 describes a three-parameter Weibull family, which includes a location parameter τ, representing the lower bound.

In other books, it is common to use the symbol β to represent the Weibull shape parameter, which is α here. Use of α to represent the shape parameter is consistent with customary notation for gamma and other distribution families. In this book, the symbol β represents scale parameters for gamma, two-parameter exponential, extreme value and Weibull distributions. Be careful to avoid confusing the Weibull parameters.

The following example illustrates a scientific answer to a common question: to burn in or not to burn in? Many companies exercise their products before shipment in an attempt to screen out infant mortality failures before their customers find them. This expensive "burn in" process is a frequent target of cost reduction pressure, even if the process is successful in finding product failures.

The Weibull model provides tools to answer two important questions about managing burn-in processes:

- *Is it helpful to burn in products?* If the time to failure distribution has a decreasing hazard function, then the burn in process identifies product defects more quickly at first then it would later. It is helpful to prevent unhappy customers by finding these early failures before shipment. However, if there is no strong evidence that the hazard function is decreasing, then the burn in process is not serving any useful purpose. Testing whether the Weibull shape parameter α is equal to 1 provides a quick way to answer this question. If the shape parameter $\alpha < 1$, then failures experienced by customers would increase if the burn in process were eliminated.
- *How long should products be burned in?* The burn in time should last until the hazard function decreases to a rate of defects that is tolerable for customers to experience. After fitting the Weibull parameters, the formula for the hazard function can answer this question directly.

Example 28.1

Pilar is a reliability engineer at Rambutan Computers. On a new computer assembly line, each computer is burned in for 24 hours prior to shipment. As part of a Six Sigma project, Pilar is investigating whether the burn-in process serves any useful business purpose, and if so, should the burn-in time be adjusted?

For 1000 recently shipped computers, 12 computers failed during the 24-hour burn in process. These were the failure times, in hours:

0.01   0.12   1.47   3.11   4.55   4.76   7.85   14.32   14.34   15.10

21.61   22.83

The remaining 988 computers did not fail in the 24 hour burn in. For the purpose of reliability analysis, these 988 computers are censored at 24 hours.

In MINITAB, Pilar performed a Weibull analysis of this data, using the Stat ⇨ Reliability/Survival ⇨ Distribution Analysis (Right Censoring) ⇨ Parametric Distribution Analysis menu option. In the Test submenu, Pilar asked MINITAB to test whether the Weibull shape parameter is equal to 1. This test will determine whether the burn in process is useful or not.

Figure 28-3 is the probability plot produced by the MINITAB analysis. In the MINITAB Session window, the test of whether the shape parameter equals 1 produced a P-value of 0.022. This means that if the shape parameter is actually 1, the probability is 0.022 that Pilar would observe a dataset as extreme as this one. Since this probability is small (less than 0.05), Pilar concludes that the shape parameter is not 1. In fact, MINITAB reports a 95% confidence interval for the shape parameter of (0.16, 0.87), so Pilar is 95% confident that $0.16 < \alpha < 0.87$.

Since this data provides strong evidence that the shape parameter is less than 1, this means that the burn-in process is detecting infant mortality failures. Therefore, it is effectively reducing the rate of failures suffered by customers.

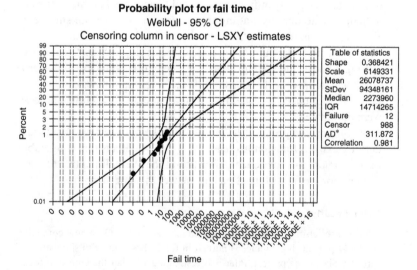

Figure 28-3   MINITAB Report of Weibull Analysis of Computer Burn-in failures.

To determine the appropriate length of burn-in, Pilar needs the hazard function. The hazard function has this form: $h_X(x) = \frac{\alpha x^{\alpha-1}}{\beta^\alpha}$. From the MINITAB report, point estimates of the parameters are $\hat{\alpha} = 0.368$ and $\hat{\beta} = 6{,}149{,}331$. Using these values, the estimated hazard function is $\hat{h}_X(x) = 0.001168 x^{-0.632}$. The reliability goal for the useful life of this product is one failure per 4000 hours of use, which is a hazard rate of 0.00025. Pilar decides that when the infant mortality failures have declined to a hazard rate of 0.0005, twice the long-term reliability goal, then burn-in should stop. Solving the equation $\hat{h}_X(x) = 0.0005$ leads to a solution of $x = 3.83$ hours.

Based on this analysis, Pilar recommends that the burn-in continue, but that its duration be reduced to 4 hours.

The above example does not consider economic issues. The cost of burn-in and the cost of shipping products which will soon fail are important factors in management decisions about burn-in. By analyzing the available data and combining Weibull probability models with the economic considerations, one can calculate what length of burn-in minimizes total cost.

Parameters: The Weibull distribution family has two parameters:

- α is a shape parameter, which may be any positive number. Special cases of the Weibull distribution include the exponential distribution, when $\alpha = 1$, and the Rayleigh distribution, when $\alpha = 2$.
- β is a scale parameter, which may be any positive number. Another name for β is "characteristic life," because of the property that $F_X(\beta) = 1 - e^{-1} \cong 0.632$. That is, in a reliability application, 63.2% of the population has failed before time β, regardless of the value of the shape parameter α.

Representation: $X \sim Weibull(\alpha, \beta)$

Support: $[0, \infty)$

Relationships to Other Distributions:

- Exponential and Weibull distributions are power transformations of each other. If $X \sim Weibull(\alpha, \beta)$, then $X^\alpha \sim Exp(\beta^\alpha)$, where β^α represents the mean parameter, and $\left(\frac{X}{\beta}\right)^\alpha \sim Exp(1)$. Also, if $Y \sim Exp(\mu)$, then $\beta\left(\frac{X}{\mu}\right)^{1/\alpha} \sim Weibull(\alpha, \beta)$.

- When the shape parameter $\alpha = 1$, a Weibull random variable is also an exponential distribution. If $X \sim Weibull(1, \beta)$, then $X \sim Exp(\beta)$, where β represents the mean parameter.

- The minimum of a set of independent observations from a Weibull distribution is also a Weibull distribution. When $X_1, \ldots,$ $X_n \overset{iid}{\sim} Weibull(\alpha, \beta)$, then $Min\{X_i\} \sim Weibull(\alpha, \beta/n)$.
- When the shape parameter $\alpha = 2$, a Weibull random variable is also a Rayleigh distribution. If $X \sim Weibull(2, \beta)$, then $X \sim Rayleigh(\beta)$
- The natural log of a Weibull random variable is a smallest extreme value random variable. If $X \sim Weibull(\alpha, \beta)$, then $\alpha\ln\left(\frac{X}{\beta}\right) \sim SEV(0, 1)$. Also, $-\alpha\ln\left(\frac{X}{\beta}\right) \sim LEV(0, 1)$.

Normalizing Transformation: In theory, a power transformation in the Box-Cox family can transform a scaled Weibull distribution into an approximately normal distribution. If the parameters α and β are known, then $\left(\frac{X}{\beta}\right)^{\alpha}$ has an exponential distribution with mean $= 1$, and $\left(\frac{X}{\beta}\right)^{0.2654\alpha}$ has an approximately normal distribution.

In practice, applying this transformation requires values for the parameters α and β, or at least for the shape parameter α. If these parameters are estimated from a sample, the uncertainty in these estimates may make the transformation unsatisfactory. However, if the transformation works, any of the wide range of methods available for exponential or normal distributions may be used on the transformed Weibull data. Attempts to transform Weibull data by Box-Cox or Johnson transformation should be assessed by probability plots and goodness-of-fit test statistics.

Process Control Tools: If the parameters α and β have been estimated by software, an individual X chart can be constructed for Weibull process data, using the formulas listed below. These formulas require a false alarm rate ε, which is generally 0.0027 for parity with standard Shewhart control charts. The typical symbol for false alarm rate is α, but since α is a parameter of the Weibull distribution, the false alarm rate is ε here.

$$UCL = \beta[-\ln(\varepsilon/2)]^{1/\alpha}$$
$$CL = \beta[-\ln 0.5]^{1/\alpha}$$
$$LCL = \beta[-\ln(1 - \varepsilon/2)]^{1/\alpha}$$

An alternate process control approach is to transform the Weibull data into a normal distribution as described above, and apply normal-based Shewhart control charts.

Be cautious about any control chart for Weibull data, if the process has not demonstrated long-term stability in its distribution. If the process is unstable, this would dramatically affect the parameter estimates, and control charts will not perform well. If the Weibull family was selected using probability plots, this choice could also be a result of process instability.

The detection of instability is often not simple. Methods of checking a process for stability include probability plots or histograms of several samples taken over a long time, and normal-based Shewhart charts like $\overline{X}$, s charts with a large subgroup size. If the choice of the Weibull distribution relies on theoretical arguments, not just the observed distribution of the data, Weibull control charts can be very effective, and they will accurately control the false alarm rate.

Estimating Parameter Values: When both parameters are unknown, there are no simple formulas to estimate them. In general, maximum likelihood estimation requires the iterative, numerical solution of two complex formulas. In certain situations, particularly when the shape parameter is small, many alternate estimation methods have been proposed. Section 4 of Chapter 21 of Johnson *et al* (1994) describes many of these methods, and research continues in this area.

For Six Sigma practitioners, the best advice is to use a major statistical package such as MINITAB or STATGRAPHICS, or one of the many specialized programs for reliability engineers designed to fit Weibull distributions. A very common complication is censoring of the dataset, which happens when a value is not known for some units in the sample. There are many types of censoring, some of which are described in Chapter 11 on exponential random variables. Censored data always calls for estimation by software algorithms.

Capability Metrics: For a process producing a Weibull distribution, calculate equivalent capability metrics using the Weibull CDF. Here are the formulas:

$$\text{Equivalent } P_{PU}^{\%} = \frac{-\Phi^{-1}\left(\exp\left[-\left(\frac{UTL}{\beta}\right)^{\alpha}\right]\right)}{3}$$

$$\text{Equivalent } P_{PL}^{\%} = \frac{-\Phi^{-1}\left(1 - \exp\left[-\left(\frac{LTL}{\beta}\right)^{\alpha}\right]\right)}{3}$$

$$\text{Equivalent } P_{P}^{\%} = \frac{\text{Equivalent } P_{PU}^{\%} + \text{Equivalent } P_{PL}^{\%}}{2}$$

$$\text{Equivalent } P_{PK}^{\%} = Min\{\text{Equivalent } P_{PU}^{\%}, \text{Equivalent } P_{PL}^{\%}\}$$

$\Phi^{-1}(p)$ is the standard normal inverse CDF, evaluated in Excel with the =NORMSINV(p) function. Table 28-1 lists selected values of equivalent $P^{\%}_{PL}$ and $P^{\%}_{PU}$ for selected Weibull random variables.

Each tolerance limit in Table 28-1 is listed as a multiple of the scale parameter β. The values of the scale parameter α represented in the table, 0.5, 1, 2, and 3.6, are the same values used for Table 14-1 on the gamma distribution. Comparing these tables provides some insight into differences between gamma and Weibull families. One characteristic of the gamma family is that the mean value is $\alpha\beta$, the product of both parameters. In the Weibull family, the "characteristic life" β represents the 63.2% probability point, regardless of the value of α.

The Weibull random variable with shape parameter $\alpha = 3.6$, on the right side of the table, is often cited as having a nearly normal distribution. For this Weibull random variable, the mean is 0.901β, and the standard deviation is 0.278β. Therefore, the bottom line in the table represents a lower tolerance limit at 3.2 standard deviations below the mean and an upper tolerance limit at 13 standard deviations above the mean. The defect rates with tolerances set at these points are similar to a normal distribution with the tolerances set at 6 standard deviations away from the mean. This fact illustrates how any similarity between this Weibull distribution and the normal distribution breaks down in the tails of the distribution.

Probability Density Function:

$$
f_{Weibull(\alpha,\beta)}(x) = \begin{cases} \dfrac{\alpha x^{\alpha-1}}{\beta^{\alpha}} \exp\left(-\left[\dfrac{x}{\beta}\right]^{\alpha}\right) & x \geq 0 \\ \\ 0 & x < 0 \end{cases}
$$

To calculate the Weibull PDF in an Excel worksheet, use the formula =WEIBULL(x,α,β,FALSE).

Cumulative Distribution Function:

$$
F_{Weibull(\alpha,\beta)}(x) = \begin{cases} 1 - \exp\left(-\left[\dfrac{x}{\beta}\right]^{\alpha}\right) & x \geq 0 \\ \\ 0 & x < 0 \end{cases}
$$

To calculate the Weibull CDF in an Excel worksheet, use the formula =WEIBULL(x,α,β,TRUE).

Table 28-1   Selected Capability Values for Weibull Random Variables

Shape = 0.5		Shape = 1 (Exponential)		Shape = 2 (Rayleigh)				Shape = 3.6			
UTL	Equiv. $P\%_{PU}$	UTL	Equiv. $P\%_{PU}$	LTL	Equiv. $P\%_{PL}$	UTL	Equiv. $P\%_{PU}$	LTL	Equiv. $P\%_{PL}$	UTL	Equiv. $P\%_{PU}$
4β	0.367	3β	0.549	0.5β	0.256	1.33β	0.317	0.6β	0.350	1.33β	0.515
7β	0.490	5β	0.824	0.4β	0.349	1.5β	0.417	0.5β	0.470	1.5β	0.417
14β	0.661	7β	1.039	0.3β	0.455	1.66β	0.508	0.4β	0.599	1.66β	0.508
26β	0.835	9β	1.222	0.25β	0.517	2β	0.697	0.3β	0.742	1.83β	0.603
44β	1.003	11β	1.383	0.2β	0.587	2.5β	0.963	0.2β	0.914	2β	0.697
70β	1.167	13β	1.529	0.1β	0.776	3β	1.222	0.1β	1.160	2.5β	0.963
108β	1.336	15β	1.662	0.01β	1.240	3.5β	1.476	0.05β	1.366	3β	1.222
159β	1.501	17β	1.787	0.001β	1.584	4β	1.726	0.02β	1.602	3.5β	1.476
225β	1.662	19β	1.904	0.0001β	1.871	4.5β	1.973	0.01β	1.761	4β	1.726
431β	2.001	21β	2.014	0.00001β	2.120	5β	2.219	0.003β	2.010	4.6β	2.023

Inverse Cumulative Distribution Function:

$$F^{-1}_{Weibull(\alpha,\beta)}(prob) = \beta[-\ln(1 - prob)]^{1/\alpha}$$

To calculate the Weibull quantiles using the inverse CDF in an Excel worksheet, use the formula =β*((-LN(1-*prob*))^(1/α))

Random Number Generation:

To generate Weibull random numbers in an Excel worksheet, use the formula =β*((-LN(RAND()))^(1/α)).

To generate Weibull random numbers with Crystal Ball and Excel software, use the formula =CB.Weibull(0,β,α).

Survival Function:

$$R_{Weibull(\alpha,\beta)}(x) = \begin{cases} \exp\left(-\left[\dfrac{x}{\beta}\right]^{\alpha}\right) & x \geq 0 \\ \\ 1 & x < 0 \end{cases}$$

Hazard Function:

$$h_{Weibull(\alpha,\beta)}(x) = \frac{\alpha x^{\alpha-1}}{\beta^{\alpha}}$$

Depending on the value of α, the hazard function can be a decreasing, constant, or increasing function of x.

Figure 28-4 illustrates the PDF, CDF, hazard, and survival functions, with various shape parameters ranging from α = 0.5 to α = 10. All five Weibull random variables illustrated have the same scale parameter, β = 1. Observe that the PDF and survival functions intersect where the value of the random variable equals the scale parameter. This point, called the "characteristic life" in reliability applications, represents the time before which $1 - e^{-1} \cong 0.632$ of the units have failed, regardless of the shape parameter.

In the formulas below, $\Gamma(x)$ represents the gamma function, which can be evaluated in Excel by =EXP(GAMMALN(x)).

Mean (Expected Value): $E[\,Weibull(\alpha, \beta)] = \beta\Gamma\left(\frac{\alpha + 1}{\alpha}\right)$

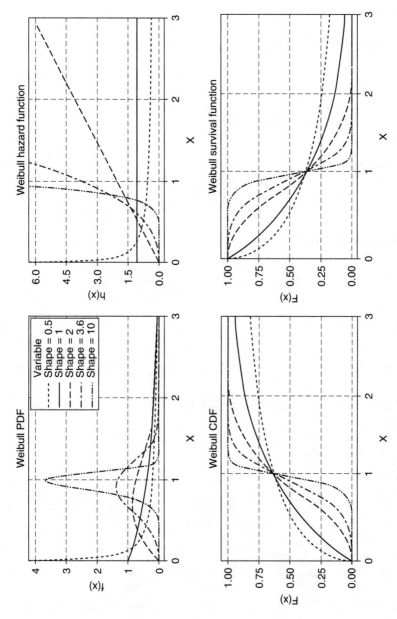

Figure 28-4 PDF, CDF, Hazard, and Survival Functions of Weibull Random Variables, with Various Shapes. All Random Variables Have a Scale Parameter β = 1.

Median: $\beta(\ln 2)^{1/\alpha}$

Mode:

The mode is 0 if $\alpha \le 1$, or $\beta\left(1 - \dfrac{1}{\alpha}\right)^{1/\alpha}$ if $\alpha \ge 1$

Standard Deviation:

$$SD[\boldsymbol{Weibull}(\alpha, \beta)] = \beta\sqrt{\Gamma\left(\frac{\alpha + 2}{\alpha}\right) - \left[\Gamma\left(\frac{\alpha + 1}{\alpha}\right)\right]^2}$$

Variance:

$$V[\boldsymbol{Weibull}(\alpha, \beta)] = \beta^2\left[\Gamma\left(\frac{\alpha + 2}{\alpha}\right) - \left[\Gamma\left(\frac{\alpha + 1}{\alpha}\right)\right]^2\right]$$

Coefficient of Variation:

$$CV[\boldsymbol{Weibull}(\alpha, \beta)] = \sqrt{\frac{\Gamma\left(\frac{\alpha + 2}{\alpha}\right)}{\left[\Gamma\left(\frac{\alpha + 1}{\alpha}\right)\right]^2} - 1}$$

Coefficient of Skewness:

$$Skew[\boldsymbol{Weibull}(\alpha, \beta)] = \frac{\Gamma\left(\frac{\alpha + 3}{\alpha}\right) - 3\Gamma\left(\frac{\alpha + 2}{\alpha}\right)\Gamma\left(\frac{\alpha + 1}{\alpha}\right) + 2\left[\Gamma\left(\frac{\alpha + 1}{\alpha}\right)\right]^3}{(SD[\boldsymbol{Weibull}(\alpha, \beta)])^3}$$

A Weibull random variable is skewed to the right for $\alpha < 3.6$, and skewed to the left for $\alpha > 3.6$.

Coefficient of Kurtosis:

$Kurt[\boldsymbol{Weibull}(\alpha, \beta)]$

$$= \frac{\Gamma\left(\frac{\alpha + 4}{\alpha}\right) - 4\Gamma\left(\frac{\alpha + 3}{\alpha}\right)\Gamma\left(\frac{\alpha + 1}{\alpha}\right) + 6\Gamma\left(\frac{\alpha + 2}{\alpha}\right)\left[\Gamma\left(\frac{\alpha + 1}{\alpha}\right)\right]^2 - 3\left[\Gamma\left(\frac{\alpha + 1}{\alpha}\right)\right]^4}{(SD[\boldsymbol{Weibull}(\alpha, \beta)])^4}$$

Coefficient of Excess Kurtosis:
$ExKurt[\boldsymbol{Weibull}(\alpha, \beta)] = Kurt[\boldsymbol{Weibull}(\alpha, \beta)] - 3$

A Weibull random variable is almost always leptokurtic, except for a region near $\alpha = 3.6$, in which it become slightly platykurtic. Figure 28-5 is a graph of the coefficients of skewness and kurtosis versus the shape parameter α.

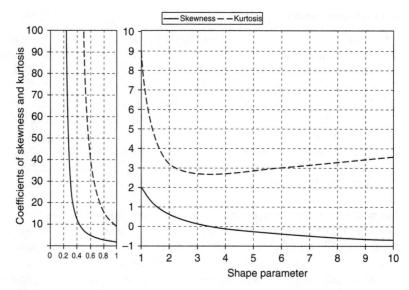

Figure 28-5   Weibull Coefficients of Skewness and Kurtosis Versus the Shape Parameter.

28.1   Three-Parameter Weibull Distribution Family

A three-parameter Weibull random variable is a continuous random variable with a lower bound τ. A two-parameter Weibull random variable plus a constant τ is a three-parameter Weibull random variable. The two-parameter exponential distribution is a special case of the three-parameter Weibull distribution.

Parameters: The three-parameter Weibull distribution family has these parameters:

- α is a shape parameter, which may be any positive number.
- β is a scale parameter, which may be any positive number. Another name for β is "characteristic life," because of the property that $F_X(\tau + \beta) = 1 - e^{-1} \cong 0.632$. That is, in a reliability application, 63.2% of the population has failed before time $\tau + \beta$.

- τ is a threshold or location parameter, which may be any number. τ represents the lower bound for the three-parameter Weibull random variable.

Representation: $X \sim Weibull(\alpha, \beta, \tau)$

Support: $[\tau, \infty)$

Relationships to Other Distributions:

- When $\tau = 0$, a three-parameter Weibull random variable is also a two-parameter Weibull random variable. $\textbf{\textit{Weibull}}(\alpha, \beta, 0) \sim \textbf{\textit{Weibull}}(\alpha, \beta)$. Also, any two-parameter Weibull random variable plus a constant τ is a three-parameter Weibull random variable.
- When the shape parameter $\alpha = 1$, a three-paremeter Weibull random variable is also a two-parameter exponential random variable. If $X \sim \textbf{\textit{Weibull}}(1, \beta, \tau)$, then $X \sim \textbf{\textit{Exp}}(\beta, \tau)$.
- When the shape parameter $\alpha = 2$, a three-parameter Weibull random variable is also a Rayleigh random variable. If $X \sim \textbf{\textit{Weibull}}(2, \beta, \tau)$, then $X \sim \textbf{\textit{Rayleigh}}(\beta, \tau)$.

Normalizing Transformation: In theory, a power transformation in the Box-Cox family can transform a translated and scaled three-parameter Weibull distribution into an exponential distribution, and from the exponential distribution into a normal distribution. If the parameters α, β and τ are known, then $\left(\frac{X - \tau}{\beta}\right)^{\alpha}$ has an exponential distribution with mean $= 1$, and $\left(\frac{X - \tau}{\beta}\right)^{0.2654\alpha}$ has an approximately normal distribution.

In practice, applying this transformation requires values for the parameters α, β and τ. If these parameters are estimated from a sample, the uncertainty in these estimates may make the transformation unsatisfactory. This lack of precision in estimates of these parameters is more significant for the three-parameter Weibull distribution than for the two-parameter Weibull distribution. However, if the transformation works, any of the wide range of methods available for exponential or normal distributions may be used on the transformed Weibull data. Attempts to transform Weibull data by Box-Cox or Johnson transformation should be assessed by probability plots and goodness-of-fit test statistics. If the threshold parameter τ is known from theoretical knowledge about the process, then the problem reduces to a two-parameter estimation problem, and the precision of estimates improves.

Process Control Tools: If the parameters α, β and τ have been estimated by software, an individual X chart can be constructed for three-parameter Weibull process data, using the formulas listed below. These formulas require a false alarm rate ε, which is generally 0.0027 for parity with standard Shewhart control charts. The typical symbol for false alarm rate is α, but since α is a parameter of the Weibull distribution, the false alarm rate is ε here.

$$UCL = \tau + \beta[-\ln(\varepsilon/2)]^{1/\alpha}$$
$$CL = \tau + \beta[-\ln 0.5]^{1/\alpha}$$
$$LCL = \tau + \beta[-\ln(1 - \varepsilon/2)]^{1/\alpha}$$

An alternate process control approach is to transform the Weibull data into a normal distribution as described above, and apply normal-based Shewhart control charts.

Estimating Parameter Values: Parameter estimation for three-parameter Weibull data is a task best left to a major statistical package or to one of the many specialized programs for reliability engineers designed to fit Weibull distributions. Section 4 of Chapter 21 of Johnson *et al* (1994) describes Weibull estimation methods, and research continues in this area.

Capability Metrics: For a process producing a three-parameter Weibull distribution, calculate equivalent capability metrics using the Weibull CDF. Here are the formulas:

$$\text{Equivalent } P_{PU}^{\%} = \frac{-\Phi^{-1}\left(\exp\left[-\left(\frac{UTL - \tau}{\beta}\right)^{\alpha}\right]\right)}{3}$$

$$\text{Equivalent } P_{PL}^{\%} = \frac{-\Phi^{-1}\left(1 - \exp\left[-\left(\frac{LTL - \tau}{\beta}\right)^{\alpha}\right]\right)}{3}$$

$$\text{Equivalent } P_{P}^{\%} = \frac{\text{Equivalent } P_{PU}^{\%} + \text{Equivalent } P_{PL}^{\%}}{2}$$

$$\text{Equivalent } P_{PK}^{\%} = Min\{\text{Equivalent } P_{PU}^{\%}, \text{Equivalent } P_{PL}^{\%}\}$$

$\Phi^{-1}(p)$ is the standard normal inverse CDF, evaluated in Excel with the =NORMSINV(p) function.

Probability Density Function:

$$f_{Weibull(\alpha,\beta,\tau)}(x) = \begin{cases} \dfrac{\alpha(x - \tau)^{\alpha-1}}{\beta^{\alpha}} \exp\left(-\left[\dfrac{x - \tau}{\beta}\right]^{\alpha}\right) & x \geq \tau \\ \\ 0 & x < \tau \end{cases}$$

To calculate the Weibull PDF in an Excel worksheet, use the formula =WEIBULL(x-τ,α,β,FALSE).

Cumulative Distribution Function:

$$F_{Weibull(\alpha,\beta,\tau)}(x) = \begin{cases} 1 - \exp\left(-\left[\dfrac{x - \tau}{\beta}\right]^{\alpha}\right) & x \geq \tau \\ \\ 0 & x < \tau \end{cases}$$

To calculate the Weibull CDF in an Excel worksheet, use the formula =WEIBULL(x-τ,α,β,TRUE).

Inverse Cumulative Distribution Function:

$$F^{-1}_{Weibull(\alpha,\beta,\tau)}(prob) = \tau + \beta[-\ln(1 - prob)]^{1/\alpha}$$

To calculate the Weibull quantiles using the inverse CDF in an Excel worksheet, use =τ+β*((-LN(1-$prob$))^(1/α))

Random Number Generation:

To generate Weibull random numbers in an Excel worksheet, use the formula =τ+β*((-LN(RAND()))^(1/α)).

To generate Weibull random numbers with Crystal Ball and Excel software, use the formula =CB.Weibull(τ,β,α).

Survival Function:

$$R_{Weibull(\alpha,\beta,\tau)}(x) = \begin{cases} \exp\left(-\left[\dfrac{x - \tau}{\beta}\right]^{\alpha}\right) & x \geq \tau \\ 1 & x < \tau \end{cases}$$

Hazard Function:

$$h_{Weibull(\alpha,\beta,\tau)}(x) = \frac{\alpha(x - \tau)^{\alpha-1}}{\beta^{\alpha}}$$

Depending on the value of α, the hazard function can be a decreasing, constant, or increasing function of x.

In the formulas below, $\Gamma(x)$ represents the gamma function, which can be evaluated in Excel by =EXP(GAMMALN(x)).

Mean (Expected Value):

$$E[Weibull(\alpha, \beta, \tau)] = \tau + \beta\Gamma\left(\frac{\alpha + 1}{\alpha}\right)$$

Median: $\tau + \beta(\ln2)^{1/\alpha}$

Mode: The mode is τ if $\alpha \leq 1$, or $\tau + \beta\left(1 - \frac{1}{\alpha}\right)^{1/\alpha}$ if $\alpha \geq 1$

Standard Deviation:

$$SD[Weibull(\alpha, \beta, \tau)] = \beta\sqrt{\Gamma\left(\frac{\alpha + 2}{\alpha}\right) - \left[\Gamma\left(\frac{\alpha + 1}{\alpha}\right)\right]^2}$$

Variance:

$$V[Weibull(\alpha, \beta, \tau)] = \beta^2\left[\Gamma\left(\frac{\alpha + 2}{\alpha}\right) - \left[\Gamma\left(\frac{\alpha + 1}{\alpha}\right)\right]^2\right]$$

Coefficient of Variation:

$$CV[Weibull(\alpha, \beta, \tau)] = \frac{\beta\sqrt{\Gamma\left(\frac{\alpha + 2}{\alpha}\right) - \left[\Gamma\left(\frac{\alpha + 1}{\alpha}\right)\right]^2}}{\tau + \beta\Gamma\left(\frac{\alpha + 1}{\alpha}\right)}$$

Coefficient of Skewness:

$$Skew[Weibull(\alpha, \beta, \tau)] = \frac{\Gamma\left(\frac{\alpha + 3}{\alpha}\right) - 3\Gamma\left(\frac{\alpha + 2}{\alpha}\right)\Gamma\left(\frac{\alpha + 1}{\alpha}\right) + 2\left[\Gamma\left(\frac{\alpha + 1}{\alpha}\right)\right]^3}{(SD[Weibull(\alpha, \tau)])^3}$$

A Weibull random variable is skewed to the right for $\alpha < 3.6$, and skewed to the left for $\alpha > 3.6$.

Coefficient of Kurtosis:

$$Kurt[Weibull(\alpha, \beta, \tau)] =$$

$$\frac{\Gamma\left(\frac{\alpha + 4}{\alpha}\right) - 4\Gamma\left(\frac{\alpha + 3}{\alpha}\right)\Gamma\left(\frac{\alpha + 1}{\alpha}\right) + 6\Gamma\left(\frac{\alpha + 2}{\alpha}\right)\left[\Gamma\left(\frac{\alpha + 1}{\alpha}\right)\right]^2 - 3\left[\Gamma\left(\frac{\alpha + 1}{\alpha}\right)\right]^4}{(SD[Weibull(\alpha, \beta, \tau)])^4}$$

Coefficient of Excess Kurtosis:

$$ExKurt[Weibull(\alpha, \beta, \tau)] = Kurt[Weibull(\alpha, \beta, \tau)] - 3$$

A Weibull random variable is almost always leptokurtic, except for a region near $\alpha = 3.6$, in which it become slightly platykurtic.

REFERENCES

AIAG (2002), *Measurement Systems Analysis*, 3d ed., Southfield, MI: Automotive Industry Action Group, *http://www.aiag.org/*.

AIAG (2005), *Statistical Process Control (SPC) Reference Manual*, 2d ed., Southfield, MI: Automotive Industry Action Group, *http://www.aiag.org/*.

ASQ (2005), *Glossary and Tables for Statistical Quality Control*, 4th ed., American Society for Quality, Milwaukee, WI: ASQ Quality Press.

Anderson, T. W., and D. A. Darling (1952), "Asymptotic Theory of Certain Goodness-of-Fit Criteria Based on Stochastic Processes," *Ann Math. Stat*, 23, 193–212.

Anderson, T. W., and D. A. Darling (1954), "A Test for Goodness-of-Fit," *J Am Stat Assoc*, 49, 300–310.

Anscombe, F. J. (1948), "The Transformation of Poisson, Binomial, and Negative Binomial Data," *Biometrika*, 37, 358–382.

Arnold, B. C. (1983), *Pareto Distributions*, Fairland, MD: International Co-operative Publishing House.

Bailey, B. J. R. (1980), "Accurate Normalizing Transformations of a Student's t Variate" *Appl Stat*, 29(3), 304–306.

Balakrishnan, N., and S. Kocherlakota (1986), "Effects of Non-Normality on X charts: Single Assignable Cause Model," *Sânkhya*, B 48, 439–444.

Balakrishnan, N. (ed.) (1992), *Handbook of the Logistic Distribution*, New York: Marcel Dekker.

Balakrishnan, N., and V. B. Nevzorov (2003), *A Primer on Statistical Distributions*, Hoboken, NJ: Wiley.

Ball, W. W. R. (1908), *A Short Account of the History of Mathematics*, 4th ed, Mineola, New York: Dover.

Bernstein, P. L. (1998), *Against the Odds: The Remarkable Story of Risk*, New York: Wiley.

Bothe, D. R. (1992), "A Capability Study for an Entire Product," *46th ASQC Annual Quality Congress Transactions*, Nashville, TN, May 1992, 172–178.

Bothe, D. R. (1997), *Measuring Process Capability*, New York: McGraw-Hill.

Bothe, D. R. (2002), "Statistical Reason for the 1.5-Sigma Shift," *Qual Eng*, 14(3), March 2002, 479–487.

Bothe, D. R. (2002a), *Reducing Process Variation: Using the DOT-STAR Problem-Solving Strategy*, Cedarburg, WI: Landmark Publishing.

Box, G. E. P., and D. R. Cox (1964), "An Analysis of Transformations," *J R Stat Soc, Ser B*, 26, 211–243.

Bowman, K. O., and L. R. Shenton (1982), "Properties of Estimators for the Gamma Distribution," *Commun Stat Simul Comput*, 12, 697–710.

Bresciani-Turroni, C (1939), "Annual Survey of Statistical Data: Pareto's Law and Index of Inequality of Incomes," *Econometrica*, 7, 107–133.

Burr, I. W. (1967), "The Effect of Non-Normality on Constants for X and R Charts," *Ind Qual Control*, 34 563–569.

Calvin, T. W. (1983), "Quality Control Techniques for 'Zero Defects'," *IEEE Trans Comp Hyb Manu Tech*, CHMT-6 (3), September, 323–328.

Chan, L. K., K. P. Hapuarachchi, and B. D. Macpherson (1988), "Robustness of X and R Charts," *IEEE Trans Relia*, 37, 117–123.

Chapman, D. G. (1956), "Estimating the Parameters of a Truncated Gamma Distribution," *Ann Math Stat*, 27, 498–506.

Chou, Y. M., A. M. Polansky, and R. L. Mason (1998), "Transforming Non-normal Data to Normality in Statistical Process Control," *J Qual Tech*, 30(2), 133–141.

Chung, J. H., and D. B. DeLury (1950), *Confidence Limits for the Hypergeometric Distribution*, Toronto: University of Toronto Press.

Clements, J. A. (1989), "Process Capability Calculations for Non-normal Distributions," *Qual Prog*, September, 95–100.

Cramer, J. S. (1971), *Empirical Econometrics*, Amsterdam North-Holland.

D'Agostino, R. B. and M. A. Stephens (1986), *Goodness-of-Fit Techniques*, New York: Marcel Dekker.

Decisioneering (2006), *Crystal Ball 7.x Software Validation Documentation*, Denver, CO: Decisioneering, Inc.

Efron, B., and R. J. Tibshirani (1994), *An Introduction to the Bootstrap*, Boca Raton, FL: Chapman & Hall/CRC.

Evans, D. H. (1975), "Statistical Tolerancing: The State of the Art, Part III: Shifts and Drifts," *J Qual Tech*, 7(2) April, 72–76.

Evans, M., N. Hastings, B. Peacock (2000), *Statistical Distributions*, 3rd ed, New York: Wiley.

Fisher, R. A. (1922), "On the Interpretation of Chi-Square for Contingency Tables and the Calculation of P," *J R Stat Soc*, 85.

Fisher, R. A. (1924), "On A Distribution Yielding The Error Functions Of Several Well-Known Statistics," *Proceedings of the International Mathematical Congress*, Toronto.

Fisher, R. A. (1925), "Applications of 'Student's' Distribution," *Metron*, 5, 90–104.

Fisher, R. A., and L. H. C. Tippett (1928), "Limiting Forms of the Frequency Distribution of the Largest or Smallest Number of a Sample," *Proc Cambridge Phil Soc*, 24, 180.

Fréchet, M. (1927), "Sur la Loi de Probabilité de l'Écart Maximum," *Annales de la Société Polonaise de Mathematique*, Cracovie, 6, 93–116.

Furry, W. H. (1937), "On Fluctuation Phenomena in the Passage of High Energy Electrons Through Lead," *Phys Rev*, 52, 569–581.

Glushkovsky, E. A. (1994), " 'On-line' G-control chart for attribute data," *Qual Relia Eng Intl*, 10, 217–227.

Gnedenko, B. (1943), "Sur la Distribution Limite du Terme Maximum d'une Serie Aleatoire," *Ann Math*, 2nd Series, 44(3), July, 423–453.

Good, P. I., and P. Good (2001), *Resampling Methods*, Basel: Birkhäuser.

Gumbel, E. J. (1958), *Statistics of Extremes*, New York: Columbia University Press; republished 2004 Minneola, New York: Dover.

Hallowell, D. (2006), "Six Sigma Software Metrics," Parts 1 and 2, *http://software.isixsigma.com*.

Harry, M. (2003), *Resolving the Mysteries of Six Sigma: Statistical Constructs and Engineering Rationale*, Phoenix, AZ: Palladyne Publishing.

Hsaing, T. C. and Taguchi, G. (1985), "A Tutorial on Quality Control and Assurance—The Taguchi Methods," *ASA Annual Meeting*, Las Vegas, NV.

Huber-Carol, C., N. Balakrishnan, M. Mikulin, and M. Mesbah (ed) (2002), *Goodness-of-Fit Tests and Model Validity*, Boston, MA: Birkhauser.

Johnson, N. L. (1949), "Systems of Frequency Curves Generated by Methods of Translation," *Biometrika*, 73, 149–176.

Johnson, N. L., Samuel Kotz, N. Balakrishnan (1994), *Continuous Univariate Distributions*, Volume 1, 2d ed, New York: Wiley.

Johnson, N. L., Samuel Kotz, N. Balakrishnan (1995), *Continuous Univariate Distributions*, Volume 2, 2d ed, New York: Wiley.

Johnson, N. L., Adrienne W. Kemp, Samuel Kotz (2005), *Univariate Discrete Distributions*, 3d ed, New York: Wiley.

Juran, J. M. (1974), *Juran's Quality Control Handbook*, 3d ed., New York: McGraw-Hill.

Juran, J. M., and A. B. Godfrey (1999), *Juran's Quality Handbook*, 5th ed, New York: McGraw-Hill.

Kan, S. A. (2003), *Methods and Models in Software Quality Engineering*, 2d ed, Boston, MA: Addison-Wesley.

Kane, V. E. (1986), "Process Capability Indices," *J Qual Tech*, 18, 41–52.

Kececioglu, D. (1991), *Reliability Engineering Handbook*, Prentice-Hall.

Kelton, W. D., and A. Law (1991), *Simulation Modeling & Analysis*, New York: McGraw-Hill.

Knüsel, L. (1998), "On the Accuracy of Statistical Distributions in Microsoft Excel 97," *Comput Stat Data Anal*, 26(3), 375–377.

Knüsel, L. (2005), "On the Accuracy of Statistical Distributions in Microsoft Excel 2003," *Comput Stat Data Anal*, 48, 445–449.

Kotz, S., and C. R. Lovelace (1998), *Process Capability Indices in Theory and Practice*, London: Arnold.

Kotz, S., and N. L. Johnson (2002), "Process Capability Indices: A Review 1992–2000," *J Qual Tech*, 34(1), 2–19.

Kotz, S., and J. R. van Dorp (2004), *Beyond Beta: Other Continuous Families with Bounded Support and Applications*, Singapore: World Scientific Publishing.

Kolmogorov, A. (1933), "Sulla Determinazione Empirica di una Legge de Distributione," *Giornale dell'Instituto Italiano degli Attuari*, 4, 1–11.

Laguna, M (1997), "Metaheuristic Optimizationn with Evolver, Genocop and OptQuest," Graduate School of Business, University of Colorado, *http://crystalball.com/optquest/comparisons.html*.

Laguna, M (1997a), "Optimization of Complex Systems with OptQuest," Graduate School of Business, University of Colorado, *http://www.crystalball.com/optquest/complexsystems.html*.

Laplace, P. S. (1774), "Mémoire sur la Probabilité des Causes par les Évènemens," *Mémoires de Mathématique et de Physique*, 6, 621–656.

L'Ecuyer, P. (1994), "Uniform Random Number Generation," *Ann Oper Res*, 53, 77–120.

Luceño, A (1996), "A Process Capability Index with Reliable Confidence Intervals," *Commun Stat Simul*, 25(1), 235–245.

McCullough, B. D., and B. Wilson (1999), "On the Accuracy of Statistical Procedures in Microsoft Excel 97," *Comput Stat Data Anal*, 31, 27–37.

McCullough, B. D., and B. Wilson (2005), "On the Accuracy of Statistical Procedures in Microsoft Excel 2003," *Comput Stat Data Anal*, 49, 1244–1252.

Montgomery, D. C. (2005), *Introduction to Statistical Quality Control*, 5th ed., New York: Wiley.

Moore, P. G. (1954), "A Note on the Truncated Poisson Distributions," *Biometrics*, 10, 402–406.

Nelson, W. B. (2004), *Applied Life Data Analysis*, Hoboken, NJ: Wiley.

Olshen, A. C. (1937), "Transformation of the Pearson Type III Distributions," *Ann Math Stat*, 6, 176–200.

Ord, J. K. (1968), "Approximations to Distribution Functions Which are Hypergeometric Series," *Biometrika*, 55, 243–248.

Owen, D. B. (1962), *Handbook of Statistical Tables*, Reading, MA: Addison-Wesley.

Pareto, V. (1897), *Cours d'Economie Politique*, Paris: Rouge et Cie.

Pearson, K. (1900), "On the Criterion that a Given System of Deviations from the Probable in the Case of a Correlated System of Variables is such that it can be Reasonably Supposed to have Arisen from Random Sampling," reprinted in Pearson (1948).

Pearson, K. (1948), *Karl Pearson's Early Statistical Papers*, edited by E. S. Pearson, London: Cambridge University Press.

Pearson, E. S., R. L. Plackett, and G. A. Barnard (ed) (1990), *'Student', A Statistical Biography of William Sealy Gosset*, Oxford: University Press.

Pearn, W. L., S. Kotz, and N. L. Johnson (1992), "Distributional and Inferential Properties of Process Capability Indices," *J Qual Tech*, 24(4), 216–231.

Peizer, D. B., and J. W. Pratt (1968), "A Normal Approximation for Binomial, F, Beta and Other Common Related Tail Probabilities, I," *J Am Stat Assoc*, 63(324), December, 1416–1456.

Lilliefors, H., (1967), "On the Kolmogorov-Smirnov Test for Normality with Mean and Variance Unknown," *J Am Stat Assoc*, 62(318), June, 399–402.

Luceño, A. (1996), "A Process Capability Index with Reliable Confidence Intervals," *Commun Stat Simul*, 25(1), 235– 245.

Massey, F. J. (1951), "The Kolmogorov-Smirnov Test for Goodness of Fit," *J Am Stat Assoc*, 46(253), March 1951, 68–78.

NIST/SEMATECH, *NIST/SEMATECH e-Handbook of Statistical Methods*, *http://www.itl.nist.gov/div898/handbook/*

Quesenberry, C. P. (1995), "Geometric Q-charts for High Quality Processes," *J Qual Tech*, 27, 304–315.

Rodriguez, R. N. (1992), "Recent Developments in Process Capability Analysis," *J Qual Tech*, 24(4), 176–187.

Schilling, E. G. and P. R. Nelson (1976), "The Effect of Non-normality on the Control Limits of X Charts," *J Qual Tech*, 8, 183–188.

Shapiro, S. S. and M. B. Wilk (1965), "An Analysis of Variance Test for Normality (Complete Samples)," *Biometrika*, 52(3–4), 591–611.

Shore, H. (2000a), "General Control Charts for Attributes," *IIE Trans (QualReliab Eng)*, 32(12), 634–637.

Shore, H. (2000b), "General Control Charts for Variables," *Intl J Prod Res,* 38(8), 1875–1897.

Shore, H. (2001), "Note: General Control Charts for Variables," *Intl J Prod Res*, 39(9), 2063–2064.

Shore, H. (2005), *Response Modeling Methodology (RMM): Empirical Modeling for Engineering and the Sciences*, Singapore: World Scientific Publishing.

Sleeper, A (2006), *Design for Six Sigma Statistics: 59 Tools for Diagnosing and Solving Problems in DFSS Initiatives*, New York: McGraw-Hill.

Smirnov, H. (1939), "Sur les Écarts de la Courbe de Distribution Empirique," *Recueil Mathématique (Matematiceskii Sbornik)*, N. S. 6 (1939), 3–26.

Student (Gossett, W. S.) (1908), "The Probable Error of a Mean," *Biometrika*, 6(1), 1–25.

Stephens, M.A. (1974), "EDF Statistics for Goodness of Fit and Some Comparisons," *J Am Stat Assoc*, 69(347), Sept. 1974, 730–737.

Sugiura, N., and A. Gomi, (1985), "Pearson Diagrams for Truncated Normal and Truncated Weibull Distributions," *Biometrika*, 72, 219–222.

U. S. (1950), *President's Water Resources Policy Commision*, Washington, DC.

Vännman, K. (1995), "A Unified Approach to Capability Indices," *Statistica Sinica*, 5, 805–820.

Weibull, W. (1939), "A Statistical Theory of the Strength of Material," Report No. 151, Ingeniörs Vetenskaps Akademiens Handligar, Stockholm.

Weibull, W. (1939a), "The Phenomenon of Rupture in Solids," Report No. 153, Ingeniörs Vetenskaps Akademiens Handligar, Stockholm.

Weibull, W. (1951), "A Statistical Distribution of Wide Applicability," *J Applied Mechanics*, 18, 293–297.

Wheeler, D. J. (2003), *Making Sense of Data*, Knoxville, TN: SPC Press.

Wilson, E. B., and M. M. Hilferty (1931), "The Distribution of Chi-Square," *Proc Natl Acad Sci*, 17, 684–688.

Xie, M., T. N. Goh, and X. Y. Tang (2000), "Data Transformation for Geometrically Distributed Quality Characteristics," *Qual Relia Eng Intl*, 16, 9–15.

Yang, Z., and M. Xie (2000), "Process Monitoring of Exponentially Distributed Characteristics Through an Optimal Normalizing Transformation," *J Appl Stat*, 27(8), 1051–1063.

Yeh, A. B., and S. Bhattacharya (1998), "A Robust Process Capability Index," *Commun Stat Simul Comput*, 27(2) 565–589.

INDEX

Continuous Distribution Models with a Lower Bound

Distribution Family	PDF	Skew	Typical Application
Exponential Chapter 11, page 203		Right	Time or distance between independent events in reliability or process modeling. Lack of memory property. Simplest continuous model.
Gamma (γ) Chapter 14, page 245 **Chi-squared** (χ^2) Chapter 9, page 179		Right	Sum of exponentials. Sum of squared normals. χ^2, exponential, and Erlang are special cases of gamma family.
Weibull Chapter 28, page 391		Right or Left	Time to failure in reliability applications. May have decreasing, constant or increasing hazard rate. Exponential and Rayleigh are special cases.
Lognormal Chapter 19, page 299		Right	Reliability applications. Log of a lognormal is normal.
Loglogistic Section 18.1, page 294		Right	Reliability applications. Log of a loglogistic is logistic.
Half-Normal Section 21.1, page 328		Right	Absolute value of a normal.
Pareto Chapter 22, page 337		Right	Economics applications. Extremely heavy and long tail.
Rayleigh Chapter 24, page 357		Right	Reliability and project management applications.